NEUROPHARMACOLOGY METHODS in EPILEPSY RESEARCH

CRC Press
METHODS IN THE LIFE SCIENCES

Gerald D. Fasman - Advisory Editor
Brandeis University

Series Overview

Methods in Biochemistry
John Hershey
Department of Biological Chemistry
University of California

Cellular and Molecular Neuropharmacology
Joan M. Lakoski
Department of Pharmacology
Penn State University

Research Methods for Inbred Laboratory Mice
John P. Sundberg
The Jackson Laboratory
Bar Harbor, Maine

Methods in Neuroscience
Sidney A. Simon
Department of Neurobiology
Duke University

Joseph M. Corless
Department of Cell Biology,
Neurobiology and Ophthalmology
Duke University

Methods in Pharmacology
John H. McNeill
Professor and Dean
Faculty of Pharmaceutical Science
The University of British Columbia

Methods in Signal Transduction
Joseph Eichberg, Jr.
Department of Biochemical and Biophysical Sciences
University of Houston

Methods in Toxicology
Edward J. Massaro
Senior Research Scientist
National Health and Environmental Effects Research Laboratory
Research Triangle Park, North Carolina

CRC Press
CELLULAR AND MOLECULAR
NEUROPHARMACOLOGY

Joan M. Lakoski, *Advisory Editor*

The CRC Press *Cellular and Molecular Neuropharmacology Series* provides the reader with state-of-the-art research methods that address the cellular and molecular mechanisms of the neuropharmacology of brain function in a clear and concise format. Topics covering all aspects of neuropharmacology are being reviewed for publication.

Published Titles

Molecular Regulation of Arousal States, Ralph Lydic

Neuropharmacology Methods in Epilepsy Research, Steven L. Peterson and Timothy E. Albertson

Forthcoming Title

Methods in Neuroendocrinology, Louis D. Van De Kar

NEUROPHARMACOLOGY METHODS in EPILEPSY RESEARCH

Edited by

Steven L. Peterson, Ph.D.
College of Pharmacy
University of New Mexico
Albuquerque, New Mexico

and

Timothy E. Albertson, M.D., Ph.D.
Department of Medical Pharmacology and Toxicology
School of Medicine
University of California Davis
Davis, California

CRC Press

Boca Raton   Boston   London   New York   Washington, D.C.

Front cover art drawn by Tara L. Peterson.

Acquiring Editor:	Paul Petralia
Project Editor:	Joanne Blake
Marketing Manager:	Becky McEldowney
Cover design:	Denise Craig
PrePress:	Kevin Luong

Library of Congress Cataloging-in-Publication Data

Neuropharmacology methods in epilepsy research / edited by Steven L.
Peterson and Timothy E. Albertson.
     p. cm. — (CRC Press methods in the life sciences. Cellular
and molecular neuropharmacology)
     Includes bibliographical references and index.
     ISBN 0-8493-3362-8 (alk. paper)
     1. Epilepsy—Research—Methodology. 2. Epilepsy—Research—Animal
models. 3. Neuropharmacology—Research—Methodology. I. Peterson,
Steven Lloyd. II. Albertson, Timothy Eugene. III. Series.
     [DNLM: 1. Epilepsy—drug therapy. 2. Disease Models, Animal.
3. Anticonvulsants—pharmacology. 4. Neuropharmacology—methods.
WL 385 N493392 1998]
RC372.N39 1998
616.8′53027—dc21
DNLM/DLC
for Library of Congress
     98-9863
     CIP

The Editors

Steven L. Peterson, Ph.D., is an Associate Professor in the College of Pharmacy at the University of New Mexico. Dr. Peterson received his B.S. degree from the University of California, Davis in 1975 with a major in Animal Science. In 1980 he earned a Ph.D. in Pharmacology and Toxicology from the Department of Pharmacology in the University of California, Davis, School of Medicine.

After two years as a postdoctoral fellow in the Department of Pharmacology in the Texas Tech University College of Medicine, he became an Assistant Professor at the Texas A&M University College of Medicine. While at Texas A&M, Dr. Peterson was twice awarded the Distinguished Teaching Award for the College of Medicine and was recognized as a Scholar by the Texas A&M University Center for Teaching Excellence. Dr. Peterson was promoted to Professor before moving to his present position.

Dr. Peterson is a member of the Society for Neuroscience and the Western Pharmacology Society. He is the recipient of grants from the National Institute of Neurological Disease and Stroke. He has authored more than 40 papers. His current research interests include the study of brainstem substrates that contribute to the pharmacological activity of anticonvulsant drugs.

Dr. Timothy E. Albertson, M.D., Ph.D., is a Professor of Medicine and Medical Pharmacology and Toxicology for the Departments of Internal Medicine and Medical Pharmacology and Toxicology in the University of California, Davis, School of Medicine. He serves as the Chief of the Division of Pulmonary and Critical Care Medicine, Director of Clinical Pharmacology for the Department of Internal Medicine, and Medical Director of the Davis Division of the University of California Poison Control Center.

Dr. Albertson received his B.A. degree from the University of California, San Diego in 1973 with majors in Biology and Psychology. He was awarded an M.S. in Pharmacology and Toxicology in 1976 from the University of California, Davis. Dr. Albertson received his M.D. degree in 1977 and his Ph.D. in Pharmacology and Toxicology in 1980 from the University of California, Davis, School of Medicine.

Dr. Albertson completed a two-year fellowship program in Pulmonary and Critical Care Medicine in 1983 at the University of California, Davis, Medical

Center. That year he became an Assistant Professor, section of Critical Care Medicine in the Divisions of Pulmonary Medicine and of Emergency Medicine and Clinical Toxicology at the University of California, Davis, School of Medicine. Dr. Albertson was promoted to Professor of Medicine and Pharmacology and Clinical Toxicology in 1993 at the University of California, Davis, School of Medicine.

Dr. Albertson is an investigator on numerous clinical studies. He has been the recipient of many awards, including Outstanding Faculty Teacher and Best Attending Physician. He is the author of over 150 book chapters and peer reviewed articles. His current research interests include mechanisms of neurotoxicities of pesticides.

Dedication

This book is dedicated to the memory of
Robert M. Joy, Ph.D.,
teacher, mentor, and friend.

Contributors

Timothy E. Albertson, M.D., Ph.D.
Department of Medical Pharmacology
and Toxicology
School of Medicine
University of California Davis
Davis, CA

Thomas H. Champney, Ph.D.
Department of Human Anatomy and
Medical Neurobiology
College of Medicine
Texas A&M University
College Station, TX

Charles R. Craig, Ph.D.
Department of Pharmacology and
Toxicology
West Virginia University Health
Science Center
Morgantown, WV

John W. Dailey, Ph.D.
Department of Biomedical and
Therapeutic Science
College of Medicine
University of Illinois
Peoria, IL

Jeffrey H. Goodman, Ph.D.
Neurology Research Center
Helen Hayes Hospital
West Haverstraw, NY

Mary Ellen Kelly, Ph.D.
Department of Pharmacology
Dalhousie University
Halifax, Nova Scotia, Canada

Wolfgang Löscher, Ph.D., D.V.M.
Department of Pharmacology,
Toxicology and Pharmacy
School of Veterinary Medicine
Hannover, Germany

Pravin K. Mishra, Ph.D.
Department of Biomedical and
Therapeutic Science
College of Medicine
University of Illinois
Peoria, IL

Steven L. Peterson, Ph.D.
College of Pharmacy
University of New Mexico
Albuquerque, NM

Charles E. Reigel, Ph.D.
Department of Pharmacology
Texas Tech University Health Science
  Center
Lubbock, TX

Larry G. Stark, Ph.D.
Department of Medical Pharmacology
  and Toxicology
School of Medicine
University of California Davis
Davis, CA

Janet L. Stringer, M.D., Ph.D.
Department of Pharmacology
Baylor College of Medicine
Houston, TX

Laurence H. Tecott, M.D., Ph.D.
Department of Psychiatry
University of California San Francisco
San Francisco, CA

H. Steve White, Ph.D.
Department of Pharmacology and
  Toxicology
University of Utah
Salt Lake City, UT

Piotr Wláz, Ph.D.
Department of Pharmacology
Faculty of Veterinary Medicine
Agricultural University
Lublin, Poland

Preface

Having worked with whole animal models of epilepsy for over 20 years we noticed a gradual shift in the methodological standards and interpretation of data in epilepsy research. Some investigators employed hybrid convulsive scales, others used stimulus paradigms that resulted in inconclusive results, and some attempted to characterize the activity of anticonvulsant drugs using inappropriate seizure models. More recently, otherwise capable molecular biologists produced genetically altered animals that exhibited convulsions, but the investigators seemed uncertain as to how to characterize the seizure phenotype. The situation has been made even more difficult by the absence of a comprehensive and detailed text concerning methods in epilepsy research since *Experimental Models of Epilepsy* in 1972. As we experienced difficulty in even locating copies of that text, we decided that perhaps there was a void in the recent epilepsy research literature that needed to be filled. So it is in that spirit that we offer this text as a comprehensive and detailed description of methodology that can be used in epilepsy research.

The text describes the fundamental methodology and procedures employed in the modern study of experimental models of epilepsy. All chapters are written by authors with extensive experience using the techniques described and who actively employ the technology in their own laboratories. Techniques covered include today's use of classic models of epilepsy, such as electroshock, chemoconvulsions, kindling, audiogenic seizures, focal seizures, and brain slice preparations. The book also describes more recently developed seizure models, including models of status epilepticus and massed trial stimulations. The influence of circadian and diurnal rhythms on convulsive activity is considered, as is the evaluation of behavioral and cognitive deficits associated with anticonvulsant drug testing. The use of gene knockout technology in the study of epilepsy is also presented. Each chapter contains the basic steps required for the technique and describes how the results of the experiments should be interpreted so that they contribute to the understanding of epilepsy.

Steven Peterson
Timothy Albertson
February 1998

Contents

Chapter **1**

Electroshock

Steven L. Peterson

Contents

0-8493-3362-8/98/$0.00+$.50
© 1998 by CRC Press LLC

I.   Introduction

Electroshock is the generalized electrical stimulation of the brain. Typically an electrical current of less than a second's duration is passed from one side of the head to the other. Originally this involved passing the electrical current through trephined holes in the skull[1] or from the roof of the mouth to the top of the skull.[3] Experimental techniques commonly used in rodents today pass the current between the eyes or the ears, and such techniques are the subject of this chapter. Electroconvulsive therapy (ECT) is widely used in psychiatry for the treatment of severe depression and may involve bilateral stimulation from one side of the head to the other or unilateral stimulation with both electrodes applied to the same side of the head.[4]

Electroshock stimulates large portions of the brain. The generalized stimulation induces neurons to fire repetitively and synchronously, which is the hallmark of epileptic neurons and epilepsy.[5] Most investigators consider the peripheral manifestations or convulsive responses that are induced by this aberrant neuronal activity to be tonic-clonic convulsions. However, other convulsive responses can be induced, depending on the strength of the stimulating current. The various convulsive responses induced by electroshock that are most commonly used in research today are discussed in this chapter.

Although electroshock was first demonstrated in animals over 120 years ago,[1] the potential value was not fully realized until 1939, when Merritt and Putnam used the technique to establish the selective antiepileptic activity of phenytoin.[3] Since then, maximal electroshock has become a critical tool for detecting potential antiepileptic drugs effective against generalized tonic-clonic (grand mal) seizures.[6-8] However, maximal electroshock is more than an empirical model of epilepsy in that the convulsions are common to mammalian species and have many of the features of generalized tonic-clonic (grand mal) seizures. Such stereotyped responses indicate common neural substrates for generalized tonic-clonic convulsions[9] and suggest that study of these convulsions in rodents is applicable to the human epileptic condition. Indeed, the rat and human electroshock responses to antiepileptic drugs are similar.[10,11] Given the current hypothesis concerning the important role of the brainstem in generalized tonic-clonic convulsions,[12] it may be expected that continued study of the electroshock response will be central to the understanding of the basic mechanisms of the epilepsies.

II.   Methodology

A.   Electroshock Induced by Corneal Electrodes

The following itemization provides a description of the relatively simple technique for administering electroshock convulsions using corneal electrodes in rodents. This chapter deals solely with the use of rodents as they are the most commonly used due to their low cost and because they are a reliable model of human seizures.[6-8,10,11]

While both convenient and inexpensive, electroshock convulsions can be complicated because they are expressed in a variety of forms that are dependent on the
rodent strain, the strength of the electrical current used, and the placement of the
stimulating electrodes. The chapter sections following the itemization provide a more
in-depth consideration of the specific details of electroshock techniques, including
the rationale for commonly used procedures, electrode placement, electrical current-
related variations in the convulsive response, interpretation of the results, and other
factors that influence the convulsive response.

The electroshock is best performed by a single person. Reproducible results are
most reliably achieved if a single investigator holds the animal, applies the shock,
and measures the duration of the convulsive phases. With practice, consistent results
can be readily obtained. Prior training in rodent handling skills will assure proper
and humane treatment of the animals. It is highly recommended that a leather
gardening glove be worn on the hand that restrains the animal during the administration of the electroshock current. Some animals struggle while being restrained
for the electroshock and may bite. After the electroshock-induced convulsion, rats
often exhibit an exaggerated startle response and may bite aggressively. A leather
glove is especially important in handling a rat in the first 5 to 10 min after an
electroshock-induced convulsion.

The electroshock procedure described involves the use of a Wahlquist stimulator
designed specifically for electroshock. Although no longer available, many departments of pharmacology still retain Wahlquist stimulators in storage. Alternative
sources for commercial electroshock stimulators are discussed later in the chapter.

The procedure for corneal electroshock in rats or mice may be itemized as
follows:

1. Place the electroshock stimulator on a nonconducting table or counter top. Clear a 4
    to 5 ft² space in front of the stimulator for handling the animal. The stimulus initiation
    switch is usually a foot pedal that is placed on the floor directly underneath the cleared
    workspace. The stimulator is activated by stepping down on the pedal and releasing it.
    The stimulus is induced when the pedal is released, thereby eliminating the need to
    "hunt" for the pedal while restraining the animal.

2. Remove the animal from the cage and place it on the cleared area. Using the gloved,
    nondominant hand, restrain the animal by cupping the palm of the hand over the
    animal's back with the middle and index fingers on each side of the neck (Figure 1.1).

3. When using mice, place a drop of 0.5% tetracaine in 0.9% saline (from any commer-
    cially available source) in each eye of the animal. This procedure is proposed to reduce
    the incidence of electroshock-induced death and corneal pain.[8]

4. While continuously restraining the rat or mouse using the gloved hand, place the saline-
    soaked, cotton-covered corneal electrodes over the eyes, using the dominate hand
    (Figure 1.1). The corneal electroshock stimulus is administered through electrodes
    mounted in a nonconducting acrylic handle (Figure 1.1). The electrodes that are pro-
    vided with the stimulator are rigid 20 gauge wires that are connected to two insulated
    electrical wires which proceed through the handle and are connected to the stimulator
    by standard banana plugs (Figure 1.1). The heads of the electrodes are the contact
    points with the animal and they are wrapped with cotton that is tied with surgical
    suture. The cotton enhances the comfort for the animal and is soaked in 0.9% saline

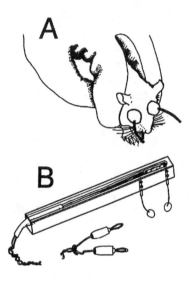

FIGURE 1.1

Panel A depicts an artist's conception of the handling technique for administering corneal electroshock. The animal is restrained by cupping the palm of the nondominate hand over the back of the animal and placing the middle and index fingers on each side of the neck. Mice may be held in the position shown to apply the drops of 0.5% tetracaine to the eyes. As illustrated, the saline-soaked, cotton-covered corneal electrodes are placed directly over the eyes. The electrodes must be held firmly in place for the entire duration of the stimulus (usually 0.2 s) to ensure that the intended stimulation is delivered. The technique may be used in rats or mice. It is recommended that a leather gardening glove be worn on the hand that restrains the animal. Panel B depicts the electroshock stimulus delivery handle as manufactured by Wahlquist, Inc. As illustrated, the handle is constructed of clear, nonconducting acrylic, through which the components are clearly visible. The electrodes are attached by connectors mounted in the acrylic. The stimulus conducting wires proceed internally through a channel in the handle. A cover is permanently fixed over the channel to internalize the conducting wires. The stimulus conducting wires are interfaced with the stimulator using standard banana plugs.

to facilitate electrical conductivity. The electrodes should have been previously adjusted to fit snugly over the eyes. The electrodes are held firmly against the animal throughout the stimulus to ensure constant contact.

5. Pass a 60 Hz, 0.2 msec electrical current of variable amplitude (18 to 500 mA in rats as discussed below) through the electrodes. The electrodes must be held to the animal's eyes for the entire duration of the stimulus to prevent arcing. Arcing is the spark resulting from the electrical current passing through the air between the electrode and the animal. When an arc or spark occurs the animal may not have received the intended electrical stimulus. The strength and duration of the stimulating current may vary according to the desired response and can be set by the controls on the stimulator. The convulsive responses evoked by the various electrical stimulation currents are described in Table 1.1.

6. Just after the stimulus, quickly roll the animal onto its side with the feet toward the investigator so that the evoked convulsion may be observed in its entirety (Figure 1.2). The convulsive phases are described in detail below.

7. A battery of timers that typically are included with the electroshock stimulator are triggered by the stimulus and may be used to time the duration of the various convulsive

TABLE 1.1
Continuum of Convulsive Responses Induced by Increasing Corneal Stimulation Currents

Response to corneal electroshock	Alternative names	Stimulus current (mA)		Quantified convulsion components	Brain region activated
		Rats	Mice		
Subconvulsive response	Stun, rage	<18	<5	No convulsive response	No epileptiform activity
Face and forelimb clonus	Minimal clonic seizure, minimal electroshock	18–20	5	Amygdala kindling scale, clonic spasm	Forebrain, limbic seizures
Running-bouncing	Wild running	20–21	10	Occurrence of running episode	Minimal activation of brainstem
Tonic-flexion	Flexion, opisthotonus	21–22	11	Occurrence of tonic flexion	Minimal activation of brainstem
Threshold tonic-clonic	Threshold tonic extension (TTE), threshold for maximal seizures (MEST), threshold electroshock	22–50	12–30	Occurrence of tonic hindlimb extension (THE), duration of tonic extension, F/E ratio (maximal)	Submaximal activation of brainstem
Maximal tonic-clonic	Maximal electroshock (MES)	150	50	Occurrence of tonic hindlimb extension (THE), duration of tonic extension, F/E ratio (minimal)	Maximal activation of brainstem

Note: The currents indicated for inducing the convulsive responses in rats and mice are approximate only and may be expected to vary greatly. It is recommended that investigators perform pilot studies to determine the exact electrical current ranges for their specific animals and laboratory conditions.

phases. Some investigators mount a video camera above the electroshock table and record both the convulsive activity and the timers, which allows easy and accurate timing of the convulsive phases.[13]

8. The animal should be returned to the home cage within 60 s of the end of the convulsion. After that time the rodents exhibit exaggerated startle responses and can be difficult to handle. Quiet, gentle handling will reduce the incidence of aggressive postictal behaviors.

B.   Convulsive Response

Electroshock induces a continuum of motor convulsions that are dependent on the intensity of the electrical stimulation current. The convulsive continuum as it occurs in rodents is indicated in Table 1.1 and is based on corneal stimulation (also known as transcorneal stimulation) using a 0.2 s duration, 60 Hz stimulus. Lower currents (below 20 mA in rats and 5 mA in mice) activate the forebrain selectively and produce only clonic convulsions. The clonic convulsions are useful

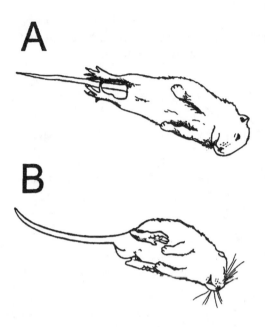

FIGURE 1.2
Tonic extension in the rat. Note that the animal has been rolled onto the side so that the convulsive phases may be observed. Depicted in panel A is a rat in tonic extension with tonic hindlimb extension (THE). This is the most severe electroshock response and may last up to 15 s. The rat depicted in part B is in tonic extension without THE. Abolition of THE is an indication of antiepileptic drug activity in generalized tonic-clonic (grand mal) seizures. Inhibition of THE is considered to have occurred when the legs do not extend beyond a 90° angle to the torso.[29,30]

for detecting anticonvulsant activity of drugs effective against absence (petit mal) seizures. Repetition of such low current electroshock stimuli on a daily basis induces limbic kindling, as described below. Higher corneal electroshock currents activate brainstem mechanisms that produce tonic-clonic convulsions. The minimal electrical currents that are just sufficient to induce tonic-clonic convulsions are called threshold currents. The threshold currents induce tonic-clonic convulsions that are predictive of general anticonvulsant drug activity. Supramaximal currents that are five to seven times threshold induce tonic-clonic convulsions that are predictive of anticonvulsant drug activity against generalized tonic-clonic (grand mal) seizures. Thus, depending on the intensity of the stimulus a number of preclinical models of epilepsy are possible using corneal electroshock.

The convulsive continuum associated with increasing current strength of corneal electroshock, as outlined in Table 1.1, is detailed in the following sections. It should be noted that the electrical stimulation current strengths (in milliamps) shown to evoke each convulsive response in rats and mice as depicted in Table 1.1 are approximate and should be expected to vary, sometimes greatly, depending on the strain of animal.[1,7,8,14,15] When initiating studies that require use of convulsive thresholds, pilot studies should always be performed to determine the electrical current thresholds for a given strain of animal in a particular laboratory setting.

1. Subconvulsive Response

Corneal electroshock in rats at currents below approximately 18 mA induces no convulsive activity (Table 1.1). Instead the animals appear irritated and display hyperactivity as well as escape and pain behaviors.[1] The animals may exhibit a catatonic or "stun" response.[1]

2. Face and Forelimb Clonus

Slightly higher currents, in the 18 to 20 mA range in the rat, induce face and forelimb clonic activity[1,14,15] that may include a 5 s period of clonic spasms which involve facial clonus as well as clonic activity of all four limbs (Table 1.1).[8,15] These convulsions are also termed minimal clonic seizures or minimal electroshock.[1,15] The currents as administered by corneal electrodes are thought to activate forebrain areas and induce limbic seizure activity.[12] The face and forelimb clonus induced by the initial stimulations only last for a few seconds.[8,16] However, minimal electroshock repeated once or twice daily with corneal electrodes induces limbic kindling, with the forebrain convulsions becoming progressively longer in duration and greater in severity.[8,16] The duration of the stimulus for this kindling effect is 4.0 s rather than the 0.2 s duration used in most electroshock procedures.[8] This means that the investigator applying the stimulus must hold the corneal electrodes tightly to the animal for the entire 4 s period. Such corneal kindling induces convulsions that are similar to those observed in amygdala kindling[17] and have a similar anticonvulsant profile.[18]

Besides serving as a model of limbic kindling that does not involve the placement of intracranial electrodes,[18] the forebrain convulsions induced by corneal electroshock are proposed to represent a threshold for minimum seizures.[15] The 0.2 s duration electrical current required to induce these convulsions represents the threshold current or the minimal stimulus needed to induce seizure activity in 97% to 100% of the animals.[15] Drugs that raise this threshold prevent the occurrence of face and forelimb clonus or clonic spasms and are active against absence (petit mal) seizures.[15] For example, valproate and ethosuximide are active in both the minimal clonic seizures induced by electroshock and generalized absence seizures.[15] The absolute change in threshold induced by anticonvulsant drugs can be determined by the method of Litchfield and Wilcoxon,[19] in which multiple groups of animals are tested with various stimulation currents until a CC_{50} (convulsant current in 50% of the animals) is determined.[1] Alternatively, fewer animals and time are required to determine changes in threshold using the "up-down" or "staircase" methods in which the stimulation current for each animal is determined by the response of the previous animal.[1,20-22]

Minimal clonic seizures are also produced by low doses of the chemical convulsants pentylenetetrazol, bicuculline, and picrotoxin,[8,12,15] which are presented in Chapter 2. The face and forelimb clonus produced by these chemical convulsants is analogous to that produced by corneal stimulation in that both identify drugs that raise seizure threshold and that are effective against absence seizures.[6,8,15]

The animals are considered to be unconscious during the minimal clonic seizures.[1] This is because the seizure activity generalizes to, or involves, the entire forebrain and disrupts conscious awareness. Since seizure generalization is a feature common to all electroshock-induced convulsions, it is likely that all animals are unconscious during electroshock-induced minimal clonic seizures, as well as during more severe convulsions induced by higher electroshock currents.[1]

3.   Running-Bouncing Clonus; Tonic Flexion

The next two successive convulsive responses evoked by progressively higher corneal electroshock stimulating currents are running-bouncing clonus and tonic flexion (Table 1.1). These occur in relatively limited current ranges and are therefore difficult to observe in isolation using corneal electroshock. The running-bouncing clonic convulsions have also been referred to as wild running and represent a response to the initial, minimal activation of the brainstem nuclei that mediate tonic-clonic convulsions.[12] The running-bouncing clonus or wild running convulsions induced by electroshock correspond to the convulsions that are reported as rank 1 seizures (audiogenic response score of 1) or minimal audiogenic seizures in genetically epilepsy-prone rats,[23] as described in Chapter 6, and the initial convulsive response seen with seizures induced by localized electrical stimulation of the inferior colliculus.[24]

Tonic flexion represents the first stage of the flexion-extension sequence that is characteristic of tonic-clonic convulsions. This convulsion is the result of contraction of the flexor muscles with the back of the animal arched and the limbs directed forward in a flexed position. This convulsion corresponds to the clonic phase of the running-bouncing clonus described as ranks 2 and 3 seizures (audiogenic response scores of 2 and 3) in genetically epilepsy-prone rats[23] and also has been described as opisthotonus.[12] Tonic flexion represents slightly greater activation of the brainstem nuclei that mediate tonic-clonic convulsions than the activation that induces running-bouncing clonus.[1] Although it is possible to induce flexion-only convulsions with electroshock, such convulsions typically are not used to test for antiepileptic drugs, in part because the narrow current range for inducing the convulsions makes for unreliable seizure responses.

4.   Tonic-Clonic Convulsions

Further increases in corneal electroshock stimulation current elicit convulsions that include both tonic flexion and tonic extension. These are commonly known as tonic-clonic convulsions (Table 1.1). While tonic-clonic convulsions are classified as threshold tonic-clonic or maximal tonic-clonic (Table 1.1), the convulsions are qualitatively the same in overall appearance. The convulsions are observed in the rat as a 2 to 3 s period of tonic flexion followed by a tonic extension phase lasting a period of approximately 10 to 12 s. The durations of the flexion and extension phases are slightly shorter in mice. The tonic extension is observed as a "wave" of muscular contraction that passes down the body from the head to the tail (rostro-caudal). The forelimbs extend first and this occurs reliably.[25] Tonic hindlimb extension (Figure 1.2) represents the maximal convulsive response, but may not always occur reliably as described below. The body remains in a rigid extension for most of the 10 to 15 s period before relaxing in reverse order from tail to head (caudorostral).

The tonic-clonic convulsion may or may not include a terminal clonic phase.[26] The terminal clonus does not occur reliably and is rarely used in seizure analysis.

a. Tonic Extension

Tonic extension is a manifestation of tonic contraction of the musculature. Although both the flexor and extensor muscles contract during the extension phase, extension dominates because the antigravity extensor muscles are stronger than the corresponding flexor muscles. Support for this hypothesis includes the observation that in the sloth, which hangs upside down and uses the flexors as antigravity muscles, tonic flexion is displayed after electroshock.[27]

b. Tonic Hindlimb Extension

By far the most frequently used endpoint in the quantification of tonic-clonic convulsions is tonic hindlimb extension (THE). Other methods for quantifying tonic-clonic convulsions (described below) are typically used only when THE cannot be reliably induced. THE is the last event to occur in the wave of contraction that progresses down the body and is observed as the hindlimbs projecting straight back behind the body (Figure 1.2). THE may be induced by electroshock or high doses of chemical convulsants that induce tonic-clonic convulsions.[15,28] It is proposed that electroshock or the high-dose chemical convulsants act directly on the neural substrates in the brainstem that mediate tonic-clonic convulsions and THE.[12] Drugs that inhibit THE in rats are also effective against generalized tonic-clonic (grand mal) seizures.[6,8] Animals treated with effective antiepileptic drugs actually still display tonic extension but without the THE component (Figure 1.2). The endpoint for anticonvulsant protection is usually considered to be the failure of the hindlimbs to extend beyond a 90° angle to the torso.[6,29,30] Mice reliably respond to electroshock with THE.[1] Unfortunately, as indicated in Table 1.2, not all Sprague-Dawley rats reliably respond with THE and the failure rate may be as high as 50%.[14,31-35] This is discussed below in Section II.G, although it should be noted that Table 1.2 provides sources of rats that respond reliably with THE. It is recommended that rats be tested (screened) for THE before use in experimental studies involving the THE component of maximal electroshock (MES) and animals that do not respond with THE should not be used.[8,36] In the event that a large percentage of the rats screened do not display THE, the flexion/extension (F/E) ratio may be used to quantify the seizure response. Alternatively, the duration of the tonic extension phase may be measured and a decrease in duration considered a decrease in seizure severity.[35,37,38] These other methods for quantifying the tonic-clonic convulsive response are described next.

c. Flexion/Extension Ratio and Duration of Extension

The flexion-extension (tonic-clonic) convulsions represent a greater electroshock activation of the brainstem mechanisms that mediate generalized tonic-clonic convulsions than occurs when running-bouncing clonus or tonic flexion is induced.[12] At electroshock currents just above threshold for tonic-clonic convulsions, the flexion phase is still prominent. With increasing stimulation current the seizure response becomes more severe as reflected by a decrease in duration of the tonic flexion phase

TABLE 1.2
The Tonic Hindlimb Extension (THE) Response to Electroshock as Reported in Various Rat Strains

Rat strain and source	% with THE	Stimulus parameters	Ref.
Sprague-Dawley albino (male), (source unknown)	90	Transcorneal, 150 mA, 0.2 s	29, 33
Sprague-Dawley albino (male), (source unknown)	100	Transcorneal, 150 mA, 0.2 s	8, 31
Sprague-Dawley albino (male), (source unknown)	50	Transcorneal, 150 mA, 0.2 s	14
Sprague-Dawley albino (male), (source unknown)	90	Transauricular, 150 mA, 0.2 s	14
Sprague-Dawley (male), Zivic-Miller, Allison Park, PA	50	Transcorneal, 50 mA, 0.2 s	35
Wistar (female), Harlan Windelmann, Borchen, F.R.G.	100	Transcorneal and transauricular, 150 mA, 0.2 s	7
Wistar (male), Harlan Sprague-Dawley Inc.	100	Transcorneal, 150 mA, 0.2 s	30
Wistar (male), Charles River, Sulzfeld	100	Transcorneal, 150 mA, 0.2 s	71

Note: The last three listings provide sources for rats that reliably respond with THE. However, it is recommended that all rats be seizure tested (screened) and the nonextenders discarded prior to use.

and an increase in duration of the extension phase. This may be quantified by the flexion/extension or F/E ratio.[1,31,39] The duration of the tonic flexion and tonic extension phases can be measured in a number of ways,[14,31-34,40-43] but typically involve flexion being considered as the period from the electroshock stimulus to the fullest extent of the tonic extension and extension being the period from the fullest extent of the tonic extension until some point in the relaxation of the tonic phase. Whatever convulsion endpoints are chosen for the phases they must be applied consistently. Tonic-clonic convulsions evoked by low currents are less severe, as reflected by a short extension phase or a high F/E ratio. Antiepileptic drugs that inhibit generalized tonic-clonic seizures also reduce the severity of flexion-extension seizures which is observed as an increased F/E ratio as compared to untreated control animals.[1,31,39] For convenience the reciprocal of the F/E ratio, the extension/flexion (E/F) ratio, may be used because it allows the use of numbers greater than unity and thus an *increase* in the E/F ratio is associated with an *increase* in the seizure severity.[41]

It should be reemphasized that the critical standard for quantifying generalized tonic-clonic convulsions is the occurrence of THE. The F/E ratio has not been systematically evaluated as a parameter for screening anticonvulsant drug activity, since it was originally shown to correlate with the effectiveness of phenytoin and phenobarbital.[26] Although a reduced duration of the tonic extension phase is corre-lated with anticonvulsant activity,[35,37,44] the F/E ratio has not been systematically tested against all clinically effective antiepileptic drugs. Perhaps future studies might evaluate these convulsive parameters in addition to THE and determine their relative utility in detecting anticonvulsant drug activity.

d.   Seizure Spread

It has been hypothesized that tonic-clonic convulsions represent the spread of seizure activity through the brain and drugs that inhibit tonic-clonic convulsions inhibit seizure spread.[15,45] The effectiveness of antiepileptic drugs is reflected by a change in seizure pattern, such as an increase in the duration of tonic flexion and a decrease in tonic extension or as an inhibition of THE.[8,15,26,31] Therefore, drugs that alter the threshold or maximal electroshock seizure pattern are proposed to inhibit seizure spread.[1,45] This concept was originally based on the theory of an "oscillator" or a minimal collection of neurons that must discharge in an epileptiform manner to produce the seizure.[45] The theoretical "oscillator" projects or spreads the aberrant epileptiform neuronal activity to other parts of the brain, resulting in the observed convulsions.[45] The maximal tonic-clonic convulsions involve spread of seizure activity to the entire brain, resulting in the tonic flexion and tonic extension response.[45] The greater the stimulating current, the greater the seizure spread, which results in a shorter period of tonic flexion and a greater period of tonic extension.[31] Antiepileptic drugs are proposed to suppress the seizure spread and thereby alter the seizure patterns.[1] The inhibition of THE is the alteration of seizure pattern that is most commonly used to evaluate anticonvulsant drug activity in electroshock[7,8] and is thought to represent a reduction in seizure spread.[15,26,31] The increased F/E ratio is also a change in seizure pattern that is interpreted as anticonvulsant drug activity,[1,26,31] as is the reduction in duration of tonic extension phase that is induced by anticonvulsant drugs.[35,37,38,44]

e.   Seizure Threshold

Tonic-clonic convulsions induced by low corneal stimulation currents represent a convulsive threshold; in other words, the minimal current required to initiate the flexion-extension convulsion in 97% to 100% of the animals.[15] For this reason these low current or submaximal threshold seizures are known as the threshold tonic extension (TTE) test,[6] maximal threshold seizures,[15] or threshold for maximal electroshock seizures (MEST) (Table 1.1).[7] The quantal all or none response of THE or minimal clonic seizures can be used to establish the current threshold for tonic-clonic convulsions,[1,7,15] either by the Litchfield and Wilcoxon[19] procedure or the "up-down" or "staircase" method.[1,20-22] Although threshold currents are proposed to activate the "oscillator" sufficiently to initiate seizure activity and the seizure activity spreads to other areas of the brain, the spread is considered minimal compared to that induced by higher (supramaximal) currents.[15,26] Although considered minimal, the seizure spread is still sufficient to induce a tonic-clonic convulsion (Table 1.1).[15,26] Thus, the maximal threshold seizures or MEST represent seizure mechanisms associated with both the threshold for seizure induction and the spread of the seizure activity to other brain areas.[15] Further, MEST convulsions respond to antiepileptic drugs that raise seizure threshold, inhibit seizure spread, or both.[15] In a classic experiment, Piredda et al.[15] demonstrated this relationship by showing that ethosuximide (which increases convulsive thresholds), phenobarbital (which inhibits seizure spread), and valproate (which increases threshold and inhibits spread) all are effective in MEST. Those authors concluded that MEST is capable of detecting anticonvulsant activity, but is nonspecific in differentiating spread and threshold mechanisms of action. However, Löscher et al.[7] have shown that MEST is capable of detecting specific anticonvulsant drug activity where electroshock stimulation using higher currents (maximal electroshock) does not. In studies with mice, Löscher demonstrated that both primidone and clonazepam are effective anticonvulsants in MEST but not maximal electroshock.[7] The Anticonvulsant Screening Project of the Epilepsy Branch of the National Institute of Neurological Disorders and Stroke has used MEST to identify compounds that are effective only in MEST convulsions; however, the clinical utility of such agents is debatable.[8]

5. Maximal Electroshock

The maximal tonic-clonic convulsion is the most severe electroshock convulsive response and it is induced by supramaximal electrical stimulation currents,[15] more commonly known as maximal electroshock (MES) (Table 1.1). MES is typically induced in rats and mice using supramaximal corneal electroshock currents of 150 mA and 50 mA, respectively.[1,8] Such currents are approximately five to seven times the threshold current and are considered supramaximal or suprathreshold. At the supramaximal currents the tonic-clonic seizures are more severe than those induced by threshold currents in that the tonic flexion phase is reduced in duration while the tonic extension phase is increased.[31] This change in phase duration is reflected in a minimal F/E ratio (Table 1.1). In a rat population that does not uniformly respond to MES with THE, the percentage of animals responding with THE increases as the current is increased from threshold to supramaximal.[31]

The most common endpoint for anticonvulsant drug activity in MES is the inhibition of THE (Figure 1.2).[6-8] Drugs that inhibit the THE component of MES are effective antiepileptics against generalized tonic-clonic (grand mal) seizures.[6-8] Due to the suprathreshold current used to induce MES it is not possible to determine if the drugs raise convulsive threshold,[15] but drugs that inhibit THE do alter the seizure pattern and therefore are proposed to reduce the spread of seizure activity.[15,26,45] Antiepileptic drugs also reduce the severity of seizures, as observed by an increase in the duration of tonic flexion and a decrease in the duration of tonic extension resulting in an increased F/E ratio.[26,31,39] The reduction in the duration of the tonic extension phase has been used as a measure of anticonvulsant drug activity.[35,37,38,44]

Electroshock stimulation currents greater than supramaximal (150 mA in rats, 50 mA in mice) do not produce significantly greater convulsive responses. Currents up to 10 times suprathreshold result in no significant change in the F/E ratio, but may increase the incidence of hindlimb paralysis after the seizure.[31] Hindlimb paralysis is the result of vertebral fractures that occur during the course of the tonic-clonic convulsions. The animals do not recover from the paralysis and it is recommended that the animal be euthanized as quickly as possible after evidence of a vertebral fracture is recognized.

C. Corneal vs. Transauricular Electroshock

The preceding section dealt with electroshock convulsions induced by corneal (or transcorneal) electroshock, that is, the placement of saline-soaked electrodes over the eyes. When electroshock techniques were initially developed, concave corneal cups were attached to the end of the electrodes and the cornea was stimulated directly because this method offered the least resistance to the electrical current.[26,46] The corneal cups have been largely replaced by saline soaked, cotton-tipped electrodes that are designed to provide snug placement over the eye sockets. Electroshock convulsions can also be induced by transauricular stimulation involving the passage of electrical current from ear to ear. The transauricular stimulation may be induced using toothless alligator clips that are connected to the stimulator via flexible wires (Figure 1.3).[14] The clamps of the alligator clips should be bent such that they maintain

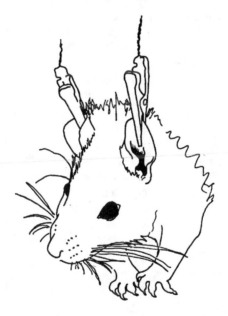

FIGURE 1.3

Diagram of electrode placement for the administration of transauricular electroshock. Toothless alligator clips are attached to the ears and the animal is freely moving at the time of the stimulus. The clamps of the alligator clips should be adjusted by bending such that contact with the ear is maintained without significant discomfort to the animal. The stimulus conducting wires should be flexible and lightweight such that they do not impede movement.

contact with the ear but are not an obvious irritation to the animal as manifested by vocalization or immediate attempts to remove the clips. Felt pads may be attached to the clamps to improve comfort to the animal. The felt pads should be soaked in saline to improve conductivity. Since the investigator must directly handle the conductive alligator clip electrodes to attach them to the animal, it is recommended that nonconducting rubber gloves be worn to prevent accidental shock to the investigator.

Wiping the ears with an ethanol solution will remove dirt and oils and improve conductivity. Once attached to the clips, and therefore the stimulator, the animal may be placed in an observation cage. The observation cage should be placed near the stimulator in order that the electrical leads connecting the animal to the stimulator are kept short and the electrical resistance induced by the wire kept to a minimum. The animal is freely moving at the time of transauricular stimulation, in contrast to corneal electroshock in which the animal is lightly restrained by a gloved hand.

As with corneal electroshock, transauricular stimulation induces running-bouncing clonus, tonic flexion, and tonic-clonic convulsions (both threshold and maximal tonic-clonic) by a current-dependent electrical activation of the brainstem nuclei that mediate generalized tonic-clonic convulsions.[12,14] In contrast to corneal stimulation, transauricular stimulation does not induce the minimal clonic seizures that involve face and forelimb clonus.[12,14] Transauricular stimulation has been hypothesized *not* to activate the forebrain mechanisms that mediate the face and forelimb clonus or clonic spasms, as explained in greater detail in Section III of this chapter.

Transauricular stimulation produces electroshock convulsions at lower, more consistent thresholds than corneal stimulation.[7,14] The threshold for tonic-clonic convulsions in mice is reported to be 4.0 mA with transauricular stimulation vs. 7.98 mA for corneal stimulation.[7] Similarly the transauricular threshold for tonic-clonic convulsions is lower in rats, being reported as 17.4 vs. 22.9 mA in one study[7] and 30 vs. 40 mA in another.[14] Transauricular stimulation induces more severe convulsions, as the incidence of hindlimb paralysis[7,14] and of animals exhibiting THE (in those populations that do not uniformly respond with THE)[14] are greater than with corneal electroshock. It has been hypothesized that transauricular convulsions may be more difficult to control with antiepileptic drugs.[14] In that regard, phenobarbital and primidone have been shown to be more potent in mouse MEST induced by corneal stimulation than by transauricular stimulation.[7] These data indicate that the less severe corneal electroshock stimulus may detect anticonvulsant activity in experimental drugs more readily than transauricular stimulation.[7] In addition, Löscher et al.[7] have suggested that because the electrical stimulus must pass from the eyes to the brainstem, corneal electroshock may be a better representative of seizure spread mechanisms than transauricular stimulation.

D. Seizure Repetition

Multiple threshold or maximal electroshock stimuli on the same day reduce the tonic-clonic convulsion response.[26] Rats will respond to a second electroshock stimulus with tonic-clonic convulsions within a matter of 6 to 12 min, but the recovery time is dependent on the initial electroshock stimulus intensity[31] and the convulsion pattern of the second seizure is altered.[26] Recovery of the normal tonic-clonic convulsive response to maximal electroshock is only 90% 2 h after an electroshock stimulus in rats.[26] Readministration of an electroshock stimulus at 10-min intervals in rabbits results in a progressive increase in the stimulus-induced duration of tonic flexion and a decrease in duration of the tonic extension resulting in an increase in the F/E ratio.[26] The change in seizure pattern (increased F/E ratio) induced by the repetitive stimulation is similar to that induced by antiepileptic drugs.[26] In addition to a decreased severity of the convulsion patterns, multiple electroshock stimuli in a single day increase the seizure threshold.[25,47] These data would indicate that anticonvulsant drug tests should not involve more than one electroshock stimulus per day as multiple daily stimulations by themselves will reduce the convulsive response. One maximal electroshock stimulus per day induces no apparent effect on seizure severity,[25] but when testing anticonvulsant drug activity in the same animal it is recommended that 1 week be allowed between seizure tests to allow for complete elimination of the administered drugs between tests.[8]

E.   Kindling of Tonic-Clonic Convulsions

Maximal electroshock stimulation repeated at 3-d intervals induces an intensification of tonic-clonic convulsions in rats, as manifested by an increased duration of hind-limb extension.[25] This is an apparent kindling effect in that the seizure response is progressively enhanced in response to the same convulsive stimulus. It is important to note that the kindled tonic-clonic convulsions differ from the face and forelimb convulsions induced by corneal kindling.[8,25] Corneal kindling refers to limbic seizures resulting from the specific activation of the forebrain[18,48] and induces convulsions similar to those described for amygdala kindling.[17] Kindling of tonic-clonic convulsions presumably is the result of activation of brainstem mechanisms that mediate the tonic-clonic convulsions.[48] Thus, maximal electroshock repeated at 3-d intervals induces a kindling of the generalized tonic-clonic response although the effect is slower to develop and is not as obvious an increase in convulsive response as observed in limbic kindling.

F.   Spinal Cord Convulsions

Frequently overlooked is the fact that the spinal cord contains all the neuronal mechanisms necessary to produce tonic-clonic convulsions independent of supraspinal structures. Experiments performed early in the development of the electroshock technique demonstrated that direct electrical stimulation of the spinal cord in a decapitated rat produces the entire tonic-clonic convulsion sequence of maximal electroshock seizures.[49] The spinal cord is also capable of self-sustained discharge that maintains the tonic-clonic activity after termination of the electrical stimulus.[49] This indicates that the spinal cord is capable of afterdischarge, resulting in a convulsion that is virtually identical to that produced by afterdischarge in supraspinal regions. Because section of the pyramidal tracts does not affect the maximal electroshock response,[50] it would appear that electroshock convulsions are a stereotyped response of the spinal cord to stimulation by nonspecific cerebrospinal tracts.[9,51] In effect, the brain serves as a spinal cord stimulator in the intact animal to activate the convulsive response mediated by the spinal cord and it has been suggested that tonic-clonic convulsions result from generalized, nonspecific supraspinal activation of the spinal cord.[45,49]

Several lines of evidence support the contention that supraspinal structures activate the spinal cord-mediated tonic-clonic convulsions. Doses of phenobarbital or phenytoin that abolish THE in the intact animal have little or no effect on spinal cord convulsions,[49] suggesting that the action of these drugs is not on the spinal cord. Rats that respond to MES with THE emit exactly the same tonic-clonic convulsion (including THE) to spinal cord-induced convulsions as rats that do not respond to MES with THE.[36] This suggests that the difference between the extenders and nonextenders is based on supraspinal mechanisms.

G. Choice of Experimental Subjects

As described previously and indicated in Table 1.2, not all rats respond reliably to either maximal threshold or supramaximal electroshock stimulation with THE.[1,14,31-35] Table 1.2 also provides a listing of sources for rats that respond reliably to electroshock with THE. All rats should be screened for THE response before use and nonextenders are discarded in an approved ethical manner. If THE is not induced in a significant percentage of rats then the F/E ratio[1,31] or the duration of the tonic extension phase[35,37,38] are useful alternatives for quantification of the seizure response. Tonic forelimb extension has been suggested as an alternative for quantifying electroshock response[22] because it occurs reliably;[25] however, this approach has only been used infrequently.[38]

Gender differences should be considered when using rats in electroshock. Females eliminate drugs less rapidly than males, which could be an advantage in anticonvulsant potency testing.[7] Females can also be housed in larger groups, which may be cost effective.[7] However, female rats have been shown to emit enhanced electroshock responses, including longer flexion phases, shorter extension phases, and lower thresholds to electroshock.[52] The estrus cycle also significantly alters the electroshock convulsive response[42] and some investigators only use male rats to avoid such fluctuations in the seizure response. Whatever the choice of gender it should be used consistently throughout a series of experiments.

Mice exhibit THE to electroshock stimuli much more reliably and screening for seizure response probably is not warranted,[1] although vehicle control groups that demonstrate complete THE response in all mice tested should be used. Table 1.3 provides a listing of sources for mice that respond reliably to electroshock with THE.

Numerous factors are known to affect the electroshock seizure response. Animals of the same age, weight, sex, strain, and that are housed in the same environment should be used in any given series of experiments to eliminate biological variability. Other considerations include diet, hydration, temperature, blood gases, endocrine state, acid-base balance, stress, blood glucose, diurnal and circadian factors. To achieve reproducible results it is necessary to control as many of these factors as possible. The reader is referred to the excellent reviews by Swinyard,[1] Woodbury,[53] Maynert,[54] and Browning[55] for further details.

TABLE 1.3
Commercial Sources of Mice that Respond Reliably (100%) to Threshold or Maximal Electroshock with Tonic Hindlimb Extension (THE)

Mouse strain and source	Ref.
CF No.1 albino (male), Charles River, Wilmington, MA	8
NMRI (male), Winkelmann, Versuchstierzucht GmbH, Borchen, F.R.G.	7
CFW-1 (male), Winkelmann, Versuchstierzucht GmbH, Borchen, F.R.G.	7
Swiss (male), Mus Rattus, Brunnthal, F.R.G.	7
C57BL/6J (male), Jackson Laboratory, Bar Harbor, ME	7
CDI (male), Charles River, Wilmington, MA	7

Note: Although seizure testing (screening) is not considered necessary in mice,[1] vehicle control groups demonstrating 100% THE in drug naïve mice are recommended.

H. Stimulators

Most electroshock convulsions are induced using stimulators based on the design originally described by Woodbury and Davenport.[56] This involves passing an alternating current of 50 to 60 Hz through a large resistance in series with the animal such that the approximately 5 kΩ resistance of the animal causes an insignificant change in the current delivered.[56,57] This ensures that the intended current is reliably delivered to the subject. For many years the standards of electroshock stimulators were those produced by Wahlquist (Wahlquist Instrument Co., Salt Lake City, UT) although such stimulators are no longer produced. The only commercial electroshock stimulators that are currently available that the author is aware of are those produced by Ugo Basil and distributed by Stoelting (Wood Dale, IL). The basic stimulator produced by Ugo Basil produces constant currents up to 99 mA, which is suitable for mice. The stimulators are equipped with transauricular electrodes and a push button to induce the stimulus. Stimulators with currents up to 198 mA and that are appropriate for rats are available on request, as are corneal electrodes and a pedal for stimulus induction.

Dangerous voltages are capable of being induced by electroshock stimulators. Extreme care must be taken to ensure that investigators do not shock themselves when using the equipment, especially when the gardening glove worn on the restraining hand during transcorneal stimulation becomes wet. It is recommended that rubber gloves be worn when handling transauricle electrodes. Assistance from an electronics professional is highly recommended when building a custom constant-current electroshock stimulator or attempting to convert a conventional electrophysiology stimulator to an electroshock stimulator.

III. Interpretation

A. Sites of Seizure Origin

Over the last decade a hypothesis has been developed by Browning[12] suggesting that there are two sites of seizure origin in the rat: the forebrain and the brainstem (Figure 1.4).[12,58,59] Convulsions mediated by the forebrain involve face and forelimb clonus

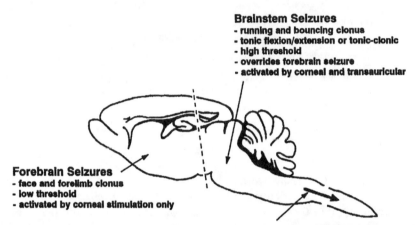

Brainstem Seizures
- running and bouncing clonus
- tonic flexion/extension or tonic-clonic
- high threshold
- overrides forebrain seizure
- activated by corneal and transauricular

Forebrain Seizures
- face and forelimb clonus
- low threshold
- activated by corneal stimulation only

Generalized activation of spinal cord by brainstem structures induces tonic-clonic (flexion/extension) sequence.

FIGURE 1.4

A schematic depiction of the two seizure origin sites in the rat as proposed by Browning.[12] Although the activation indicated is for electroshock, both forebrain and brainstem seizures may be induced by chemical convulsants. Animals with a complete brain transection at the precollicular level, indicated by the dotted line, still respond to corneal electroshock with tonic-clonic convulsions.[62] Similarly, rats with such precollicular transections also respond to systemic chemical convulsant administration with appropriate forebrain electrographic seizures.[58] In seizure naïve animals the two seizure origin sites function independently, but with repeated seizures (kindling) each area begins to influence the seizure response of the other. The brainstem structures activate the spinal cord which contains all the neuronal mechanisms necessary to generate tonic-clonic convulsions.[49]

and may include rearing and falling in the case of kindled limbic seizures.[12] Forebrain convulsions can occur independently of the brainstem as intact electrographic forebrain seizures can be induced after complete forebrain transection.[58] While forebrain convulsions may be induced by low, systemic doses of chemical convulsants or low electrical stimulation currents applied directly to the limbic system through depth electrodes, the most sensitive site for activation is the deep prepyriform cortex or the area tempestas.[59] At low stimulation currents corneal electroshock also activates the forebrain selectively inducing minimal clonic seizures (Table 1.1). Such minimal clonic seizures are considered an experimental model of absence (petit mal) seizures.[6,8,15] Repeated corneal electroshock stimulation with low currents induces corneal kindling which may serve as a model of complex partial seizures[18,55] as does amygdala kindling, as described in Chapter 3. In contrast, transauricular stimulation does not activate the forebrain and does not induce face and forelimb clonus.[14,18,55]

The brainstem is the proposed site of origin for tonic-clonic convulsions.[12] Although direct electrical stimulation of the reticular formation produces tonic-clonic convulsions,[60,61] the most convincing evidence is that rats with complete precollicular brain transections at the level indicated in Figure 1.4 still respond to maximal electroshock stimuli with generalized tonic-clonic convulsions.[62] The pontine reticular formation is the site considered most likely to generate the tonic-clonic convulsions[12,59] and the nucleus reticularis pontis oralis is proposed to be essential for the occurrence of THE as induced by electroshock, audiogenic seizures, or pentylenetetrazol.[40,48,63-65] High doses of chemical convulsants or high currents of electrical stimulation are proposed to activate directly the neuronal substrates in the brainstem that mediate tonic-clonic convulsions.[12,62] The wild running and generalized tonic-clonic convulsions associated with audiogenic seizures in genetically epilepsy-prone rats are also proposed to involve only the brainstem site of seizure origin.[12] Repeated maximal electroshock at 3-d intervals using high currents induces a kindling effect in that the tonic-clonic convulsion response becomes increasingly severe.[25]

The forebrain and brainstem possess separate mechanisms for seizure origin with separate thresholds for seizure initiation. Either seizure origin site may generate seizures in the absence of the other site.[58,62] Although the brainstem threshold is higher, once initiated it overrides forebrain seizures and only the tonic-clonic convulsions are observed (Figure 1.4).[12] Initially, in seizure-naïve, normal experimental animals the forebrain and brainstem mechanisms function separately. However, with repeated seizures the separation of the two seizure origin sites erodes. For example, in kindled amygdala seizures the incidence of THE is significantly enhanced in comparison to nonkindled rats.[65] Genetically epilepsy-prone rats given repeated audiogenic seizures that exhibit generalized tonic-clonic convulsions solely by way of brainstem mechanisms eventually also exhibit face and forelimb clonus as mediated by forebrain mechanisms.[66] When subjected to amygdala kindling, genetically epilepsy-prone rats develop spontaneous generalized tonic-clonic convulsions.[67] These studies indicate that the interaction between forebrain and brainstem seizure sites are facilitated in genetically epilepsy-prone rats. Thus, not only do both seizure origin sites demonstrate kindling,[8,16,25] but the kindling influences seizure activity mediated by the other site.[65-67]

Once activated by an electroshock stimulus the brainstem neural substrate that mediates generalized tonic-clonic convulsions is proposed to induce a generalized activation of the spinal cord (Figure 1.4).[45,49] Although all of the neuronal mechanisms necessary to emit tonic-clonic convulsions are in the spinal cord it is the supraspinal brainstem neural substrate that activates the spinal cord[51] and it would appear that supraspinal brainstem nuclei and not the spinal cord are the site of action of antiepileptic drugs.[33,49] The hypothesis of two separate origins of seizure activity indicates that the antiepileptic drugs act on the individual seizure origin sites,[55] rather than the spread of seizure activity from the "oscillator" as originally proposed.[26,31] This raises the interesting possibility that the site of action of drugs in tonic-clonic convulsions is the brainstem.[68]

B.   Evaluation of Anticonvulsant Drug Activity

Threshold or maximal electroshock are widely accepted techniques to screen compounds for anticonvulsant activity and the endpoint most commonly used to determine the effectiveness of anticonvulsant compounds is the abolition of tonic hindlimb extension (THE).[6-8] All clinically effective antiepileptic drugs active against generalized tonic-clonic (grand mal) seizures inhibit the THE component of threshold electroshock and most inhibit the THE component of maximal electroshock (Table 1.4).[6-8] As indicated in Table 1.4, the clinically effective antiepileptics phenobarbital,

TABLE 1.4
Relative Anticonvulsant Activities of Clinically Effective Antiepileptic Drugs in Threshold and Maximal Electroshock Models of Epilepsy in Mice and Rats

	Mice MEST	PI	Mice MES	PI	Rats MES	PI
Phenobarbital	++	20–21	+	3–4	+	2–3
Phenytoin	++	9–10	+	4–7	+	11–22
Carbamazepine	++	23	+	4–9	+	5–100
Valproate	++	5–6	+	2	+	2
Primidone	++	13	–	–	–	–
Diazepam	++	1–2	+	<1	++	<1
Clonazepam	+	<1	–/+	<1	–/+	<1
Ethosuximide	–	–	–	–	–	–

Note: Anticonvulsant effects are based on the abolition of THE: ++ indicates high potency; + indicates moderate potency; and – indicates no anticonvulsant activity. Protective index (PI) is defined as the TD_{50}/ED_{50} (see text). High PI values indicate minimal drug-induced neurological deficit at doses that inhibit THE. The antiepileptic drugs effective in generalized tonic-clonic seizures have the greatest potency and lowest incidence of adverse neurological effects (highest PI) when tested in threshold electroshock or threshold for maximal seizures (MEST) in mice. Primidone induces effective anticonvulsant activity only when tested in MEST.[69] The benzodiazepines have anticonvulsant activity only at doses that induce neurological deficit in all models. Ethosuximide is effective only in absence seizures and has no activity in the rodent electroshock models. Relative potency and PI values are calculated from the reviews by Löscher et al.,[7] Krall et al.,[6] White et al.,[8] Löscher and Nolting,[69] and Löscher and Schmidt.[22]

phenytoin, carbamazepine, and valproate all inhibit the THE component of MES in both rats and mice.[6-8,22] However, these drugs are most potent against the THE component of threshold electroshock (MEST) seizures in mice (Table 1.4).[6-8] The lower potency might indicate that MEST is a preferred method for screening anticonvulsant activity, but it should be remembered that MEST does not differentiate between drugs that raise threshold and those that inhibit seizure spread.[15] Although primidone is reported to be active only in MEST and this has been offered as evidence that MEST detects anticonvulsant activity that might be missed by MES,[7] the clinical value of compounds detected by MEST has been questioned.[8]

Other evidence also supports the role of threshold or maximal electroshock as preclinical models of generalized tonic-clonic epilepsy. Diazepam and clonazepam are effective against threshold or maximal electroshock only in doses that induce neurological deficit, which indicates that the benzodiazepines are not selective anticonvulsants in those models of epilepsy. Selective anticonvulsant activity indicates that a drug inhibits seizure activity without producing adverse neurological effects and is determined by the protective index (PI) (Table 1.4).[8,69] The PI is defined as the TD_{50}/ED_{50}, where the TD_{50} is the toxic dose (TD) that induces adverse effects in 50% of the animals, while the ED_{50} is the effective dose (ED) that protects 50% of the animals from the stimulus-induced convulsive response. Determination of anticonvulsant neurological deficit by evaluation of PI is described in Chapter 8, but in essence the higher the PI the more selective the anticonvulsant drug action and the lower the incidence of expected adverse clinical effects. Both diazepam and clonazepam have a PI of one or less (Table 1.4), indicating anticonvulsant activity occurs only in association with adverse neurological effects. Neither drug is useful in the treatment of generalized tonic-clonic seizures,[70] further validating the use of electroshock and the PI in detecting drugs effective in that variant of epilepsy. Ethosuximide has selective activity against absence (petit mal) seizures, but has no anticonvulsant activity in either threshold or maximal electroshock (Table 1.4). This observation is also cited as evidence that threshold and maximal electroshock identify drugs that are active in generalized tonic-clonic seizures.[6-8,15]

The potency and PI values for anticonvulsants determined by MES in rats are equivalent to those determined in mice (Table 1.4).[8,22] Anticonvulsant activity against threshold electroshock (MEST) in rats is rarely determined and such data are not included in Table 1.4. As with mice, MES in rats predicts activity of phenobarbital, phenytoin, carbamazepine, and valproate against generalized tonic-clonic seizures.[6-8] Primidone is reported to have little or no activity in MES in rats.[7,69] The benzodiazepines are effective against MES in rats only at neurotoxic doses and ethosuximide is ineffective.[6-8,15] Further discussions of the relationships between electroshock models of epilepsy in rodents and the clinical efficacy of anticonvulsant drugs may be found in reviews by Krall et al.,[6] Löscher et al.,[7] and White et al.[8]

The intense reliance on electroshock for the screening of anticonvulsant drugs has been questioned, the continued argument being that use of electroshock may select for compounds that are not of any greater efficacy than those compounds currently in use. This is especially true with the development of mechanism-based models, such as the genetic knockout technologies that may detect specific alterations in gene expression which produce epilepsy. While it is reasonable to expect that other experimental models of epilepsy will be developed that detect novel antiepileptic drugs, it also should be expected that electroshock will remain the standard against which all future antiepileptic drugs will be compared. Further, the obvious correspondence between the electroshock-induced tonic-clonic convulsions and the generalized tonic-clonic convulsions observed in epileptics suggests common neural mechanisms. Electroshock will continue to serve as a model of the neuronal networks that mediate generalized tonic-clonic seizures.

References

1. Swinyard, E. A., Electrically induced convulsions, in *Experimental Models of Epilepsy,* Purpura, D. P., Penry, J. K., Tower, D., Woodbury, D. M., and Walter, R., Eds., Raven Press, New York, 1972, 431.
2. Putnam, T. J. and Merritt, H. H., Experimental determination of the anticonvulsant properties of some phenyl derivatives, *Science,* 85, 525, 1937.
3. Merritt, H. H. and Putnam, T. J., A new series of anticonvulsant drugs tested by experiments on animals, *Arch. Neurol. Psychiat.,* 39, 1003, 1939.
4. Weiner, R. D., Electroconvulsive Therapy, in *Comprehensive Textbook of Psychiatry,* Vol. 5, Pt. 2, Kaplan, H. I. and Sadock, B. J., Eds., Williams & Wilkins, Baltimore, 1989, chap. 31.7.
5. Dichter, M. A. and Ayala G. F., Cellular mechanisms of epilepsy: a status report, *Science,* 237, 157, 1987.
6. Krall, R. L., Penry, J. K., White, B. G., Kupferberg, H. J., and Swinyard, E. A., Antiepileptic drug development. II. Anticonvulsant drug screening, *Epilepsia,* 19, 409, 1978.
7. Löscher, W., Fassbender, C. P., and Nolting, B., The role of technical, biological and pharmacological factors in the laboratory evaluation of anticonvulsant drugs. II. Maximal electroshock seizure models, *Epilepsy Res.,* 8, 79, 1991.
8. White, H. S., Woodhead, J. H., Franklin, M. R., Swinyard, E. A., and Wolf, H. H., Experimental selection, quantification, and evaluation of antiepileptic drugs, in *Antiepileptic Drugs,* Levy, R. H. and Mattson, R. H., Eds., B. S. Meldrum Raven Press, New York, 1995, 99.
9. Van Der Kooy, D., The reticular core of the brain-stem and its descending pathways: anatomy and function, in *Epilepsy and the Reticular Formation: The Role of the Reticular Core in Convulsive Seizures,* Fromm, G. H., Faingold, C. L., Browning, R. A., and Burnham, W. M., Eds., Alan R. Liss, New York, 1987, 9.
10. Toman, J. E. P., Loewe, S., and Goodman, L. S., Physiology and therapy of convulsive disorders, *Arch. Neurol. Psychiat.,* 58, 312, 1947.
11. Swinyard, E. A., Laboratory evaluation of antiepileptic drugs, *Epilepsia,* 10, 107, 1969.
12. Browning, R. A., Effect of lesions on seizures in experimental animals, in *Epilepsy and the Reticular Formation: The Role of the Reticular Core in Convulsive Seizures,* Fromm, G. H., Faingold, C. L., Browning, R. A., and Burnham, W. M., Eds., Alan R. Liss, New York, 1987, 137.
13. Jobe, P. C., Lasley, S. M., Burger, R. L., Bettendorf, A. F., Mishra, P. K., and Dailey, J. W., Absence of an effect of aspartame on seizures induced by electroshock in epileptic and non-epileptic rats, *Amino Acids,* 3, 155, 1992.
14. Browning, R. A. and Nelson, D. K., Variation in threshold and pattern of electroshock-induced seizures in rats depending on site of stimulation, *Life Sci.,* 37, 2205, 1985.
15. Piredda, S. G., Woodhead, J. H., and Swinyard, E. A., Effect of stimulus intensity on the profile of anticonvulsant activity of phenytoin, ethosuximide and valproate, *J. Pharmacol. Exp. Ther.,* 232, 741, 1985.
16. Sangdee, P., Turkanis, S. A., and Karler, R., Kindling-like effect induced by repeated corneal electroshock in mice, *Epilepsia,* 23, 471, 1982.

17. Racine, R. J., Modification of seizure activity by electrical stimulation. II. Motor seizure, *Electroenceph. Clin. Neurophysiol.,* 32, 281, 1972.
18. Swinyard, E. A., Wolf, H. H., White, H. S., Skeen, G. A., Stark, L. G., Albertson, T., Pong, S. F., and Drust, E. G., Characterization of the anticonvulsant properties of F-721, *Epilepsy Res.,* 15, 35, 1993.
19. Litchfield, J. T. and Wilcoxon, F., A simplified method of evaluating dose-effect experiments, *J. Pharmacol. Exp. Ther.,* 96, 99, 1949.
20. Finney, D.J., *Probit Analysis,* Cambridge University Press, Cambridge, 1952.
21. Kimball, A.W., Burnett, W.T., and Doherty, D.G., Chemical protection against ionizing radiation. I. Sampling methods for screening compounds in radiation protection studies with mice, *Radiat. Res.,* 7, 1, 1957.
22. Löscher, W. and Schmidt, D., Which animal models should be used in the search for new antiepileptic drugs? A proposal based on experimental and clinical considerations, *Epilepsy Res.,* 2, 145, 1988.
23. Dailey, J. W. and Jobe, P. C., Anticonvulsant drugs and the genetically epilepsy-prone rat, *Fed. Proc.,* 44, 2640, 1985.
24. McCown, T. J., Greenwood, R. S., Frye, G. D., and Breese, G. R., Electrically elicited seizures from inferior colliculus: a potential site for the genesis of epilepsy, *Exp. Neurol.,* 86, 527, 1984.
25. Ramer, D. and Pinel, J. P. J., Progressive intensification of motor seizures produced by periodic electroconvulsive shock, *Exp. Neurol.,* 51, 421, 1976.
26. Toman, J. E. P., Swinyard, E. A., and Goodman, L. S., Properties of maximal seizures and their alternation by anticonvulsant drugs and other agents, *J. Neurophysiol.,* 9, 231, 1946.
27. Esplin, D. W. and Woodbury, D. M., Spinal reflexes and seizure patterns in the two-toed sloth, *Science,* 133, 1426, 1960.
28. Goodman, L. S., Grewal, M. S., Brown, W. C., and Swinyard, E. A., Comparison of maximal seizures evoked by pentylenetetrazol (metrazol) and electroshock in mice, and their modification by anticonvulsants, *J. Pharmacol. Exp.Ther.,* 108, 168, 1953.
29. Swinyard, E. A., Brown, W. C., and Goodman, L. S., Comparative assays of antiepileptic drugs in mice and rats, *J. Pharmacol. Exp. Ther.,* 106, 319, 1952.
30. Peterson, S. L., Localization of an anatomical substrate for the anticonvulsant activity induced by D-cycloserine, *Epilepsia,* 35, 933, 1994.
31. Laffan, R. J., Swinyard, E. A., and Goodman, L. S., Stimulus intensity, maximal electroshock seizures, and potency of anticonvulsants in rats, *Arch. Int. Pharmacodyn.,* 1, 60, 1957.
32. Buterbaugh, G. G., A role for serotonergic systems in the pattern and intensity of the convulsive response of rats to electroshock, *Neuropharmacology,* 16, 707, 1977.
33. Buterbaugh, G. G., Effect of drugs modifying central serotonergic function on the response of extensor and nonextensor rats to maximal electroshock, *Life Sci.,* 23, 2393, 1978.
34. Novack, G. D., Stark, L. G., and Peterson, S. L., Anticonvulsant effects of benzhydryl piperazines on maximal electroshock seizures in rats, *J. Pharmacol. Exp. Ther.,* 208, 480, 1978.
35. Berman, E. F. and Adler, M. W., The anticonvulsant effect of opioids and opioid peptides against maximal electroshock seizures in rats, *Neuropharmacology,* 23, 367, 1984.

36. Zablocka, B. and Esplin, D. W., Role of seizure spread in determining maximal convulsion pattern in rats, *Arch. Int. Pharmacodyn.,* 147, 525, 1964.

37. Iadarola, M. J. and Gale, K., Substantia nigra: site of anticonvulsant activity mediated by gamma-amino butyric acid, *Science,* 218, 1237, 1982.

38. Tortella, F. C., Ferkany, J. W., and Pontecorvo, M. J., Anticonvulsant effects of dextrorphan in rats: possible involvement in dextromethorphan-induced seizure protection, *Life Sci.,* 42, 2509, 1988.

39. Tedeschi, D. H., Swinyard, E. A., and Goodman, L. S., Effects of variations in stimulus intensity on maximal electroshock seizure pattern, recovery time, and anticonvulsant potency of phenobarbital in mice, *J. Pharmacol. Exp. Tech.,* 116, 107, 1956.

40. Browning, R. A., Simonton, R. L., and Turner, F. J., Antagonism of experimentally induced tonic seizures following a lesion of the midbrain tegmentum, *Epilepsia,* 22, 595, 1981.

41. Peterson, S. L., Trezciakowski, J. T., Frye, G. D., and Adams, H. R., Glycine potentiation of anticonvulsant drugs in maximal electroshock seizures in rats, *Neuropharmacology,* 29, 399, 1990.

42. Woolley, D. E. and Timiras, P. S., Estrous and circadian periodicity and electroshock convulsions in rats, *Am. J. Physiol.,* 202, 379, 1962.

43. Woolley, D. E. and Timiras, P. S., Gonad-brain relationship: effects of castration and testosterone on electroshock convulsions in male rats, *Endocrinology,* 71, 609, 1962.

44. McNamara, J. O., Russell, R. D., Rigsbee, L., and Bonhaus, D.W., Anticonvulsant and antiepileptogenic actions of MK-801 in the kindling and electroshock models, *Neuropharmacology,* 27, 563, 1988.

45. Woodbury, D. M. and Esplin, D. W., Neuropharmacology and neurochemistry of anticonvulsant drugs, *Res. Publ. Assoc. Nerv. Ment. Dis.,* 37, 24, 1959.

46. Speigel, E. A., Quantitative determination of the convulsive reactivity by electric stimulation of the brain with the skull intact, *J. Lab. Clin. Med.,* 22, 1274, 1937.

47. Essig, C. F. and Flanary, H. G., The importance of the convulsion in occurrence and rate of development of electroconvulsive threshold elevation, *Exp. Neurol.,* 14, 448, 1966.

48. Browning, R. A., Role of the brain-stem reticular formation in tonic-clonic seizures: lesion and pharmacological studies, *Fed. Proc.,* 44, 2425, 1985.

49. Esplin, D. W. and Freston, J. W., Physiological and pharmacological analysis of spinal cord convulsions, *J. Pharmacol. Exp. Ther.,* 130, 68, 1960.

50. Ninchoji, T., Burnham, W. M., and Livingston, K. E., Effect of lesions on cortical-generalized seizures in the kindled rat: spinal transections, *Exp. Neurol.,* 73, 642, 1981.

51. Burnham, W. M. and Browning, R. A., The reticular core and generalized convulsions: a unified hypothesis, in *Epilepsy and the Reticular Formation: The Role of the Reticular Core in Convulsive Seizures,* Fromm, G. H., Faingold, C. L., Browning, R. A., and Burnham, W. M., Eds., Alan R. Liss, New York, 1987, 193.

52. Woolley, D. E., Timiras, P. S., Rosenzweig, M. R., Krech, D., and Bennett E. L., Sex and strain differences in electroshock convulsions of the rat, *Nature,* 190, 515, 1961.

53. Woodbury, D. M., Role of pharmacological factors in the evaluation of anticonvulsant drugs, *Epilepsia,* 10, 121, 1969.

54. Maynert, E. W., The role of biochemical and neurohumoral factors in the laboratory evaluation of antiepileptic drugs, *Epilepsia,* 10, 145, 1969.

55. Browning, R. A., The electroshock-model, neuronal networks, and antiepileptic drugs, in *Drugs for Control of Epilepsy,* Faingold, C. L. and Fromm, G. H., Eds., CRC Press, Boca Raton, FL, 1987, 195.

56. Woodbury, L. A. and Davenport, V. D., Design and use of new electroshock seizure apparatus, and analysis of factors altering seizure threshold and pattern, *Arch. Int. Pharmacodyn.,* 92, 97, 1952.

57. Barany, E. H. and Stein-Jensen, E., The mode of action of anticonvulsant drugs on electrically induced convulsions in the rabbit, *Arch. Int. Pharmacodyn.,* 1-2, 1, 1946.

58. Browning, R., Maggio, R., Sahibzada, N., and Gale, K., Role of brainstem structures initiated from the deep prepiriform cortex of rats, *Epilepsia,* 34, 393, 1993.

59. Gale, K. and Browning, R. A., Anatomical and neurochemical substrates of clonic and tonic seizures, in *Mechanisms of Epileptogenesis,* Dichter, M. A., Eds., Plenum Press, New York, 1988, 111.

60. Kreindler, A., Zuckerman, E., Steriade, M., and Chimion, D., Electrochemical features of convulsions induced by stimulation of brainstem, *J. Neurophysiol.,* 21, 430, 1958.

61. Chiu, P. and Burnham, W. M., The effect of anticonvulsant drugs on convulsions triggered by direct stimulation of the brainstem, *Neuropharmacology,* 21, 355, 1982.

62. Browning, R. A. and Nelson, D. K., Modification of electroshock and pentylenetetrazol seizure patterns in rats after precollicular transection, *Exp. Neurol.,* 93, 546, 1986.

63. Browning, R. A., Turner, F. J., Simonton, R. L., and Bundman, M. C., Effect of midbrain and pontine tegmental lesions on the maximal electroshock seizure pattern in rats, *Epilepsia,* 22, 583, 1981.

64. Browning, R. A., Nelson, D. K., Mogharrenban, N., Jobe, P. C., and Laird H. E., II, Effect of midbrain and pontine tegmental lesions on audiogenic seizures in genetically epilepsy-prone rats, *Epilepsia,* 26, 175, 1985.

65. Applegate, C. D., Samoriski, G. M., and Burchfiel, J. L., Evidence for the interaction of brainstem systems mediating seizures in kindling and electroconvulsive shock seizure models, *Epilepsy Res.,* 10, 142, 1991.

66. Naritoku, D. K., Mecozzi, L. B., Aiello, M. T., and Faingold, C. L., Repetition of audiogenic seizures in genetically epilepsy-prone rats induces cortical epileptiform activity and additional seizure behaviors, *Exp. Neurol.,* 115, 317, 1992.

67. Coffey, L. L., Reith, M. E. A., Cheu, N., Mishra, P. K., and Jobe, P. C., Amygdala kindling of forebrain seizures and the occurrence of brainstem seizures in genetically epilepsy-prone rats, *Epilepsia,* 37, 188, 1996.

68. Fromm, G. H., The brain-stem and seizures: summary and synthesis, in *Epilepsy and the Reticular Formation: The Role of the Reticular Core in Convulsive Seizures,* Fromm, G. H., Faingold, C. L., Browning, R. A., and Burnham, W. M., Eds., Alan R. Liss, New York, 1987, 203.

69. Löscher, W. and Nolting, B., The role of technical, biological and pharmacological factors in the laboratory evaluation of anticonvulsant drugs. IV. Protective indices, *Epilepsy Res.,* 9, 1, 1991.

70. McNamara, J. O., Drugs effective in the treatment of the epilepsies, in *The Pharma-cological Basis of Therapeutics,* Hardman, J. G., Limbird, L. G., Molinoff, P. B., and Ruddun, R. W., Eds., McGraw-Hill, New York, 1996, 461.

71. Rostock, A., Tober, C., Rundfeldt, C., Bartsch, R., Jürgen, E., Polymeropoulos, E. E., Kutscher, B., Löscher, W., Hönack, D., White, H. S., and Wolf, H. H., D-23129: a new anticonvulsant with a broad spectrum of activity in animal models of epileptic seizures, *Epilepsy Res.,* 23, 211, 1996.

Chapter  2

Chemoconvulsants

H. Steve White

Contents

I. Introduction

The use of the chemoconvulsant pentylenetetrazol (PTZ) for the discovery of new antiepileptic drugs (AEDs) began in 1944 when Everett and Richards[1] demonstrated that trimethadione and phenobarbital, but not phenytoin, could block PTZ-induced seizures. A year later, Goodman et al.[2] demonstrated that phenytoin and phenobarbital, but not trimethadione, could block tonic-extension seizures induced by maximal electroshock (MES). The critical link between human seizure disorders and experimental seizure models was made by Lennox[3] when he demonstrated that trimethadione was effective in the treatment of human "petit mal" attacks and was ineffective toward, or worsened, "grand mal" attacks. Today, results obtained from the subcutaneous PTZ (s.c. PTZ) test and the MES test direct the flow of an investigational AED through the testing protocol of the National Institute of Health (NIH)-sponsored Anticonvulsant Screening Project which examines approximately 800 to 1000 investigational AEDs each year.[4]

With minor exceptions, the s.c. PTZ test, when used appropriately, is a reliable predictor of a drug's ability to elevate seizure threshold and its potential activity against human myoclonic jerks and spike-wave seizures. Investigational AEDs are most often tested for their ability to block a clonic seizure induced by a subcutaneous dose of PTZ sufficient to induce a seizure in 97% of the animals tested, otherwise referred to as the convulsive dose 97 (CD97). The s.c. PTZ test provides a quantal evaluation of the ability of a drug to block a defined seizure endpoint. Since it is known that the seizure threshold can vary among various rodent strains, it becomes important that each individual laboratory establish its own CD97 values in the particular mouse or rat strain that they intend to employ.

In addition to the s.c. PTZ test, the intravenous PTZ seizure threshold test (IV PTZ) is useful for assessing the effect of an AED on seizure threshold. The IV PTZ test is an extremely sensitive test that provides a quantitated estimate of individual seizure thresholds within a relatively small population of mice.[5] As such, this test is employed to determine the effect of drug treatment on seizure threshold. In addition, the IV PTZ test can be used to estimate the effect of nonpharmacological treatments on seizure threshold (e.g., lesions, hyperthermia, gene knockouts of a specific receptor, ion channel, or second messenger system, etc.).

The remainder of this chapter focuses primarily on the methods employed for conducting both the s.c. and IV PTZ seizure tests, the anatomical substrate for PTZ-induced seizures, the potential advantages and disadvantages of these models as tools for AED discovery, and factors contributing to the interpretation of results obtained. In addition, a brief comparison between the PTZ test and other animal models of generalized absence seizures will be made. However, it is beyond the scope of the current chapter to provide an in-depth description of these other tests and the reader is referred to excellent reviews by Snead,[6] Löscher et al.,[7] Marescaux et al.,[8] and Marescaux and Vergnes[9] for a more extensive review.

II. Methodology

A. Experimental Animals

The following description of the s.c. and IV PTZ tests is based for the most part on the current methodology of the Anticonvulsant Screening Project (ASP) at the University of Utah in Salt Lake City. For the PTZ tests to be described, the ASP employs adult male Carworth Farms No. 1 (CF1) albino mice (Charles River, Wilmington, MA) and Sprague-Dawley albino rats (Simonsen Laboratories, Gilroy,

CA) weighing between 18–25 g and 100–150 g, respectively. With the exception of the short time that the animals are removed from their cages for testing, all animals are maintained on an adequate diet and allowed free access to food and water. This is an important point, since dehydration and insufficient nutrition can modify seizure threshold. In this regard, animals newly received in the laboratory are permitted at least 4 to 7 d to acclimate to their environment and adjust to any food and water restriction that they may have incurred during transit.

All animals are group housed (8 to 16 mice per cage) in a temperature- and humidity-controlled animal facility accredited by the American Association for Accreditation of Animal Care (AAALAC), and maintained on a 12-hour-on/12-hour-off light schedule. Animals are housed, fed, and handled in a manner consistent with the recommendations set forth in the Department of Health, Education and Welfare (HEW) publication (NIH) No. 7423, "Guide for the Care and Use of Laboratory Animals." Typically, groups of four animals are sufficient for preliminary identification studies to estimate the potential efficacy of a compound in the s.c. PTZ test. Groups of 8 to 10 mice per time point and dose are used for definitive studies to determine the median effective dose (ED_{50}) and 95% confidence interval (see below).

B. Convulsive Response

As mentioned above, the s.c. PTZ test is a pharmacological model that provides a quantal estimation of a drug's ability to block a specific seizure endpoint. As will be discussed below, choosing an inappropriate seizure endpoint can turn a rather discriminating seizure model into a nondiscriminating seizure test. Thus, depending on the dose administered, the seizure phenotype can vary from: (1) single to repeated myoclonic jerks; (2) a brief episode of minimal clonic seizures characterized by clonus of the jaws and/or vibrissae, forelimbs, and/or hindlimbs without loss of righting reflex; (3) clonic forelimb and hindlimb seizures with loss of righting reflex; and (4) tonic extension of forelimbs and hindlimbs.[7] Since the PTZ seizure can be defined by multiple seizure endpoints, it becomes critical that the PTZ dose employed be sufficient to induce only the desired endpoint. For the Anticonvulsant Screening Project, efficacy in the s.c. PTZ test is defined by the ability of an investigational AED to block a minimal clonic seizure of the forelimbs and hindlimbs. As discussed below, the pharmacology associated with abolition of the "minimal clonic seizure" is most predictive of efficacy against generalized absence seizures.

In addition to meeting the appropriate criteria to be considered a pharmacological model of absence, the s.c. PTZ test also meets many of the criteria as defined by Snead[6] for a model of absence seizures. Specifically,

1. PTZ produces an electrographic and behavioral seizure that is consistent with human seizures
2. Seizures produced by PTZ are reproducible and predictable
3. The incidence of seizures can be quantitated
4. Seizures possess a pharmacological profile consistent with the human condition (i.e., blocked by anti-absence AEDs and potentiated by phenytoin and GABAergic drugs)

5. Seizures display a unique developmental profile
6. Seizures evolve from thalamocortical structures

All of these criteria have been reviewed and discussed for a number of chemo-convulsants.[6]

C. The Subcutaneous Pentylenetetrazol Test

Initially, the s.c. PTZ test is conducted in groups of mice (n = 4) at two different time points (0.5 and 4 h) after the intraperitoneal (i.p.) administration of a wide range of doses (e.g., 30, 100, and 300 mg/kg) of the investigational AED. These doses and times provide a preliminary estimate of the efficacy of a drug against s.c. PTZ-induced seizures. Positive results are confirmed in more extensive time- and dose-response studies (see below). All tests are conducted between 0800 and 1400 h.

1. At the time of the test, PTZ (85 mg/kg) is injected s.c. into a loose fold of skin in the midline of the neck. This dose of PTZ represents the average CD97 for CF1 mice in the ASP laboratory. Each laboratory should determine its own CD97 (see below). It is important that the site of injection be massaged in order to distribute the PTZ solution throughout the subcutaneous tissue. If the PTZ solution is injected into adipose tissue and care is not taken to distribute it by massage, the likelihood of false positives is increased. PTZ is administered as a solution of 0.85% in a volume of 0.01 ml/g body weight.

2. Mice are then placed in individual wire-mesh isolation cages (7 cm wide by 10 cm deep by 10 cm high) and observed over the course of the next 30 min for the presence or absence of a minimal clonic seizure of the vibrissae or forelimbs (Figure 2.1). Since a brief episode may persist for only 3 to 5 s, it is imperative that the investigator pay extremely close attention to each animal.

3. Animals not displaying a minimal clonic seizure are considered protected.

It is important to physically isolate individual mice because when aggregated they will likely display a different response than when tested individually.[10] For example, when placed in a group, a seizing mouse may lower the threshold of the remaining mice. Thus, a relatively small drug-induced elevation of seizure threshold could be masked by such an aggregate effect. Likewise, a drug-induced lowering of seizure threshold could be artificially enhanced by an aggregate effect.

In the event that no protection is observed at any of the prescribed doses or times by the investigational AED, no further testing is undertaken. However, in the event that protection is noted at a given dose and time, a more extensive time course (0.25, 0.5, 1, 2, and 4 h) is conducted to determine the correct time of peak effect (TPE) of the test substance. Subsequent testing at the TPE in groups of 8 to 10 mice using multiple doses of the test substance is then performed to quantitate the dose-response relationship of the drug.

FIGURE 2.1

Sketch representing a mouse placed into the middle compartment of a wire-mesh cage for observation. The mouse is shown displaying a minimal clonic seizure of the right forelimb after receiving the CD97 of PTZ (85 mg/kg, s.c.). The minimal clonic seizure depicted is usually very brief in nature, lasting between 3 and 5 s. As such, continued vigilance by the investigator is required to ensure not "missing" a seizure. At higher doses of PTZ, mice and rats will also display tonic extension seizures similar in phenotype to that depicted in Figure 1.2A in Chapter 1. (Sketch kindly provided by Erik Daniel White.)

For the quantification studies, the dose of the test substance is varied until a minimum of three dose levels has been tested, with at least two points lying between the limits of 0% and 100% protection. The medium effective dose (ED_{50}) and 95% confidence interval are then calculated by probit analysis.[11] Although the above description applies to mice, identical studies can be undertaken with rats (n = 8 to 10 rats per group), provided that the appropriate CD97 of PTZ (e.g., 70 to 90 mg/kg) has been employed. The CD97 is estimated by injecting various doses of PTZ into groups of eight mice. Individual mice are observed for the presence or absence of a minimal clonic seizure; the number of mice in each group displaying a seizure is recorded. The CD97 and 95% confidence interval can be calculated by probit analysis. As discussed below, choosing a different seizure endpoint is not without consequences when attempting to interpret the results obtained from the s.c. PTZ test. For example, the pharmacological profile associated with PTZ-induced "minimal" clonic seizures is more discriminatory for drugs active against generalized absence and myoclonic seizures, whereas the profile associated with PTZ-induced tonic extension is nondiscriminatory with respect to clinical classes of AEDs (see discussion below).

D. The Intravenous Pentylenetetrazol Seizure Threshold Test

The IV PTZ test attempts to answer a different question from that asked by the s.c. PTZ test. As described above, the results obtained in the s.c. PTZ test are purely quantal, i.e., the drug either prevented the seizure endpoint or it did not. In contrast, the IV PTZ seizure threshold test provides an extremely sensitive parametric method for assessing seizure threshold.[5,12] In the IV PTZ test, a drug does not have to completely prevent the occurrence of a seizure but only delay its appearance. When compared to the s.c. PTZ test, it has the added advantage that a quantifiable endpoint can be obtained with a minimal number of animals (e.g., 8 to 10 per treatment group). Furthermore, this test can be used to assess both the ability of a drug to modify seizure threshold and the effect of a nonpharmacological manipulation on seizure threshold (hydration, dehydration, hyperthermia, hypoxia, CO_2, hyponotremia, etc.).

The ASP employs the IV PTZ test to determine whether an investigational AED possesses the ability to increase or decrease (proconvulsant) seizure threshold. Two doses of the test compound are usually employed in this test. The first dose corresponds to the ED_{50} for protection against either MES seizures (see Chapter 1) or s.c. PTZ-induced clonic seizures. The second dose employed usually corresponds to the median toxic dose (TD_{50}) or that dose that produces rotorod impairment in 50% of the mice tested. The specific details of the IV PTZ seizure threshold test are as described below.

1. Thirty mice (25 to 30 g) are randomized into three separate groups: vehicle control, MES ED_{50} (or s.c. PTZ ED_{50}), and TD_{50}. One mouse from each group is then injected i.p. with either vehicle or one of the two drug doses; injections are two minutes apart. The same injection process is maintained until all 30 mice have been injected.

2. At the previously determined TPE, a lateral tail vein of a treated mouse is catheterized with a 27 gauge stainless steel needle attached to a length of No. 20 polyethylene (PE) tubing which is secured to the tail by a narrow piece of adhesive tape. The PE tubing (approximately 75 cm in length) is attached to a 10 ml plastic syringe containing the PTZ solution (5 mg/ml PTZ in 0.9% saline containing 10 USP units/ml of heparin sodium) which is mounted into a syringe pump. The PE tubing is clamped with a hemostat to prevent backflow of blood in the PE tubing and to prevent the PTZ solution from leaking into the tail vein. The PTZ solution is subsequently infused into the tail vein of a freely moving mouse at a constant rate of 0.34 ml/min.

3. Immediately after releasing the hemostat, two stopwatches are started simultaneously with the infusion pump. Times (in seconds) from the start of the infusion to the appearance of the "first twitch" (stopwatch no. 1) and to the onset of sustained clonus (stopwatch no. 2) are recorded for each mouse.

4. The times to each endpoint are then converted to mg/kg PTZ for each mouse as follows:

$$\text{mg/kg PTZ} = \frac{\text{infusion time (T)} \times \text{rate of infusion (ml/min)} \times \text{mg PTZ/ml} \times 1000 \text{ g}}{60 \text{ s} \times \text{weight (W) of animal in g}}$$

$$= \frac{T \times 0.34 \times 5.0 \times 1000}{60 \times W}$$

$$= \frac{28.33 \times T}{W}$$

5. The mean and standard error of the mean for each of the three groups and the significance of the difference between the test groups and the control are then calculated.

An increase in the mg/kg dose of PTZ to produce a first twitch or clonic seizure suggests that the test substance possesses the ability to elevate seizure threshold, whereas a decrease is indicative of a potential proconvulsant.

The selection of an appropriate AED dose for the IV PTZ test is dependent on individual needs. In addition to assessing the effect of a reference dose (e.g., ED_{50} or TD_{50}) on PTZ threshold, Löscher et al.[7] have advocated a method that employs several doses in order to estimate the AED dose sufficient to elevate the IV threshold (threshold increasing dose, TID) by 20% (TID_{20}) or 50% (TID_{50}). Regardless of the dose employed, the investigator should be able to relate it to some biological endpoint (e.g., the ED_{50} or TID_{20}).

III. Other Chemoconvulsant Models

A. The Subcutaneous Bicuculline and Picrotoxin Seizure Tests

In addition to PTZ, two other commonly employed chemoconvulsants include biculline and picrotoxin. The seizures that result from the s.c. administration of a CD97 of either chemoconvulsant results in a behavioral seizure that is for the most part indistinguishable from that produced by s.c. PTZ. The CD97 of bicuculline (Bic; 2.70 mg/kg) or picrotoxin (Pic; 3.15 mg/kg) is injected s.c. at the TPE for the test substance. Individual mice are then placed into isolation cages and observed for either 30 min (Bic) or 45 min (Pic) for the presence or absence of minimal clonic seizure. Pic-treated mice are observed for a longer period of time because of the slower absorption associated with Pic. Animals not displaying a clonic seizure within the prescribed time frame are considered protected. The activity of an AED can be quantitated in a similar manner to that described above for s.c. PTZ.

Results obtained from these two tests provide additional information that helps to establish the overall anticonvulsant profile of an investigational AED. Despite the similarity in the seizure phenotype associated with PTZ, Bic, and Pic administration,

TABLE 2.1

**Profile of Anticonvulsant Activity and Minimal Toxicity of
Prototype Anticonvulsants in Mice**

Substance	TD_{50}	TD_{50} or ED_{50} (mg/kg, i.p.) and PI[a]		
		s.c. PTZ	s.c. Bic	s.c. Pic
Carbamazepine[b]	47.8	>50	>60	28.9
	(39.2–59.2)			(23.9–41.6)
				PI 1.7
Clonazepam[b]	0.27	0.017	0.008	0.05
	(0.14–0.43)	(0.012–0.025)	(0.005–0.012)	(0.03–0.07)
		PI 16	PI 34	PI 5.4
Ethosuximide[b]	323	128	365	211
	(279–379)	(101–163)	(284–483)	(170–266)
		PI 2.5	PI 0.9	PI 1.53
Felbamate[c]	816	148	>300	156
	(590–1024)	(121–171)		(122–202)
		PI 5.5		PI 5.2
Gabapentin[c]	>500	47.5	>500	>500
		(17.9–86.2)		
		PI >11		
Lamotrigine[c]	48.0	>60	>50	>50
	(38.7–57.7)			
Phenobarbital[c]	69.2	13.3	37.7	27.5
	(56.1–81.6)	(10.5–18.0)	(26.5–47.4)	(20.9–34.8)
		PI 5.2	PI 1.8	PI 2.5
Phenytoin[b]	42.8	>50	>60	>60
	(36.4–47.5)			
Valproic acid[b]	483	209	437	311
	(412–571)	(176–249)	(369–563)	(203–438)
		PI 2.3	PI 1.1	PI 1.6

Note: Values in parentheses are 95% confidence interval.

[a] Protective index (PI) = TD_{50}/ED_{50}.

[b] Data from White et al.[12]

[c] Unpublished data on file with the ASP, University of Utah.

marked differences between their pharmacological profiles have been observed
(Table 2.1). For example, valproic acid, ethosuximide, and clonazepam are all
effective against clonic seizures induced by all three convulsants. In contrast, carbamazepine is effective against clonic seizures induced by s.c. Pic but not s.c. PTZ
or s.c. Bic, whereas felbamate is effective against s.c. PTZ- and s.c. Pic- but not s.c.
Bic-induced clonic seizures. Furthermore, gabapentin is effective against s.c. PTZ-
but not s.c. Bic- or s.c. Pic-induced seizures. These results clearly demonstrate that
efficacy against one chemoconvulsant does not guarantee activity against another,

even though all three chemoconvulsants are thought to exert their effect through an action at the $GABA_A$ receptor ionophore. Nonetheless, differentiating an investigational AED in all three chemoconvulsant seizure models provides useful information concerning the overall spectrum of activity of a new AED.

B. Gamma-hydroxybutyrate (GHB) Seizures

Since an extensive discussion of the GHB model of absence is beyond the scope of this chapter, the reader is referred to a review by Snead[6] for a more in-depth discussion of this electrographic model of absence. The GABA metabolite GHB occurs naturally in the mammalian brain and when injected into animals produces an electrographic and behavioral seizure that is similar in many respects to that observed in human generalized absence seizures.[6] Gamma-butyrolactone (GBL) is often used as a pro-drug to GHB because it produces a consistent and reproducible 7- to 9-Hz spike-wave electrographic discharge that is accompanied by behavioral arrest, facial myoclonus, and vibrissal twitching similar to the human condition.[6] The pharmacological profile of the GBL model is consistent with other animal models of absence. Thus, seizures induced by GBL are blocked by anti-absence drugs ethosuximide, trimethadione, and valproate. Furthermore, GBL-induced spike-wave seizures are prolonged by phenytoin and drugs that enhance GABAergic tone (e.g., direct GABA agonists, the GABA transaminase inhibitor vigabatrin). The GBL model fulfills all of the criteria outlined by Snead[6] to be considered an experimental and pharmacological model of human absence seizures and should be considered as an excellent *in vivo* animal model for screening potential anti-absence drugs.

IV. Interpretation

A. Anatomical Substrate of Pentylenetetrazol-Induced Seizures

As described above, a sufficiently high dose of PTZ can produce a continuum of seizure activity that progresses from mild myoclonic jerks to clonic seizures of the vibrissae, forelimbs, and hindlimbs without loss of righting reflex, to clonic seizures of the limbs with loss of righting reflex, to full tonic extension of both forelimbs and hindlimbs.[7] As discussed in Chapter 1, a given electrical current administered via corneal electrodes can evoke very similar seizure endpoints. However, unlike PTZ seizures, the seizure phenotype is stimulus dependent and does not represent a continuum from minimal clonic through tonic extension.

Studies by Browning[13] have suggested that seizures originate from two primary brain regions, i.e. the forebrain and the brainstem (see Chapter 1 for discussion and references). These studies suggest that seizures characterized by forelimb clonus originate from forebrain structures such as the deep prepiriform cortex or the area tempestas,[14] whereas tonic-clonic seizures are thought to originate from brainstem structures that include the pontine reticular formation and the nucleus reticularis pontis oralis. Interestingly, forebrain and brainstem seizures can occur independently of each other.[15] For example, precollicular transections in rats prevented both PTZ- and electroshock-induced minimal seizures characterized by forelimb clonus, but did not prevent tonic flexion-extension seizures. Conversely, lesions of the pontis reticular formation attenuated tonic seizures associated with both seizure stimuli but did not prevent the clonic seizure associated with minimal electroshock, PTZ, or the convulsant fluorothyl. Thus, PTZ, depending on the dose administered, can produce both forebrain and brainstem seizures. For the most part, the minimal clonic seizure associated with lower PTZ doses is most likely of forebrain origin. Considering that seizure phenotypes between low-dose PTZ and limbic kindling are similar and that projections emanating from the deep prepiriform cortex innervate other limbic structures, it is not surprising that chronic administration of subconvulsive doses of PTZ can result in a chemical kindling that is indistinguishable behaviorally from that seen with electrical kindling of the amygdala and hippocampus.[16-18]

B.   Evaluation of Anticonvulsant Drug Activity

The established and newer AEDs available today were brought to the clinic on the basis of their *in vivo* anticonvulsant profile in one or more of the animal seizure models discussed in this text. The dependence in the past on animal screening has resulted from the clinical predictability of these animal models. Of the multitude of animal models, the MES and s.c. PTZ tests are perhaps the most widely validated for AED screening.[19] For example, drugs effective against tonic extension seizures induced by MES are likely to be active against generalized tonic-clonic seizures in human patients, whereas drugs effective against threshold PTZ seizures are more likely to be effective against generalized absence seizures. However, the pharmacological profile of the PTZ tests described above would argue against such a generalization. For example, ethosuximide, valproate, the benzodiazepines, and phenobarbital have all been found to be effective in nontoxic doses against minimal clonic seizures induced by s.c. PTZ (Table 2.2). Of these AEDs, only ethosuximide, valproate, and the benzodiazepines possess clinical efficacy against generalized absence seizures. Phenobarbital and other GABAergic compounds, on the other hand, have been associated with a worsening of absence seizures.[6] In contrast, all of these AEDs possess clinical activity against human myoclonic seizures. On this basis, it has been suggested that the s.c. PTZ test is perhaps more predictive of a drug's efficacy against myoclonic vs. absence seizures.[7] Thus, the results obtained from the s.c. PTZ test should be used only as a guide for estimating the potential clinical utility of an investigational AED. Additional studies in other more discriminating absence animal models should be conducted to further differentiate the potential clinical utility of a candidate substance for management of absence seizures.

In contrast to the clonic seizure associated with PTZ administration, the PTZ-induced tonic extension seizure can be blocked by drugs effective against absence[7,20] as well as AEDs effective against MES-induced tonic extension in animals (Table

TABLE 2.2
Comparative Anticonvulsant and Minimal Toxicity Profile of Prototype Anticonvulsants in Mice and Rats

| | TD_{50} or ED_{50} (mg/kg) and PI[a] | | | | | |
| | Mouse, i.p. | | | Rat, p.o. | | |
Substance	TD_{50}	MES	s.c. PTZ	TD_{50}	MES	s.c. PTZ
Carbamazepine[b]	47.8	9.85	>50	361	3.57	>250
	(39.2–59.2)	(8.77–10.7)		(319–402)	(2.41–4.72)	
		PI 4.9			PI 101	
Clonazepam[b]	0.27	23.8	0.017	1.99	2.41	0.77
	(0.14–0.43)	(16.4–31.7)	(0.012–0.025)	(1.71–2.32)	(1.95–2.81)	(0.26–1.52)
		PI 0.01	PI 16		PI 0.8	PI 2.6
Ethosuximide[b]	323	>350	128	>500	>250	204.2
	(279–379)		(101–163)			(160–264)
			PI 2.5			PI >2.5
Phenobarbital[c]	69.2	12.8	13.3	61.1	9.14	11.5
	(56.1–81.6)	(11.1–13.9)	(10.5–18.0)	(43.7–95.8)	(7.58–11.9)	(7.74–15.0)
		PI 5.4	PI 5.2		PI 6.7	PI 5.3
Phenytoin[b]	42.8	6.48	>50	>500	23.2	>250
	(36.4–47.5)	(5.65–7.24)			(21.4–25.4)	
		PI 6.6			PI >22	
Valproic acid[b]	483	287	209	859	395	620
	(412–571)	(237–359)	(176–249)	(719–1,148)	(332–441)	(469–985)
		PI 1.7	PI 2.3		PI 2.2	PI 1.4

Note: Values in parentheses are 95% confidence interval.

[a] Protective index (PI) = TD_{50}/ED_{50}.

[b] Data from White et al.[12]

[c] Unpublished data on file with the ASP, University of Utah.

2.2) and human generalized tonic-clonic seizures (e.g., phenytoin, carbamazepine, and phenobarbital).[7,20] In this respect, PTZ-induced tonic extension has little predictive utility for estimating the clinical potential of an investigational AED.[7]

The above example exemplifies the importance of not only choosing the appropriate seizure endpoint but also providing a complete pharmacological profile of the seizure test before drawing any definitive conclusions regarding the potential clinical utility of a particular AED. The ideal model of epilepsy should predict the clinical utility of a newly identified AED. Unfortunately, no single animal seizure model will, in and of itself, provide a complete assessment of the overall clinical utility of an AED or predict efficacy in a particular patient population. Only when a drug has undergone rigorous clinical testing will its full potential be realized. Nonetheless, animal models are clearly important to the drug discovery process, since no drug is likely to become a clinical candidate without demonstrating adequate activity in one or more animal seizure models.

C. Other Factors

It is important to note that there are a number of factors that can contribute to whether a drug will be efficacious in the s.c. PTZ test that are unrelated to its inherent pharmacodynamic properties.[7] In addition to accurately assessing the TPE of an individual AED, the investigator should take appropriate measures to assure that any apparent lack of efficacy is not the result of inadequate absorption, inability to cross the blood-brain-barrier, and/or species- and sex-dependent differences in drug metabolism and distribution. Inadequate appreciation of any one of these factors can lead to "missing" a novel AED substance or "underestimating" its full potential. A relevant example of species differences is provided by gabapentin; it is effective in mice and ineffective in rats against s.c. PTZ-induced clonus (Table 2.3). Furthermore, vigabatrin was found to be inactive against both MES- and s.c. PTZ-induced seizures when screened at 0.5 and 4 h after i.p. administration to mice. However, when appropriate consideration was given to its proposed mechanism of action (i.e., inhibition of GABA transaminase) and it was tested at its peak effect (24 to 48 h), marked activity against MES was observed. This long time to peak effect presumably reflects the slow, yet irreversible inhibition of GABA transaminase and the subsequent elevation of brain GABA concentrations. The interested reader is referred to a discussion of these and other important factors by Löscher et al.[7]

TABLE 2.3
Comparative Anticonvulsant and Minimal Toxicity Profile of Newer Anticonvulsants in Mice and Rats

| | TD_{50} or ED_{50} (mg/kg) and PI[a] | | | | | |
| | Mouse, i.p. | | | Rat, p.o. | | |
Substance	TD_{50}	MES	s.c. PTZ	TD_{50}	MES	s.c. PTZ
Felbamate	816	50.1	148	>3,000	47.8	238
	(590–1,024)	(35.6–61.7)	(121–171)		(41.0–57.3)	(132–549)
		PI 16	PI 5.5		PI >63	PI >13
Gabapentin	>500	78.2	47.5	52.4	9.13	>100
		(46.6–127)	(17.9–86.2)	(35.2–76.2)	(4.83–14.4)	
		PI >6.4	PI >11		PI 5.7	
Lamotrigine	48.0	7.20	>60	325	3.21	>250
	(38.7–57.7)	(6.10–8.45)		(256–419)	(2.60–3.69)	
		PI 6.7			PI 101	

Note: Values in parentheses are 95% confidence interval.

[a] Protective index (PI) = TD_{50}/ED_{50}. Unpublished data on file with the ASP, University of Utah.

In summary, efficacy in the GBL and s.c. PTZ tests provides a reasonable estimate of the potential efficacy of an investigational AED against generalized absence seizures. However, the genetic rat of Strasbourg[9,21] and the lethargic (lh/lh) mouse[22] represent two genetic nonconvulsive animal models that display electrographic seizures that closely resemble the human condition and display a pharmacological profile consistent with the therapeutic management of absence seizures. As such, they represent excellent alternatives to the chemoconvulsants PTZ and GBL.

References

1. Everett, G. M. and Richards, R. K., Comparative anticonvulsive action of 3,5,5-trimethyloxazolidine-2,4-dione (Tridione), Dilantin and phenobarbital, *J. Pharmacol. Exp. Ther.*, 81, 402, 1944.

2. Goodman, L. S., Swinyard, E. A., and Toman, J. E. P., Laboratory techniques for the identification and evaluation of potentially antiepileptic drugs, *Proc. Am. Fed. Clin. Res.*, 2, 100, 1945.

3. Lennox, W. G., The petit mal epilepsies. Their treatment with Tridione, *JAMA*, 129, 1069, 1945.

4. White, H. S., Wolf, H. H., Woodhead, J. H., and Kupferberg, H. J., The National Institutes of Health Anticonvulsant Drug Development Program: Screening for efficacy, in *Antiepileptic Drug Development. Advances in Neurology*, French, J., Leppik, I., and Dichter, M. A., Eds., Lippincott-Raven Publishers, Philadelphia, 1997, 29.

5. Orlof, M. J., Williams, H. L., and Pfeiffer, C. C., Timed intravenous infusion of Metrazol and strychnine for testing anticonvulsant drugs, *Proc. Soc. Exp. Biol. Med.*, 70, 254, 1949.

6. Snead, O. C., Pharmacological models of generalized absence seizures in rodents, *J. Neural Transm.*, 35, 7, 1992.

7. Löscher, W., Honack, D., Fassbender, C. P., and Nolting, B., The role of technical, biological and pharmacological factors in the laboratory evaluation of anticonvulsant drugs. III. Pentylenetetrazole seizure models, *Epilepsy Res.*, 8, 171, 1991.

8. Marescaux, C., Micheletti, G., Vergnes, M., Depaulis, A., Rumbach, L., and Warter, J. M., A model of chronic spontaneous petit mal-like seizures in the rat: comparison with pentylenetetrazole-induced seizures, *Epilepsia*, 25, 326, 1984.

9. Marescaux, C. and Vergnes, M., Genetic absence epilepsy in rats from Strasbourg (GAERS), *Ital. J. Neurol. Sci.*, 16, 113, 1995.

10. George, D. J. and Wolf, H. H., Dose-lethality curves for d-amphetamine in isolated and aggregated mice, *Life Sci.*, 5, 1583, 1966.

11. Finney, D. J., *Probit Analysis*, Cambridge University Press, London, 1971.

12. White, H. S., Woodhead, J. H., Franklin, M. R., Swinyard, E. A., and Wolf, H. H., General principles: experimental selection, quantification, and evaluation of antiepileptic drugs, in *Antiepileptic Drugs,* 4th ed., Levy, R. H., Mattson, R. H., and Meldrum, B. S., Eds., Raven Press, New York, 1995, 99.

13. Browning, R. A., Effect of lesions on seizures in experimental animals, in *Epilepsy and the Reticular Formation: The Role of the Reticular Core in Convulsive Seizures*, Fromm, G. H., Faingold, C. L., Browning, R. A., and Burnham, W. M., Eds., Alan R. Liss, New York, 1987, 137.

14. Piredda, S., Lim, C. R., and Gale, K., A crucial epileptogenic site in the deep prepiriform cortex, *Nature*, 317, 623, 1985.

15. Browning, R. A. and Nelson, D. K., Modification of electroshock and pentylenetetrazol seizure patterns in rats after precollicular transections, *Exp. Neurol.*, 93, 546, 1986.

16. Ito, T., Hori, M., Yoshida, K., and Shimuzu, M., Effect of anticonvulsants on seizure developing in the course of daily administration of pentetrazol to rats, *Eur. J. Pharmacol.*, 45, 165, 1977.

17. Giorgi, O., Carboni, G., Frau, V., Orlandi, M., Valentini, V., Feldman, A., and Corda, M. G., Anticonvulsant effect of felbamate in the pentylenetetrazole kindling model of epilepsy in the rat, *Naunyn Schmiedebergs Arch. Pharmacol.*, 354, 173, 1996.

18. Rocha, L., Briones, M., Ackermann, R. F., Anton, B., Maidment, N. T., Evans, C. J., and Engel, J. J., Pentylenetetrazol-induced kindling: early involvement of excitatory and inhibitory systems, *Epilepsy Res.*, 26, 105, 1996.

19. White, H. S., Johnson, M., Wolf, H. H., and Kupferberg, H. J., The early identification of anticonvulsant activity: role of the maximal electroshock and subcutaneous pentylenetetrazol seizure models, *Ital. J. Neurol. Sci.*, 16, 73, 1995.

20. Piredda, S. G., Woodhead, J. H., and Swinyard, E. A., Effect of stimulus intensity on the profile of anticonvulsant activity of phenytoin, ethosuximide and valproate, *J. Pharmacol. Exp. Ther.*, 232, 741, 1985.

21. Marescaux, C., Micheletti, G., Vergnes, M., Rumbach, L., and Warter, J. M., Diazepam antagonizes GABAmimetics in rats with spontaneous petit mal-like epilepsy, *Eur. J. Pharmacol.*, 113, 19, 1985.

22. Hosford, D. A. and Wang, Y., Utility of the lethargic (lh/lh) mouse model of absence seizures in predicting the effects of lamotrigine, vigabatrin, tiagabine, gabapentin, and topiramate against human absence seizures, *Epilepsia*, 38, 408, 1997.

Chapter 3

The Kindling Model of Temporal Lobe Epilepsy

Mary Ellen Kelly

Contents

0-8493-3362-8/98/$0.00+$.50
© 1998 by CRC Press LLC

I.   Introduction

Animal models of human epilepsy have provided a wealth of information relevant
to understanding the processes involved in the development and/or maintenance of
this neurological condition. The existence of more than 40 different types of epilepsies and epileptic syndromes (see Beldhuis[1]) has necessitated the development of
diverse approaches with which to study this disease. To study mechanisms of human
partial epilepsy, several animal models have been employed to induce experimentally focal seizure activity that may or may not secondarily generalize. The majority
of these focal seizure models involve the application of chemoconvulsants (e.g.,
kainic acid, quinolinic acid) or specific metals (e.g., aluminum hydroxide, cobalt,
iron, as described in Chapter 7) directly onto or into the brain (for review, see Löscher
and Schmidt[2]). However, both the inability to adequately control the induction of
seizure activity and the difficulty in predicting the type of seizure that might develop
have limited the applicability of these models to the study of focal seizure development and spread.

A serendipitous discovery by Graham Goddard during the 1960s provided
researchers with a model of chronic partial epilepsy devoid of many of the problems
associated with models then in use.[3] During his time as a graduate student, Goddard
observed that some rats exposed to electrical stimulation of the amygdala on a daily
basis for 10 or more consecutive days, eventually exhibited behavioral signs of
seizure activity (see Goddard[4]). Although Goddard was not the first to observe an
increase in seizure disposition following low-intensity stimulation,[5] he was the first
to recognize that this phenomenon, later termed kindling, represented secondary
epileptogenesis and was important enough to warrant further study.[6] During the late
1960s Goddard and colleagues began a systematic investigation of the kindling
process.[7,8] These early studies, in addition to the subsequent reports by Racine,[9,10]
provided researchers with a novel approach to the study of complex partial seizures
with secondary generalization.

Kindling is described as the progressive intensification of both electrographic and behavioral seizures as a result of daily, low-intensity electrical stimulation to a particular forebrain site.[6] Since introduced, hundreds of researchers have utilized the kindling model to advance not only our understanding of mechanisms underlying epileptogenesis, but also those that underlie neural plasticity and memory.

This chapter will provide both a detailed account of the methodological steps involved in the electrical kindling paradigm, including surgical preparation of the animals, and a description of the electrophysiological and behavioral measures one should consider when utilizing this model of epileptogenesis. The issues discussed in this chapter will focus exclusively on kindling in rats. Although this phenomenon has been demonstrated in all species thus far tested, rats are usually the species of choice in most kindling experiments. Factors contributing to the popularity of the rat in kindling include (1) their relative hardiness to surgical procedures and experimental manipulations; (2) their low cost; (3) their wide use within all areas of the neuroscience field; and (4) the availability of excellent rat brain atlases.[11-13]

II. Methodology

A. Animal Preparation and Surgery

The electrical kindling model requires that animals are first surgically implanted with chronic indwelling electrodes. Though exact surgical procedures vary from laboratory to laboratory, there are several aspects of this procedure common to most. In describing the steps involved in preparing an animal for the kindling procedure, the relatively simple and inexpensive system of Molino and McIntyre[14] will be described (see Figure 3.1). Table 3.1 provides a comprehensive list of the necessary supplies and equipment.

1. Issues in Stereotaxic Surgery

Proper placement of the stimulating/recording electrode into a particular brain region is accomplished via standard stereotaxic techniques. As a number of detailed reference books on stereotaxic surgery in the rat exist,[15-18] this chapter will discuss only briefly some critical aspects of stereotaxic technique. Introduced nearly 90 years ago by Horsley and Clark,[19] the premise of stereotaxic surgery is that there is a constant and fixed relationship between the brain and two standard fixation points; the bony external auditory meatuses and the palate. With the head of the animal firmly fixed in position, the location of a particular brain structure can be determined by consulting one of many stereotaxic atlases.[11-13] The atlas used most commonly is that of Paxinos and Watson.[11] Atlases can differ from each other in two major ways: (1) whether bregma or the interaural line is used as the reference point from which to determine the stereotaxic coordinates, and (2) whether the skull surface is positioned flat or at an angle relative to the horizontal plane.

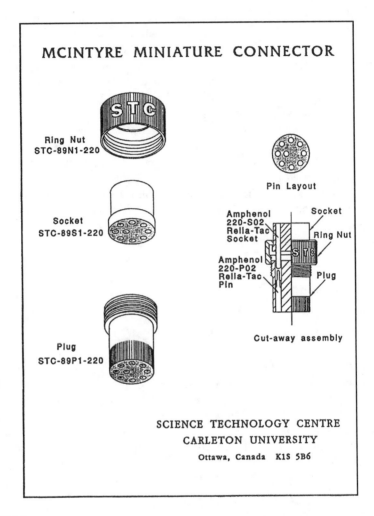

FIGURE 3.1

Schematic of the components comprising the McIntyre headcap apparatus. The plug is attached to the animal, whereas the socket is attached to the set of leads attached to the polygraph. Ordering information is located on the bottom right.

a.   Bregma vs. Interaural Line

Two systems exist for determining the relative position of a structure within the rat brain. The original system utilized the interaural line as the basis of determining the stereotaxic coordinates. The *interaural line* refers to a line through the head of the rat from one auditory meatus to the other. In the second system, *bregma* is used as the major reference. As indicated in Figure 3.2A, bregma is the point of intersection of the coronal skull suture with the midline or saggital suture. Most recent atlases provide coordinates for both systems so that the investigator can use either point of reference. Advantages to using bregma are (1) it is visible during surgery, and (2) greater accuracy is achieved when using rats of different sizes.[17,20] Whishaw et al.[20]

TABLE 3.1
**List of Necessary Supplies for Stereotaxic Implantation of
Stimulating/Recording Electrodes**

Supplies	Suggested suppliers
Large items	
Stereotaxic apparatus: stereotaxic surgical drill	Stoelting, Kopf Instruments
Drill bits (0.6 mm, 1.0 mm)	Stoelting, Small Parts, Inc.
Animal clippers	Stoelting, Harvard Apparatus
Small animal weigh scale	Stoelting, Harvard Apparatus
Hot glass bead dry sterilizer	Stoelting
Syringes and hypodermic needles (26 gauge)	
Depth electrode and ground screw supplies	
Amphenol connector pins (220-S02, 220-P02)	Amphenol Interconnect Products Corp.
Insulated wire (depth electrode and ground screw)	A-M Systems
Solder	
Voltmeter	
Jeweler's screws (approx. 1.59 mm o.d.; 3.2 mm long)	(Stoelting; #51457)
Surgical instruments and supplies	
Scalpel blades and handle	
Forceps	
Small surgical scissors	
4 curved hemostats	
Surgical marker	
Screw driver for jeweler's screws	
Needle nose hemostats (large)	
Electrode headcap assembly	Carleton University, Plastics One
Gauze pads	
Cotton swabs	
Penicillin G (topical)	
Sterile saline	
Sterile eye drops	
70% alcohol	
Petri dish	
Anesthetic	
Dental cement kit	Stoelting, Plastics One

Note: For the reader's convenience, the names of suppliers are provided for some items. The addresses
　　　are indicated below:
　　　1. A-M Systems, Inc., Carlsbourg, WA
　　　2. Amphenol Interconnect Products Corp., Endicott, NY
　　　3. Carleton University, Ottawa, Canada
　　　4. Harvard Apparatus, Holliston, MA
　　　5. Kopf Instruments, Tujanga, CA
　　　6. Plastics One, Inc., Roanoke, VA
　　　7. Small Parts, Inc., Miami Lakes, FL
　　　8. Stoelting, Wood Dale, IL

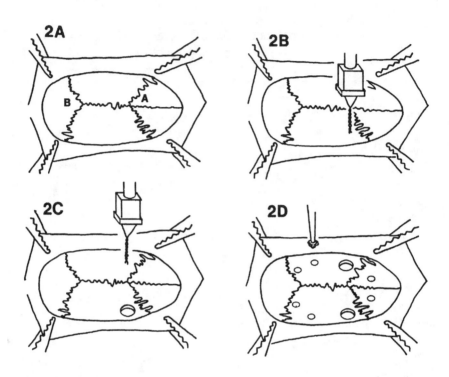

FIGURE 3.2
Series of schematic drawings illustrating steps involved in the implantation of stimulating/recording electrodes. A detailed description of each of the drawings is described in the text: 2A, steps 7 and 8; 2B, step 13; 2C, steps 13 and 14; 2D, step 15; 2E, steps 16 and 17; 2F, step 18; 2G, step 19; 2H, step 20. A and B in 2A represent bregma and lambda, respectively.

also reported greater accuracy with bregma when the anatomical structure of interest is located anterior to this reference point. Although Paxinos and Watson[11] confirmed this finding, they also noted that the interaural reference point may be better suited for localizing more posterior structures.

b.   Skull Level
A critical parameter to be aware of when using a particular stereotaxic atlas is the rostral fixation point on which the atlas is based. In some atlases the toothbar is fixed 5 mm above the interaural line,[12] whereas in others the toothbar is set so that bregma and lambda are on the same horizontal plane.[11,13] It has been reported that adjusting the position of the toothbar such that the skull is horizontal minimizes errors when rats of different sizes are used.[20] Regardless of the level at which the skull is positioned, it is imperative that it corresponds to the stereotaxic atlas being used.

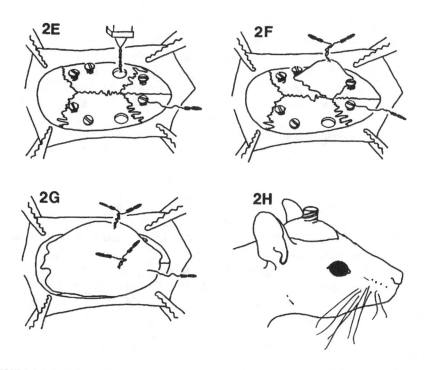

FIGURE 3.2 (continued)

2. Surgery

a. Anesthesia

The implantation of two electrodes can take between 45 and 90 min. A sufficient level of anesthesia must be maintained throughout this time. Sodium pentobarbital is a frequently used anesthetic and is administered directly into the peritoneal cavity (i.p.). The effective dose can vary depending on the strain of rat one is using. An adequate dose for Long Evans Hooded rats (between 250 and 500 g) is between 50 and 60 mg/kg. Alternatively one can use inhalant anesthetics such as halothane, which allows for better control of the depth of anesthesia. If using this latter method, it is important that an adequate ventilation or scavenger system be available as repeated exposure to low levels of halothane has been reported to have toxic effects. For assistance with doses and routes of anesthetic administration, consultation with animal care personnel is recommended.

b. Aseptic Surgery Concerns

Until recent years, rats have proved very resistant to pathogens introduced during surgical procedures. Perhaps due to changes within the breeding facilities and the requirement by many research facilities for admitting only barrier-raised, pathogen-free rats, it has been the experience of several researchers (D. C. McIntyre, personal communication) that this is no longer the case. Aseptic surgery procedures will reduce greatly the chance of introducing pathogens during surgery. Pathogens, such as *Staphylococcus epidermitis*, are known to be associated with disintegration of the skull surface and increased loss of headcaps. Given the prolonged duration of many kindling experiments, it is critical that the integrity of the skull be maintained for an extended period. Use of gloves and mask, as well as application of a penicillin G solution (personal observation) onto the cranial surface *prior* to the drilling of electrode holes are important to reduce or prevent the likelihood of postsurgical infections.

c. Electrode Implantation

The steps involved in the implantation of stimulating/recording electrode(s) are listed below. Cooley and Vanderwolf,[17] as well as Skinner,[16] have also provided a detailed step-by-step approach to stereotaxic surgery with excellent pictoral and photographic accompaniments. Those attempting stereotaxic surgery for the first time should remember that proper placement of the head into the stereotaxic apparatus is critical for proper placement of electrodes. This is sometimes difficult at first. If possible seek the guidance of experienced personnel during initial attempts. The procedure described in this chapter involves the implantation of two electrodes since many kindling experiments are designed using two or more electrode sites. However, one electrode is sufficient to kindle an animal.

1. Prior to surgery sterilize instruments, gauze pads, and cotton swabs. Maintain in sterile condition until use. The glass bead sterilizer device listed in Table 3.1 is a convenient means of sterilizing surgical instruments and meets National Institutes of Health (NIH) guidelines. During surgery, place intruments on sterile draping. If reuse is required during surgery and a glass bead sterilizer is unavailable, immerse in liquid disinfectant between procedures.

2. Prepare bipolar stimulating/recording depth electrodes and the screw/electrode that will serve as ground prior to surgery. Depth electrodes consist of a single strand of stainless steel wire with a Teflon®* coating. Typically the diameter of the wire is between 100 and 200 μm. Between 0.25 and 0.5 mm of insulation is scraped from both ends of the wire and subsequently soldered to Amphenol (220-P02) pins. The wire strand is then twisted to create the bipolar form, and cut to the desired length. Ground reference electrodes are easily constructed from insulated wire that is greater in diameter than the wire used for bipolar depth electrode construction. Again the insulation is removed from both ends (again between 0.25 and 0.5 mm of uninsulated wire should be exposed at both tips). One end is soldered to a jeweler's screw and the other end is soldered to Amphenol pins identical to those used for depth electrode construction. The viability of all electrodes can be determined using a voltmeter. Each rat will require a single depth electrode for each site one wishes to stimulate/record, one ground, and five jeweler's screws. A maximum of four bipolar electrodes can be implanted using the McIntyre connector unit.

---

* Registered trademark of E. I. Du Pont de Nemours and Company, Inc., Wilmington, Delaware.

3. Prior to surgery place cut depth electrodes, ground electrode, screws, and drill bits in a liquid disinfectant for 15 min. Remove and place on sterilized gauze. Ensure that disinfectant has been thoroughly absorbed by the gauze prior to use. Alternatively these items may be sterilized with ethylene oxide.

4. Anesthetize rat. Once the level of anesthesia is sufficient, such that the rat no longer responds to a firm tail pinch, proceed to shave the scalp with electric clippers. The shaved region should include the area extending from between the eyes to just behind the ears.

5. Position the rat in the stereotaxic apparatus. Correct positioning of the rat is crucial to accurate electrode placement. One way to determine whether the rat is placed correctly in the stereotaxic unit is to grasp the snout and assess the movement of the head from side to side. As noted by Cooley and Vanderwolf,[17] movement of the snout by more than 4 mm laterally, in either direction, indicates improper placement. First time users should confirm with more experienced users as to whether the rat is properly positioned before proceeding. To maintain a more sterile environment and to aid in maintaining body temperature, place a sterile cloth over the body of the rat.

6. Administer a few drops of sterile eye drops into each eye. Cleanse and disinfect the incision area using a combination of three different solutions (hibitaine, 70% alcohol, and bridine). The incision area is first swabbed with hibitaine, followed by 70% alcohol, and then a final swab with a small amount of bridine.

7. Make a midline incision in the scalp using firm pressure to ensure a clean cut in a single stroke. The incision should extend from the area between the eyes to the area between the ears. Do not extend your incision beyond the transverse bony ridge behind the ears.

8. So that it is possible to locate the standard reference points on the skull (see Figure 3.2A), the subcutaneous layer of tissue covering the skull (periosteum) first must be removed. This can be achieved in one of two ways. In the first, the periosteum is simply scraped back on both sides using a bone curette. The second approach involves careful removal of the periosteum by making an incision in the periosteum at the anterior point of your cut. Using your scalpel, proceed to cut the periosteum along the edges of the skull being very careful not to cut into muscle. Forceps and scissors can then be used to pull back the periosteum and cut it free from the skull. This latter method prevents temporary swelling of the periosteum which can sometimes prevent proper adherence of the dental acrylic along the entire skull surface.

9. Using hemostats, or towel clamps, expose the skull surface (see Figure 3.2A). If possible, gently pinch an edge of the remaining periosteum with the hemostats rather than the scalp.

10. Using sterile cotton swabs, wipe the skull surface several times to remove blood and other debris. Apply a few drops of a solution of penicillin G (approximately 300 to 400 IU/µl) onto the skull surface and allow to sit for 1 to 2 min. Then use a cotton swab to absorb the solution. Rinse surface of skull several times with sterile saline.

11. The next series of steps will describe how to determine the exact location at which to place the electrode, using bregma as the reference point. If using an atlas based on a flat skull position, ensure that lambda and bregma are on the same horizontal plane. This is facilitated by setting your electrode into the electrode holder of your stereotaxic apparatus. It is critical that the electrode is positioned so that the tip of the electrode is straight and perpendicular to the skull surface. One can now determine visually whether the two points are horizontal by positioning the electrode over

bregma, lowering it until it just touches the skull, and taking a depth reading off the scale. Perform a similar step, this time positioning the electrode over lambda. If the skull is horizontal, the depth readings from lambda and bregma should be identical.

12.   To obtain the coordinates for the location of your electrode(s), first determine the stereotaxic coordinates of bregma. Obtain both an anterior-posterior reference, as well as a lateral reference point, by placing the tip of your electrode directly over bregma and slowly lowering until the tip is touching the skull. Determination of coordinates is achieved by reading the vernier scales on the stereotaxic apparatus. Use these coordinates to calculate the position at which the electrode hole(s) should be drilled. Simply stated, you need to determine the distance that the electrode tip must be moved anterior (or posterior) and lateral from bregma so that it will be above the brain structure you are interested in. This distance is easily determined by referring to a stereotaxic atlas. So time is not wasted during surgery these latter coordinates should be determined prior to anesthetizing the rat.

13.   Move the electrode holder so that the tip of the electrode is directly over the area of interest. This is accomplished by adjusting the anterior-posterior scale and the lateral scale to the coordinates calculated in step 12. At your new location, lower the electrode tip so that it is just touching the skull surface (see Figure 3.2B). Mark this spot with a fine-tip marker. Double check that the mark is in the correct spot. The next step is to drill through the skull at this location using the larger drill bit (o.d. 1.0 mm). Drill into the skull gradually, checking frequently whether you have successfully passed through the skull. Extreme care should be taken not to drill into the brain. If sufficient care is taken while drilling, the dura will remain intact. Use a sterile hypodermic needle to cut the dura. This latter procedure should be performed gently so as not to pierce the brain.

14.   If implanting two or more electrodes, repeat step 13 until all electrode holes have been drilled. Continue to use the same electrode tip to determine the location of new electrode holes. The accompanying figures (Figure 3.2B and C) illustrate the implantation of two electrodes into the amygdala.

15.   Now that all electrode holes have been drilled, proceed to drill the holes for the jeweler's screws. First replace your drill bit with the smaller bit (o.d. 0.6 mm) listed in Table 3.1. Drill six holes placed at locations that will give you a broad base for your skull cap. Allow space between the jeweler's screw and the skin for dental acrylic to adhere. Figure 3.2D illustrates positioning for jeweler's screw and ground pin holes. It is important not to drill closer than 0.5 mm to the midline suture, as the sagittal sinus is located just below this line.

16.   Rinse the skull several times with sterile saline to remove bone chips and blood. Carefully insert jeweler's screws and ground into the appropriate holes. It is critical that screws and ground are inserted into the skull no more than about 1 mm in depth to avoid depressing the surface of the brain.

17.   With screws and ground in place, one can now implant the electrode(s). Reposition the stereotaxic electrode holder such that the tip of your electrode is over the first hole (Figure 3.2E). Insertion of the electrode can occur by using either the surface of the brain as the reference point from which to begin lowering your electrode or by using the surface of the skull. Only the latter method will be described, as it is difficult to visualize the surface of the brain. In determining the position of the skull surface, the fluid that has accumulated in the electrode hole is used as a reference point. After ensuring that the fluid in the hole is level with the skull, slowly lower the electrode tip

until it just breaks the surface tension. Determine the depth coordinate at this point and calculate, based on the predetermined coordinates obtained from your atlas, the depth to which the electrode must be lowered to be correctly positioned in the appropriate brain structure.

18. Once the electrode is lowered, acrylic cement is then used to fix the electrode in place. Mix a small amount of powder with the liquid solvent (methylmethacrylate) until it reaches a consistency of a thick syrup. Apply the cement to the area around your inserted electrode. If inserting more than one electrode ensure that the cement is applied in a manner such that it does not run into other electrode holes. Allow this layer of cement to dry (approximately 5 to 10 min) before proceeding. Once the electrode is securely in place, remove the electrode from the holder (see Figure 3.2F). Repeat this step if implanting more than one electrode.

19. Once all electrodes are securely fixed with dental acrylic, fill in the remaining surface of the exposed skull with dental acrylic so that it is level with the surface of the scalp. Remove hemostats once the dental acrylic begins to harden. Allow this layer of acrylic to dry completely (see Figure 3.2G).

20. Carefully insert and snap the electrodes into the McIntyre connector using the needle nose hemostats. Make sure a protective ring (see ring nut, Figure 3.1) has been placed on the connector prior to inserting electrodes. If rats are to be housed in bedding chips, protective rings can be ordered that will prevent debris from becoming lodged within the electrode holes. It is important to record which electrodes are inserted into which of the nine holes in the connector. It is best to establish a routine within a laboratory so that a standard configuration is employed by all users of the kindling apparatus; i.e. the ground pin should always be inserted into the same spot, and the two pins that comprise a single bipolar electrode should always be inserted into two holes known to represent a single electrode. This configuration can then be matched to the leads attached to the polygraph, allowing for easy recording and/or stimulating from a particular site.

21. Once the pins are inserted into the connector cap, carefully arrange and bend the wires of each electrode so that they are not touching other electrodes or ground wires. Set the cap so that it is positioned straight and as close as possible to the layer of dental acrylic. Begin applying dental acrylic until all the wires are covered and a smooth headcap is created (see Figure 3.2H).

22. Place the rat under a warming lamp until the rat becomes mobile. Analgesics such as children's Tylenol®* should be administered at this point (suppository form: approximately 7 mg/kg). To minimize the incidence of subsequent infection, a subcutaneous dose of penicillin G procaine (22,000 units/kg; i.m.) should be administered before the rat regains consciousness.

d. Postoperative Care

Rats should be allowed 10 to 14 d postoperative recovery prior to the initiation of kindling. During this time, rats should be singly housed to prevent them from grooming and damaging headcap assemblies of cage mates. It is also best if they are kept in a deep cage with a shallow cover to minimize damage to the headcap. Twenty-four to 48 h after surgery, daily handling of the rats should commence. This

---

* Registered trademark of McNeil Pharmaceuticals, Fort Washington, PA.

will minimize the stress on the rat once the kindling phase is initiated, allowing the researcher to more easily connect, and disconnect, rats during the kindling trials.

B. Kindling Procedure

1. Equipment

The equipment requirements necessary to kindle an animal can, in its simplest form, consist of a polygraph machine and a constant current stimulator. Although the level of technical sophistication between laboratories varies greatly, the minimum equipment requirements are a power source with which to stimulate the animal and a recording device with which to monitor ongoing electrographic events.

2. Protocol

Although the specifics of the kindling procedure will vary depending on the particular experimental question, a general description of the protocol would indicate that kindling involves administration of a low-intensity electrical stimulus to a particular region of the rat forebrain. The particular properties of the stimulus and the interstimulus interval must meet certain requirements in order for kindling to proceed (see below). During kindling, the electrographic seizures increase both in duration and complexity. In addition there is an orderly progression from mild to more severe forms of behavioral seizure activity with successive stimulations. In general, rats are said to be "fully kindled" when stimulation of the focus will reliably trigger a motor seizure that is characterized by bilateral forelimb clonus with rearing and falling (stage 5 seizure;[10] see below). Most kindling protocols involve the elicitation of between three to five stage 5 seizures before kindling is terminated, but again this is highly dependent on the directives of the particular study.

Details of the kindling protocol, such as stimulation parameters, interstimulus interval, seizure threshold determinations, and the manner in which behavioral seizures are traditionally classified, are outlined in the next section. In addition, measures of interest to the study of epileptogenesis, such as seizure thresholds, latency to forelimb clonus, and the duration of both the electrographic and behavioral seizure events will be discussed.

a. Stimulus Parameters

A number of researchers have investigated the influence of stimulus parameters on kindling development.[8,9,21,22] The results suggest that kindling can occur with a wide variety of stimulation parameters, provided the intensity of the stimulus is sufficient to trigger an electrographic seizure event, or afterdischarge (AD).[8,9]

Most kindling protocols involve administration of either 1-msec sine- or square-wave pulses, at a frequency of 60 Hz, for a duration of one to several seconds. The intensity of the stimulus, as noted above, must be sufficient to induce an AD. Although the rate of kindling is not affected by varying the stimulus intensity above threshold, several weeks of subthreshold (no AD triggered) stimulation provided little evidence of kindling development.[9] Determining the appropriate intensity at which to proceed with kindling can occur in one of two ways. The first involves an initial determination of the AD threshold, which is defined as the minimum stimulus intensity to provoke an AD outlasting the stimulus by 2 s or more (see the following section on afterdischarge threshold). Subsequent stimulations are then administered at this AD threshold intensity. In the second approach, the stimulus is administered, throughout the entire kindling procedure, at an intensity known to exceed the AD threshold for the particular structure. Stimulus intensities of 200 μA are well above the AD thresholds for the basolateral amygdala and hippocampus. Although this latter technique is less time consuming and provides for a constant level of stimulation across animals, the former approach provides important information relevant to the epileptogenicity of local circuitry (see Section II.D.2).

Although initial reports suggested that kindling could only occur with administration of a high-frequency stimulus (25 Hz or greater), a series of papers by Corcoran and Cain[22,23] reported successful kindling in a variety of forebrain structures with stimulations in the low-frequency range (3 Hz) using square-wave pulses. In contrast to traditional 60 Hz kindling, low-frequency kindling required relatively high-intensity stimulations applied for a relatively long duration (20 to 60 s). An important feature of this particular kindling regimen is that the rate of convulsive seizure development is very rapid as compared to kindling with conventional 60 Hz stimulation. In many rats, bilateral motor convulsions (stage 5) were triggered from the amygdala on the initial stimulation trial.[23] In contrast, 60 Hz amygdala kindling required many more stimulation trials (12.0; range 5 to 19) before a stage 5 seizure was evoked. The rapid development of low-frequency (high-intensity) limbic kindling may prove useful to the study of convulsive seizure mechanisms in experimental conditions that may not be conducive to long-term studies (example, *in vivo* microdialysis and cannulation experiments). It is important to keep in mind, however, that one of the characteristics of the kindling model that differentiates it from other models of epilepsy is the "progressive" nature of both the electrographic and behavioral seizure development. The exact parameters one decides upon when kindling are dependent on the particular question one is attempting to address. Low-frequency, high intensity stimulations may not be suitable for studying mechanisms associated with focal seizure development. Such questions would be more appropriately addressed with traditional kindling parameters in which the direct effects of the stimulus are more discretely localized to the focus.

b. Afterdischarge Threshold

An important variable, easily obtained during the kindling process, yet overlooked in many kindling reports, is the AD threshold. In combination with the duration of the afterdischarge, AD thresholds are a sensitive indicator of the level of excitability within the local circuitry. Further discussion of AD thresholds and their relevance to understanding mechanisms involved in epileptogenesis is provided in Section II.D.1. The various procedures by which AD thresholds are obtained are outlined below.

There is much variation between laboratories as to the exact methodology involved in AD threshold assessment. A common technique utilized by many in the kindling field is that developed by Racine.[9] This procedure involves stimulation of rats with a low-intensity electrical current (5 to 10 μA). Should this initial stimulus fail to elicit an AD, the subsequent stimulation is administered at an intensity that is double that of the original stimulus. This doubling of stimulus intensity continues until an electrographic seizure is observed. Following the successful elicitation of an AD, the next stimulation is then administered at an intensity that is halfway between the current intensity that produced the seizure and the previous subthreshold intensity. One limitation of this procedure is that the elicitation of an AD temporarily inhibits subsequent AD provocation. To circumvent this inhibition, Racine recommends that subjects be stimulated every second day during threshold assessment.[9] Inclusion of this interval between various threshold assessment trials may prolong significantly the length of the experimental protocol and thus may deter some investigators from assessing this variable.

A second method of threshold determination, requiring less time investment, uses a modified method of limits. With this technique only an ascending series of stimulations is administered. The initial stimulation is of a very low current intensity. If no AD is observed the intensity of the subsequent stimulus is increased slightly and administered after only a short period of time. This procedure is repeated until an AD is successfully evoked. Freeman and Jarvis[24] showed that with this particular method, because stimulation is discontinued with the appearance of an AD, the interval between subsequent stimulations can be as short as 1 min without affecting the stability of the AD threshold. However, it is important to remember that threshold trials administered at intervals of 30 s or less yielded significantly higher threshold measures. The exact increments by which the current intensity is increased between stimulations varies from study to study, but in general involve increases that are 20% to 35% of the previous intensity; for example, 15, 25, 35, 50, 75, 100, 150, 200, 250, 300, 350, 400, 450, and 500 μA peak to peak.

In most studies, thresholds are assessed at the onset of kindling and/or the experimental treatment and reassessed at the conclusion of the experiment. Daily stimulation trials proceed at the predetermined AD threshold, or a fixed intensity as set by the investigator. However, in many circumstances it may be informative to track changes in AD thresholds throughout the kindling procedure. Mohapel et al.[25] described a simple procedure with this directive in mind. Rather than administering daily kindling stimulations at the initial AD threshold, rats instead were administered daily kindling stimulations that were one intensity increment below the AD threshold of the previous day. If that stimulus intensity was insufficient to trigger an AD, then, after a 1-min delay, the intensity of the stimulus was increased by one increment, until an AD was triggered.[25] This procedure allowed for daily assessment of local seizure thresholds and provided important information as to local excitability changes within various amygdala nuclei as a consequence of kindling.

c.   Interstimulus Interval

The interval between subsequent kindling stimulations, or the interstimulus interval (ISI), is a critical variable influencing the kindling process. In the classic 1969 study Goddard et al.[8] investigated the effect of a variety of ISIs on the development of amygdala kindling. It was reported that ISIs of 20 min or greater were necessary for kindling to progress. Although there was evidence of partial kindling with short ISIs, complete kindling progression was not possible. The ISI that kindled with the fewest number of stimulations was that in which the kindling stimuli were separated by a minimum of 24 h. Administration of stimuli twice a day slightly increased the number of stimulations before a fully kindled motor seizure was observed.

A later study by Racine and colleagues[21] modified somewhat the conclusions drawn by Goddard's group.[8] In the absence of electrographic recordings Goddard and colleagues were unaware that short ISIs often increased the AD threshold and thus a number of the stimulations may have failed to elicit an AD. Racine's assessment[21] incorporated both behavioral and electrographic data and noted that the minimal ISI at which one could kindle without increasing the number of stimulations during amygdala kindling was 1 to 2 h. Consistent with the earlier report by Goddard et al.,[8] Racine's group observed a lack of kindling progression (amygdala) with ISIs of 15 min or less.[21] The inability of short ISIs to induce a fully kindled state appears to be an age-dependent phenomenon. Moshe et al.[26] attempted amygdala kindling in suckling rats with ISIs of 15 min and, unlike the adult, observed consistent prolongation of the afterdischarges and repeated generalized seizures.

Traditionally most kindling experiments are designed so that stimuli are administered either once or twice a day. In comparing the results from different kindling studies one must be aware of the particular ISI used in each experiment as there is now substantial evidence of dramatic differences in seizure response as a result of atypical intertrial intervals.[8,21,27] However, in certain instances, the experimental question may require the implementation of a paradigm incorporating a nontraditional ISI.

The "rapid hippocampal kindling" paradigm described by Lothman et al.,[28,29] presented in Chapter 4, involves a massed stimulation protocol of the hippocampus and produces generalized motor seizures within a few hours. Although this enhanced seizure response is evident when stimulations are continued on the subsequent day, the permanence of these changes has been the subject of controversy.[29,30] A recent assessment of this issue by Elmer and colleagues[30] suggests that a brief period of "rapid hippocampal kindling" can lead to a progressive, but delayed, development of the fully kindled state without additional stimulations. Although evidence of kindling was not pronounced 1 week following the stimulation regimen, rats tested 4 weeks following stimulation responded with fully generalized stage 5 seizures. As noted by Elmer et al.,[30] the rapid kindling protocol described by Lothman et al.[28,29] may be suited to the study of mechanisms regulating both plasticity and kindling acquisition. The delayed development of the "kindled state" in the absence of continued stimulation trials provides a novel approach to studying kindling-related growth processes.[30]

"Massed kindling" of the amygdala is another example of a kindling protocol that may be suited to certain experimental objectives. In this protocol, stimulation of the amygdala every 5 to 10 min induces retarded epileptogenesis.[8,21,31,32] This retardation is characterized by an initial growth in AD (10 to 15 stimulation trials) that is soon followed by a sudden truncation of AD length. Subsequent stimulations produce only brief electrographic responses with no further development of seizure activity. As noted by McIntyre et al.,[31] the suppression of amygdala kindling with massed stimulation provides a robust example of arrested seizure development. The processes responsible for this suppression may provide important clues as to the mechanisms operative during epileptogenesis.

C. Progression of Kindling

Repeated electrical stimulation of limbic structures such as the amygdala or hippocampus results in a stereotypic progression of electrographic and behavioral seizure activity. Although the profile of seizure development is similar between various limbic regions, differences in the electrographic and behavioral seizure response are apparent when kindling from different forebrain sites. Of all limbic regions, the amygdala has been the best characterized in terms of the temporal sequence of events one can expect during kindling. The following section will describe, in detail, the electrographic and behavioral changes observed with repeated stimulation of the amygdala. This will then be followed by a description of the kindling profile observed from other limbic regions, highlighting some of the subtle differences evident between amygdala kindling and other forebrain sites. The final section will compare the profile of anterior neocortical kindling to that of limbic kindling, indicating the dramatic differences in kindled seizure expression.

1. Amygdala Kindling

In the absence of electrographic records, Goddard and colleagues described the sequence of behavioral seizure development resulting from daily electrical stimulation of the amygdala in the rat.[8] It was reported that the initial stimulation trials failed to produce any obvious behavioral responses. It was not until several stimulations had been administered that behavioral automatisms, outlasting the kindling stimulus by several seconds, were observed. Such automatisms consisted of behavioral arrest, closing of the eye ipsilateral to the stimulus electrode, and chewing movements. Subsequent stimulations triggered bilateral clonic convulsions, which involved rearing with infrequent loss of balance, facial contractions, and forelimb clonus beginning in the contralateral limb and spreading to involve both forelimbs in bilateral synchrony. Racine[10] confirmed this qualitative assessment of kindling progression and developed a 5 point rating scale to describe each stage of behavioral seizure development. This rating scale, used today by most researchers to describe limbic kindling, classifies the oralimentary movements (mouth and facial twitches) and clonic head movements as stages 1 and 2, respectively. The appearance of contralateral forelimb clonus is designated as stage 3. The progression to stage 4 is characterized by bilateral forelimb clonus that is now associated with rearing. Stage 5 seizures represent the final seizure stage in most kindling paradigms, and involves bilaterally generalized motor responses with rearing and falling. Typically between 10 and 15 stimulation trials are required before a stage 5 seizure is successfully triggered from an amygdala focus. The exact rate of amygdala kindling is dependent not only on the strain of rat but also on the particular amygdala nucleus in which the electrode is situated.[25,33]

It was Racine[9] who first examined the electrographic correlates of the kindling phenomenon and observed that the changes in the amygdala electrographic responses were as dramatic as the behavioral changes described above. Initial stimulation of the amygdala triggers a brief electrographic seizure event, or AD, lasting about 10 s. These early epileptiform discharges consist of simple, biphasic spikes, occur at a frequency of about 1 Hz, and propagate only weakly to distant sites such as the contralateral amygdala. The typical behavior associated with the initial amygdala ADs is behavioral arrest. With repeated stimulations, the AD becomes more complex and increases in duration, frequency, and amplitude. Such changes are evident not only in the focus but also at distant sites. Concomitant with this electrographic seizure development is the simultaneous progression of behavioral seizures as outlined above. A series of electrographic traces recorded from a rat at various times during an amygdala kindling protocol are presented in Figure 3.3.

2.   Kindling from Other Limbic Regions

In general the development of kindling from other limbic sites proceeds in a manner similar to that described for the amygdala. Kindling of the hippocampus or piriform cortex, for example, involves progressive changes in the complexity of the ADs that mirror those described above for amygdala kindling. Associated with changes in the electrographic response is a heightening of behavioral seizure activity that eventually culminates in a stage 5 rearing and falling convulsion. Although, with the exception of higher AD thresholds in the piriform cortex, there is little difference in kindled seizure expression between the basal amygdala and the piriform cortex,[34] subtle differences are apparent when kindling from the hippocampus.

Regardless of whether one is kindling from the dorsal or ventral hippocampus, electrographic seizure discharges are almost always characterized by a primary AD (20 to 30 s), followed by a period of suppressed or isoelectric activity (10 to 40 s), and then a so-called rebound[35] or secondary AD.[10,36-39] Rarely is this type of profile observed when kindling from other limbic regions. Several behavioral responses also tend to be associated with hippocampal kindling.[37,38,40] "Wet dog" shakes (WDS), and sometimes grooming, are a characteristic behavioral response exhibited during the early stages of hippocampal kindling that tend to decline as kindling progresses. Racine's classification[9] of amygdala seizure activity is applicable to motor seizure development in the hippocampus, with only minor exceptions. As reported by Racine et al.,[40] there are two basic profiles of behavioral seizures exhibited during hippocampal kindling. The first pattern is nearly identical to that described for the amygdala (with the exception of the WDS) with the initial development of stages 1 and 2 seizures, followed in subsequent trials by contralateral forelimb clonus (stage 3). As with amygdala kindling, continued stimulation results in the eventual development of bilateral forelimb clonus with rearing and falling (stage 5). However, a subset of rats kindled from the hippocampus sometimes exhibit seizure responses more characteristic of anterior neocortical kindling (see below). This profile is characterized by a loss of postural control accompanied by forelimb

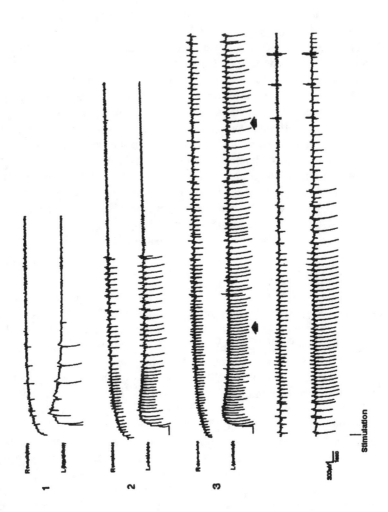

FIGURE 3.3

A series of electrographic traces taken from a rat at various times throughout amygdala kindling. This particular rat was stimulated in the left hemisphere (L). The top trace represents the initial AD trial. Note that there is minimal propagation to the contralateral hemisphere. Trace 2 represents the electrographic seizure activity observed on the fifth stimulation. Note the increased strength of propagation into the contralateral site, as well as the increase in duration and complexity of the EEG. The bottom set of traces are associated with the first stage 5 convulsion (occurring between the two arrows). Note the significant prolongation of AD duration.

clonus and mild episthotonus. A prolonged clonic response with rearing and falling often develops with subsequent hippocampal stimulations and follows the episthotonus response. In this particular group there is often little evidence of the stage 1 to 2 behavioral responses noted from other limbic sites. Although these seizure responses differ from the typical stage 5 seizure responses described for the amygdala, they are often scored as stage 5 seizures. The procedure in our laboratory is to classify such seizures according to Racine's classification system of amygdala kindling, while noting the exact nature of the motoric seizure.

As a cautionary note, it has been observed by several investigators that there is an increased aggressive response in hippocampal kindled rats immediately following evoked seizure activity.[40] Such responses are not common to amygdala kindling. This postictal aggression will dissipate if the rat is allowed to recover for 5 to 10 min before removal from the stimulation chamber.

3.   Anterior Neocortical Kindling

The most striking difference between kindling of the anterior neocortex and kindling from subcortical and cortical limbic structures is the appearance of behavioral seizure responses on the initial stimulation trial of the frontal motor regions of the rat.[41] The behavioral profile elicited by stimulation of the anterior neocortex, although somewhat variable and most likely dependent on the exact placement of the electrode within anterior neocortical areas, differs dramatically from convulsions triggered in subcortical regions. The anterior neocortical convulsion triggered on the initial trial is characterized by forelimb clonus followed by a mild tonus. These responses are often associated with exaggerated oralimentary movements similar to those classified as stage 1 and stage 2 for amygdala kindling. This series of behavioral responses is observed almost immediately following stimulation and are associated with short-duration ADs that last only 7 to 10 s. The tonic component of the seizure increases in strength with successive stimulations and eventually culminates in loss of postural control. Unlike the definitive evolution of motor seizure activity, changes in the electrographic discharges are minimal. It is not until many more stimulations before the duration of the electrographic discharge suddenly increases and approaches that observed with seizures triggered from limbic sites such as the amygdala.[42-45] In these later stages of neocortical kindling, there is an intriguing change in the convulsive response associated with such prolonged ADs. After about 25 to 30 neocortical-type seizures a late clonic component appears at the end of the typical brief neocortical seizure.[43] This latter seizure type closely resembles a generalized limbic seizure and is roughly correlated with the development of independent ADs at limbic sites.[46] This profile of neocortical kindling provides important insights into the anatomy of kindled seizure development.

D.   Dependent Measures of Interest

There are several critical measures one should consider when designing a kindling study. The following section will describe these measures and indicate their relevance to the study of both epileptogenesis and the epileptic state.

1.   AD Threshold and AD Duration

As alluded to in a previous section, the AD threshold (ADT) and associated AD duration reflect the local circuit properties of a structure and provide a relative measure of inherent excitability within a particular region.[34] Many kindling studies have failed to assess these measures, relying primarily on kindling rate data or convulsive seizure disposition as a means of characterizing the epileptogenic influences of a particular treatment or the epileptogenic potential of a limbic site. In assessing the effects of a treatment on either kindling development or the kindled state itself, the potential for erroneous conclusions is increased without the inclusion of AD thresholds as a dependent measure. For example, although it was originally believed that phenytoin was ineffective against blocking kindled seizures, more recent studies that avoided "supramaximal" stimuli and induced seizures at intensities that were only 20% above the original AD threshold, reported dramatic effects of phenytoin treatment on the AD threshold.[47-50] Following such treatment, it was necessary to increase current intensities 400% above the predetermined AD threshold before behavioral seizures could be evoked.[50] Experiments designed using only supramaximal stimulation would fail to detect such important threshold changes. According to Löscher and colleagues, the most sensitive parameter for detection of anticonvulsant activity against focal seizures is the threshold for AD induction (see Löscher and Honack[51]).

As described previously, the hippocampus has a unique electrographic seizure profile, involving both a primary electrographic event and a secondary electrographic seizure event that are separated by 20 to 40 s. Although, as stated above, AD durations are believed to reflect the degree of baseline excitability within the stimulated site, this principle may not apply to both components of the hippocampal discharge. It has been postulated that the secondary ADs observed with hippocampal stimulation originate from outside the hippocampus, and are presumably driven by circuits within the entorhinal cortex.[39,52] Thus, although the duration of primary ADs triggered from the hippocampus may represent the level of excitability within hippocampal circuits, the duration of the secondary AD may represent the excitability of extrahippocampal regions within the temporal lobe. Experimentally induced changes to the profile of the secondary electrographic response may therefore indicate important changes in the strength of seizure propagation from the hippocampus or excitability changes within other subcortical regions. This issue has been described in detail by McIntyre and Kelly.[52]

2.   Rate of Kindling Progression

The pattern and rate of behavioral seizure progression are important measures of the rate of epileptogenesis. As noted above, limbic behavioral seizures progress in a stereotypic manner and have been classified into five stages.[10] The clinical context of such stages has been described by McNamara,[53] where stages 1 and 2 are thought to mimic human complex partial seizures and reflect electrographic activity that is primarily localized to limbic regions. Given the early appearance of stages 1 and 2 seizures with stimulation of the amygdala or piriform cortex as compared to stimulation of the hippocampus or entorhinal cortex, it is thought that circuits driving such oralimentary movements reside in close proximity to the amygdala-piriform region. Stage 3 seizures are thought to represent focal (partial) seizure activity, whereas stages 4 and 5 represent secondary generalized motor seizures, involving recruitment of neuronal circuits outside of limbic regions.

In kindling studies, the measure that is usually reported is the rate of kindling. Kindling rate refers to the number of stimulation trials required to trigger a stage 5 seizure, although several investigators measure instead the cumulative duration of ADs to reach the first stage 5. The rate at which a structure kindles is thought to reflect the ability of triggered ADs to access motor structures that drive the convulsive response.[8] Differences in kindling rates between structures and/or between different experimental groups may represent (1) differential connections (or a disruption of existing connections) to the motor substrates that trigger generalized seizure events, (2) differential connections (or a disruption of existing connections) to other forebrain sites that augment the seizure discharge, or (3) differential reactivity of the stimulation sites themselves.[54] Determining which of these three possibilities could account for differences in kindling rate is aided by comparing structures (or different experimental groups) on AD thresholds and the associated AD durations. Additionally, a comparison of the rate of transition through the various seizure stages can also be used to understand the factors underlying kindling rate differences.[55,56] An example of this latter condition is described below.

Löscher and coworkers[55] measured carefully the temporal profile of behavioral seizure development at various times following electrode implantation and noted facilitated rates of kindling in groups with prolonged electrode implantation. Importantly they noted that differences in overall kindling rates could be accounted for by differences in the rate of development of stages 1 and 2 seizure activity and that the transition from stage 2 to stage 5 behavioral seizures occurred at a constant rate between the different experimental groups.[55] In combination with a decrease in the focal AD threshold, the enhanced transition through the initial stages of kindling allowed them to conclude that the pro-kindling effects of prolonged electrode implant involved local events at the site of stimulation rather than an overall increase in brain excitability.

3. Convulsive Seizure Profile

Three indices that can be used to characterize the convulsive seizure profile are, (1) latency to forelimb clonus, (2) clonus duration, and (3) total AD duration associated with the stage 5 motor seizure. Although many studies fail to report these measures, they are easily obtained and provide important information relevant to the process of secondary generalization of limbic seizures.

The latency from stimulus onset to the appearance of forelimb clonus is a measure of the rate of seizure spread from the focus and the degree to which discharges triggered in a particular site can recruit the motor regions that drive the clonic response.[2,34] It has been argued that brief latencies (1 to 2 s), such as those triggered from the perirhinal cortex, may represent direct connections from the kindled site to motor regions. What is commonly observed during kindling from some limbic sites, such as the hippocampus, is a dramatic shortening of seizure latencies with successive stage 5 seizures.[38,52] Goddard et al.[8] suggested that this latter phenomenon may reflect enhanced excitability within local circuits and their respective connections. The increased excitability of local circuitry provides for more efficient propagation of triggered discharges, allowing them greater access to circuits controlling convulsive activity.

Stage 5 motor seizures (measured from the onset to forelimb clonus until motoric forelimb responses are terminated) triggered from a variety of limbic regions are remarkably similar in duration.[8,25,34] This similarity is suggestive that triggered limbic discharges ultimately access a common motor seizure substrate to realize the motoric component of the kindled response. The anatomical routes believed to be involved in limbic seizure progression are discussed in the subsequent section.

III. Interpretation

A. Anatomical Substrate of Kindling

The proclivity for partial seizures to eventually manifest as fully generalized convulsions (stage 5) indicates that during the kindling process focal discharges activate and modify the circuitry of distant sites so that access to the neural substrates supporting motor convulsions is eventually attained. A comparison of kindling rates across different brain sites[8] was one of the first attempts to address the issue of kindled seizure circuits. The following section will discuss how such differences in kindling rates have helped to formulate a working hypothesis on the anatomy of limbic kindling. Following this, the phenomenon of "transfer" will be introduced and its contribution to our present understanding of mechanisms underlying the process of limbic kindling will be described.

1. Kindling Rate

As noted previously, kindling rate represents the number of stimulation trials necessary before a stage 5 convulsion is triggered. Structures with close anatomical proximity to the neural substrates responsible for generalizing limbic seizures should kindle quickly relative to structures anatomically "upstream", or more remote, from the mechanisms of generalization. It was evident from the original kindling study that rate of kindling development varied depending upon the site of kindling. Of the numerous sites kindled by Goddard et al.,[8] the site requiring the fewest number of stimulations to reach a stage 5 convulsion was the amygdala. In addition, a reasonable correlation was noted between the kindling rate of a structure and its anatomical distance from the amygdala. Structures closer in proximity to the amygdala, such as the septum, kindled in fewer trials than structures more anatomically remote, such as the dorsal hippocampus.[8] However, subsequent studies highlighted the potential importance of the piriform and perirhinal cortices to limbic seizure generalization. Early studies by Cain and colleagues[57,58] revealed that the piriform cortex and its primary afferent the olfactory bulb manifest stage 5 motor convulsions with fewer stimulation trials than the amygdala, while a more recent study by McIntyre et al.[34] described the fastest kindling from limbic regions to occur with stimulation of the perirhinal cortex, an area immediately adjacent to the piriform cortex. These latter studies have led several investigators to postulate that relative to other limbic regions, the piriform and/or perirhinal cortices may have preferred access, either directly or indirectly, to structures capable of supporting a motor convulsion.[54,59-65] Manifestation of kindled convulsions may therefore depend on the ability of a limbic site such as the hippocampus to access and modify the neural circuitry within these cortical regions.

2.   Transfer Phenomenon

Additional information relating to the mechanism(s) of seizure generalization is provided by a robust feature of kindling known as the "transfer" phenomenon.[8,10,66,67] It is well documented that following kindling of a limbic structure such as the amygdala, kindling of a second limbic site requires fewer stimulation trials than were kindling to only have been attempted in the secondary site. The increased seizure susceptibility of sites "remote" from the primary site of kindling is referred to as positive transfer and can be observed in structures both within (intrahemispheric) and between (interhemispheric) hemispheres.

Understanding the mechanisms of the transfer phenomenon has provided important insights into the mechanisms that may underwrite the kindling process. Initially it was not known whether the eventual appearance of convulsive responses during kindling reflected modifications of only those neurons directly adjacent to the stimulating electrode or whether neural reorganization of distant circuits was a prerequisite for kindling progression. If the former mechanism was operative then positive transfer to a secondary site might merely reflect the ability of the secondary site to activate the already kindled or modified primary site. Two approaches were taken to address this possibility.

The first involved induction of a large electrolytic lesion of the primary site immediately following kindling.[8,10] Subsequent kindling of a secondary site showed no disruption of the transfer effect, suggesting that the transfer phenomenon involves a transynaptic increase in seizure susceptibility independent of the primary site. Further confirmation of this concept was provided using a split-brain preparation. Forebrain bisections following establishment of an epileptic focus in one hippocampus did not abolish the positive transfer normally observed during secondary site kindling of the contralateral hippocampus[68] or between primary and secondary kindling sites in the amygdalae.[69] These data imply that a key feature of limbic kindling is the neuronal reorganization of distant structures. Subsequent kindling of a secondary site may utilize critical components of this previously modified seizure network resulting in facilitated kindling rates.

Studies investigating the mechanisms of "positive transfer" suggest that various limbic sites make use of the same common pathway or circuit to manifest a generalized response. As noted above, a comparison of limbic kindling rates suggested that the piriform and/or perirhinal cortical area may be pivotal to the process of seizure generalization. If this is true then certain predictions can be made concerning the degree of transfer between two limbic sites and their relative proximity to the piriform/perirhinal cortices: (1) structures in close anatomical proximity to the piriform and/or perirhinal cortex should show near complete transfer if preceded by kindling of a limbic site anatomically more remote (or upstream); and (2) primary site kindling of structures adjacent to the aforementioned cortical regions should facilitate secondary site kindling of an anatomically upstream structure by approximately the same number of stimulations that was necessary for primary site kindling. These predictions were confirmed by Burnham,[67] in an extensive study investigating transfer between numerous limbic sites. It was noted that primary site kindling of the dorsal or ventral hippocampus produced near-complete transfer to secondary site kindling of the amygdala. In agreement with the second prediction, primary site kindling of the amygdala facilitated hippocampal kindling by nearly the exact number of stimulations required to kindle the primary amygdala site. For example, dorsal hippocampal kindling normally required an average of 37.3 stimulation trials; however, if preceded by kindling of the ipsilateral amygdala (10.6 stimulations) dorsal hippocampal kindling required only 24.0 stimulation trials. These findings are consistent with the view that modification of circuits within the region of the piriform and/or perirhinal cortex may be necessary for the generalization of limbic focal seizures triggered from a variety of limbic areas.

The above discussion of kindling rates and the "transfer" phenomenon has attempted to emphasize the principle that kindling development does not proceed via random spread of seizure activity through limbic regions but rather involves the propagation of seizure discharges through specific temporal lobe circuits. A critical concept to consider in designing a kindling experiment is that the kindling of one limbic site does not necessarily induce similar modifications in distant regions. Reasonable predictions as to the temporal and spatial profile of kindling-induced changes can be made, however, by referring to several comprehensive reviews on the anatomy of kindled seizures.[54,64,65,70-72]

B. Designing a Kindling Study

1. Species

The kindling phenomenon has been demonstrated in all species tested thus far, including amphibians,[73] reptiles,[74] rodents,[8,75] felines,[76] and primates.[77] Although the majority of work pertaining to the kindling model has involved the use of rats, the experimental question may sometimes be better addressed using an alternate species. The increasing availability of transgenic mice may provide a powerful means of assessing the involvement of a particular gene(s) in the development and maintenance of the kindled state. Recently, Watanabe and colleagues[78] reported an attenuation of kindling development in homozygous *c-fos* knock-out mice. However, in deciding to use transgenic mice in a kindling paradigm it is important to monitor for evidence of spontaneous epileptiform activity. There are several reports whereby particular gene knock-outs have increased the predisposition for spontaneous seizure activity.[79,80] It would be important to determine whether any observed changes in kindling profile resulted directly from the particular gene knock-out, or was an indirect effect of uncontrolled and/or clinically undetected seizure activity.

2.   Inclusion of Proper Control Groups

A series of recent papers by Löscher and colleagues[55,81] and an earlier report by Blackwood et al.[82] emphasized the importance of including both an implanted, nonstimulated control group (surgical control) and a nonimplanted control group to assess fully the effects of intracranial stimulation. In one such study, long-lasting alterations in transmitter amino acid levels were evident in several brain regions in both kindled rats and a surgical control group as compared to levels in a nonimplanted control group.[81] Importantly, the alterations in levels of amino acids induced by electrode implantation were evident in structures remote from the site of electrode implantation. It was also noted, and confirmed in a subsequent study,[55] that prolonged implantation of electrodes into the basolateral amygdala predisposes the brain to kindling. Decreases in the prekindling AD thresholds and faster kindling rates were noted when kindling stimulations were started 4 and 8 weeks after electrode implantation as compared to those observed when kindling was initiated 1 week following the implantation of electrodes. The latter data emphasize the importance of ensuring that the duration of electrode implantation is carefully controlled.

3.   Dissociation of Transient vs. Permanent Seizure Effects

One of the most intriguing characteristics of the kindling model is that once convulsive seizure responses to focal stimulation have been established, the enhancement in seizure response is relatively permanent. Months after culmination of amygdala kindling in the rat, stage 5 convulsions can be retriggered with one or two subsequent stimulations.[8,70] This permanent increase in seizure disposition suggests that the neuronal changes underlying the kindling process must also be long lasting. Based on this premise, a number of investigators have emphasized the importance of assessing for kindling-induced changes weeks to months after completion of kindling.[54,83,84] Kindled seizures can induce a number of transitory changes that may take days or weeks to return to baseline,[71,83,85] whereas the neurochemical, electrophysiological, or molecular changes responsible for maintenance of the kindled state must be operative for the duration of the kindled state itself.[83,85] If the experimental objective is to investigate the nature of kindling permanence, then it is critical that measures be taken at appropriate intervals. However, as noted by McNamara et al.,[71] there is potential value in understanding many of the transitory changes induced by kindled seizures, as such changes may represent the endogenous inhibitory mechanisms operative in the brain that serve to suppress ongoing seizure activity.

4.   Choosing a Region of Study

Two critical variables to consider when designing an experiment are the location of the kindling stimulus and the particular region(s) one chooses to assay. Deciding on these issues is dependent not only on the experimental question one wishes to address but also requires consideration of many of the principles discussed in this chapter. As with much of epilepsy research the hippocampus has been a common "focus" in many of the recent kindling reports. Given that temporal lobe seizures in humans are most likely to originate from the hippocampus (see Engel and Cahan[86]), a statistic consistent with the low AD thresholds found with hippocampal kindling in rats,[25,34,38,41] it is not surprising that this region has received much of the investigative energy. However, it is important that the results of experiments involving hippo-campal kindling, or those that make use of hippocampal tissue to identify kindling-induced changes, are interpreted in a manner that is consistent with current theories of kindled seizure networks.

If the experimental objective is to identify kindling-induced changes that are critical to kindled seizure development, it is important to assay for such changes in tissue that is known to be modified to the extent that it is "kindled." Based on what is known of the circuits involved in kindled seizure propagation (see Section III.A), it makes little theoretical sense to kindle the amygdala and look to the hippocampus to identify the modifications critical to kindling. Undoubtedly, amygdala-triggered discharges will propagate to the hippocampus,[10,70,87] and result in a number of morphological, electrophysiological, neurochemical, and molecular changes within this latter region.[54,88-90] Importantly, however, based on the many transfer experiments, these changes are *not* sufficient to induce a kindled state within the hippocamus itself, nor should they be interpreted as being "critical" to the development of amygdala-kindled seizures. As noted in a previous section (see Section III.A.2), the hippocampus requires an additional 24 stimulation trials following amygdala kindling before a stage 5 seizure can be triggered. Thus, it is important to recognize that although the numerous changes evident in the hippo-campus following kindling from extrahippocampal areas may reflect important functional changes that can effect seizure responses in other sites, they are not necessarily associated with the development and maintenance of the "kindled state." Contrary to providing a facilitatory effect on kindling, evidence exists that certain changes evident in the hippocampus following amygdala kindling may in fact retard kindling genesis from other sites. Racine et al.[62] reported that when access to the hippocampus was blocked via bilateral knife cuts delivered posterior to the amygdala, kindling from the amygdala occurred at a facilitated rate. Thus, though amygdala kindling may induce important and interesting effects on the hippocampus, the results of such experiments should be interpreted in light of what is known about kindled seizure networks.

As indicated above the decision of where to kindle is dependent on the exper-imental question. If interested specifically in hippocampal changes critical to kin-dling development, it makes most sense to kindle the hippocampus itself. If one is concerned about confounding effects of electrode implantation on the focus, an alternative would be to kindle one hippocampus and assay the contralateral hippo-campus for kindling-induced modifications. It is well documented based on transfer studies that the kindling of one hippocampus in essence kindles both hippo-campi.[10,68,91] The decision as to where "to search" for important kindling-related changes is also directed by the experimental objective. If one is interested in focal seizure mechanism(s) important to kindling development, then restricting one's search for kindling-induced changes to the kindled site can provide important infor-mation relative to focal seizure development. However, kindling provides an oppor-tunity both to assess for changes important to focal seizure genesis as well as those involved in the secondary generalization of seizures. With respect to studies that have the latter goal in mind, it is critical that regions believed to be important to the secondary generalization of limbic discharges, such as the piriform/perirhinal cortices, are included in the analyses.[64,65]

5. Kindling Genesis vs. Kindled State

This section is particularly relevant to studies investigating the neurochemical mechanisms involved in kindling. In a comprehensive review of the neurochemistry of kindling, Peterson and Albertson[85] emphasized the important distinction between the acquisition of kindling and the kindled state. It was noted that the biochemical and physiological changes evident during kindling development, or the "dynamic phase" of kindling, are often transient and are different from those changes that define and mediate the "kindled state."[85] There are numerous studies supporting this principle, whereby pharmacological manipulations of various neurotransmitter systems affect either the rate and/or profile of kindling development or the convulsive potential in a fully kindled brain.[92-94] The involvement of noradrenaline in kindling has been well characterized and exemplifies the distinction between the "kindling process" and the "kindled state." During kindling development noradrenaline has been shown to have strong inhibitory effects on the rate of seizure spread and progression of kindling development, yet once stage 5 seizures have been fully established, pharmacological manipulations of noradrenaline have little effect (see Corcoran[72]).

The distinction noted above for kindling has a clinical corollary. As noted by Sato et al.,[54] the term "antiepileptogenic" refers to inhibition of processes underlying the development of an epileptic condition, whereas "anticonvulsant" refers to inhibition of seizures in an already epileptic state. Kindling provides a robust model that allows for the assessment of compounds with both antiepileptogenic and anticonvulsant potentials. Drugs known to be efficacious in inhibiting the development of kindling may have useful antiepiletogenic properties in humans, whereas compounds found to inhibit the manifestation of convulsions in kindled rats may be powerful anticonvulsants in the clinical realm.[54]

C. Clinical Relevance of Kindling

Kindling is a robust, highly reproducible phenomenon that satisfies most of the criteria proposed by Wada[95] as essential for a good experimental model of epilepsy. Wada[95] suggested the following as necessary features in modeling a human epileptic condition: (1) precise experimental control over the anatomy in terms of area as well as size of the epileptogenic lesion; (2) ability to create an epileptogenic site without introducing identifiable destructive pathology; (3) accurate experimental control over the initiation and development of seizures; (4) ready induction of seizure by a discrete and identifiable experimental event; (5) eventual development of a spontaneous recurrent seizure state mimicking the previously established electroclinical pattern; and (6) evidence of persistence of the epileptic state for many months. Despite satisfying most of these criteria, the relevance of kindling as a model of complex partial seizures (CPS) has been controversial.

A common criticism of kindling is that there is little documented evidence that humans kindle. For obvious reasons this concern is difficult to address. Although there exist at least two reports of kindled epileptogenesis occurring in the human brain following repeated direct stimulation with implanted intracranial electrodes,[96,97] and several instances of recurrent spontaneous convulsions in schizophrenic patients receiving repeated electroconvulsive therapy,[54] a definitive answer to this criticism is impossible. A question that can be addressed, however, is whether the phenomenon of clinical epilepsy has kindling-like properties.

There are two lines of evidence supportive of a role for a kindling-like mechanism in clinical epilepsy. First, the progressive nature of experimental kindling would suggest that if kindling-like mechanisms were operative in human epilepsy one should also observe progression or worsening of clinical symptoms. Though there is much debate as to whether epilepsy should be viewed as a progressive disorder, evidence is available indicating that seizures in some patients become more severe, more frequent and more refractory to medical treatment with time.[6] Determining the extent to which clinical seizures are indeed progressive is difficult. Most patients have already experienced one or more spontaneous epileptic events before seeking medical attention. It would be erroneous to assume that these presenting forms of seizure were not preceded by spontaneous subclinical seizures that progressed over time.

A second clinical phenomenon that many have postulated to reflect "kindling" of the human brain is posttraumatic epilepsy.[6,54,86,98] Patients with this condition often present with CPS, months to years after a serious brain insult. During the intervening "silent period" between the initial trauma and the appearance of clinical seizures, injured brain tissue, or bone fragments, may trigger intermittent subclinical epileptiform discharges. Such seizures may alter the excitability in surrounding and distant structures by mechanisms similar to those induced in kindled rats by focal stimulation. It has been hypothesized that chronic recurrent seizures in humans do not become manifest until a large area of the brain has been sufficiently altered or "kindled" by these subclinical seizures.[6]

In assessing the relevance of kindling to clinical temporal lobe epilepsy (TLE; or complex partial seizures), one is forced to make inferences based on indirect evidence. As noted by Engel and Shewmon,[6] it is difficult to derive useful information as to the natural history of human epilepsy given that patients are invariably treated with antiepileptic medications. Available evidence, though circumstantial, suggests that kindling-like mechanisms may account for the progressive worsening of partial epileptic symptoms sometimes observed with TLE.[86] However, as stated by Engel and Shewmon,[6] "this contention is by no means proved, and it is difficult to see how a definitive prospective study could be ethically designed that would convince those who remain in doubt."

D. Conclusions

Since characterized nearly 30 years ago, kindling has become a well-established model with which to study epileptogenesis and the mechanisms that maintain an epileptic state. Given the complexity of cellular events that occur with activation of a single neuron, it is unlikely we will find a single factor that is responsible for the complex network phenomenon that kindling represents. The goal of this chapter was to provide a comprehensive understanding of some of the basic principles of the kindling phenomenon. It is hoped that these fundamental principles that characterize "kindling" can be combined with the ever-growing array of techniques available in molecular biology, electrophysiology, and anatomy to understand better the processes responsible for development and maintenance of the kindled state. As noted by Goddard,[84] "no one believes that the basis of kindling can disappear without a trace, only to reappear in full force a week, a month, or a year later when the stimulus is reapplied. Some trace must exist. None has yet been found." To this end we can view the "trace" as a series of complex and intricate cellular changes that occur within and between cells. It is these changes that culminate, eventually, in the manifestation of generalized motor seizures by stimuli previously shown to produce only brief electrographic responses.

References

1. Beldhuis, H. J., Behavioural characteristics and neuronal mechanisms of amygdala kindling, *Doctoral Dissertation,* 1993, University of Groningen, The Netherlands.
2. Löscher, W. and Schmidt, D., Which animal models should be used in the search for new antiepileptic drugs? A proposal based on experimental and clinical considerations, *Epilepsy Res.,* 10, 119, 1991.
3. Morrell, F., *Kindling and Synaptic Plasticity. The Legacy of Graham Goddard,* Birkhauser, Boston, 1991.
4. Goddard, G. V., The continuing search for mechanism, in *Kindling 2,* Wada J. A., Ed., Raven Press, New York, 1981, 1.
5. Delgado, J. M. R. and Sevillan, M., Evolution of repeated hippocampal seizures in cats, *Electroenceph. Clin. Neurophysiol.,* 13, 722, 1961.
6. Engel, J. J. and Shewmon, D. A., Impact of the kindling phenomenon on clinical epileptology, in *Kindling and Synaptic Plasticity,* Morrell, F., Ed., Birkhauser, Boston, 1991, 196.
7. Goddard, G. V., Development of epileptic seizures through brain stimulation at low intensity, *Nature,* 214, 1020, 1967.
8. Goddard, G. V., McIntyre, D. C., and Leach, C. K., A permanent change in brain function resulting from daily electrical stimulation, *Exp. Neurol.,* 25, 295, 1969.
9. Racine, R. J., Modification of seizure activity by electrical stimulation. I. Afterdischarge threshold, *Electroenceph. Clin. Neurophysiol.,* 32, 269, 1972.

10. Racine, R. J., Modification of seizure activity by electrical stimulation. II: Motor seizure, *Electroenceph. Clin. Neurophysiol.*, 32, 281, 1972.

11. Paxinos, G. and Watson, C., *The Rat in Stereotaxic Coordinates*, Academic Press, New York, 1986.

12. Pellegrino, L.J., Pellegrino, A. S., and Cushman, A. J., *A Stereotaxic Atlas of the Rat Brain*, Plenum Press, New York, 1979.

13. Swanson, L. W., *Brain Maps: Structure of the Rat Brain*, Elsevier, Amsterdam, 1993.

14. Molino, A. and McIntyre, D. C., Another inexpensive headplug for the electrical recording and/or stimulation of rats, *Physiol. Behav.*, 9, 273, 1972.

15. Pellegrino, L. J. and Cushman, A. J., Use of the stereotaxic technique, in Methods in Psychobiology, Vol. 1, Myers, R. D., Ed., Academic Press, New York, 1971.

16. Skinner, J. E., *Neuroscience: A Laboratory Manual*, W. B. Saunders, Philadelphia, 1971.

17. Cooley, R. K. and Vanderwolf, C. H., *Stereotaxic Surgery in the Rat: A Photographic Series*, University of Western Ontario, London, Ontario, 1977.

18. Moore, R. Y. , Methods for selective-restrictive lesion placement in the central nervous system, in *Neuroanatomical Tract Tracing Methods*, Heimer, L. and Robards, M. J., Eds., Plenum Press, New York, 1981, chap. 2.

19. Horsley, V. and Clarke, R. H., The structure and functions of the cerebellum examined by a new method, *Brain*, 31, 45, 1908.

20. Whishaw, I. Q., Cioe, J. D. D., Previsich, N., and Kolb, B., The variability of the interaural line vs the stability of bregma in rat stereotaxic surgery, *Physiol. Behav.*, 19, 719, 1977.

21. Racine, R. J., Burnham, W. M., and Gartner, J., First trial motor seizures triggered by amygdaloid stimulation in the rat, *Electroenceph. Clin. Neurophysiol.*, 35, 487, 1973.

22. Corcoran, M. E. and Cain, D. P., Kindling of seizures with low-frequency electrical stimulation, *Brain Res.*, 196, 262, 1980.

23. Cain, D. P. and Corcoran, M. E., Kindling with low-frequency stimulation: generality, transfer, and recruiting effects, *Exp. Neurol.*, 73, 219, 1981.

24. Freeman, F. G. and Jarvis, M. F., The effect of interstimulation interval on the assessment and stability of kindled seizure thresholds, *Brain Res. Bull.*, 7, 629, 1981.

25. Mohapel, P., Dufresne, C., Kelly, M. E., and McIntyre, D. C., Differential sensitivity of various temporal lobe structures in the rat to kindling and status epilepticus induction, *Epilepsy Res.*, 23, 179, 1996.

26. Moshe, S. L., Albala, B. J., Ackermann, R. F., and Engel, J. J., Increased seizure susceptibility of the immature brain, *Brain Res.*, 283(1), 81, 1983.

27. Peterson, S. L., Albertson, T. E., and Stark, L. G., Intertrial intervals and kindled seizures, *Exp. Neurol.*, 71, 144, 1981.

28. Lothman, E. W., Hatlelid, J. M., Zorumski, C. F., Conry, J. A., Moon, P. F., and Perlin, J. B., Kindling with rapidly recurring hippocampal seizures, *Brain Res.*, 360, 83, 1985.

29. Lothman, E. W. and Williamson, J. M., Closely spaced recurrent hippocampal seizures elicit two types of heightened epileptogenesis: a rapidly developing, transient kindling and a slowly developing, enduring kindling, *Brain Res.*, 649, 71, 1994.

30. Elmer, E., Kokaia, M., Kokaia, Z., Ferencz, I., and Lindvall, O., Delayed kindling development after rapidly recurring seizures: relation to mossy fiber sprouting and neurotrophin, GAP-43 and dynorphin gene expression, *Brain Res.*, 712, 19, 1996.

31. McIntyre, D. C., Rajala, J., and Edson, N., Suppression of amygdala kindling with short interstimulus intervals: effect of norepinephrine depletion, *Exp. Neurol.*, 95, 391, 1987.
32. McIntyre, D. C., Kelly, M. E., and Dufresne, C., Suppression of amygdala kindling with massed stimulation: effect of noradrenaline antagonists, *Brain Res.*, 561, 279, 1991.
33. Le Gal La Salle, G., Amygdaloid kindling in the rat: regional differences and general properties, in *Kindling 2,* Wada, J. A., Ed., Raven Press, New York, 1981, 31.
34. McIntyre, D. C., Kelly, M. E., and Armstrong, J. N., Kindling in the perirhinal cortex, *Brain Res.*, 615, 1, 1993.
35. Dyer, R. S., Swartzwelder, H. S., Eccles, C. U., and Annau, Z., Hippocampal afterdischarges and their post-ictal sequelae in rats: a potential tool for assessment in CNS neurotoxicity, *Neurobehav. Toxicol.*, 1, 5, 1979.
36. Burnham, W. M., Primary and 'transfer' seizure development in the kindled rat, in *Kindling,* Wada, J. A., Ed., Raven Press, New York, 1976, 61.
37. Grace, G. M., Corcoran, M. E., and Skelton, R. W., Kindling with stimulation of the dentate gyrus. I. Characterization of electrographic and behavioral events, *Brain Res.*, 509, 249, 1990.
38. Lerner-Natoli, M., Rondouin, G., and Baldy-Moulinier, M., Hippocampal kindling in the rat: intrastructural differences, *J. Neurosci. Res.*, 12, 101, 1984.
39. Leung, L. S., Hippocampal electrical activity following local tetanization. I. Afterdischarges, *Brain Res.*, 419, 173, 1987.
40. Racine, R. J., Rose, P. A., and Burnham, W. M., Afterdischarge thresholds and kindling rates in the dorsal and ventral hippocampus and dentate gyrus, *Can. J. Neurol. Sci.*, 4, 273, 1977.
41. Racine, R. J., Modification of seizure activity by electrical stimulation: cortical areas, *Electroenceph. Clin. Neurophysiol.*, 38, 1, 1975.
42. Altman, I. M. and Corcoran, M. E., Facilitation of neocortical kindling by depletion of forebrain noradrenaline, *Brain Res.*, 270, 174, 1983.
43. Burnham, W. M., Cortical and limbic kindling: Similarities and differences, in *Limbic Mechanisms: The Continuing Evolution of the Limbic System Concept,* Livingston, K. E. and Hornykiewicz, O., Eds., Plenum Press, New York, 1978, 507.
44. Kelly, M. E., McIntyre, D. C., and Staines, W. A., Perirhinal connections to the orbital and frontal cortex: anatomy and kindling, *Soc. Neurosci. Abstr.*, 19, 1467, 1993.
45. McIntyre, D. C., Effects of focal vs. generalized kindled convulsions from anterior neocortex or amygdala on CER acquisition in rats, *Physiol. Behav.*, 23, 855, 1979.
46. Seidel, W. T. and Corcoran, M. E., Relations between amygdaloid and anterior neocortical kindling, *Brain Res.*, 385, 375, 1986.
47. Lothman, E. W., Williamson, J. M., and Van Landingham, K. E., Effect of phenytoin on kindled responses: influence of stimulus parameters and comparison of intravenous vs, intra-peritoneal drug administration, *Epilepsia,* 31, 632, 1990.
48. Rundfeldt, C., Honack, D., and Löscher, W., Phenytoin potently increases the threshold for focal seizures in amygdala-kindled rats, *Neuropharmacology,* 29, 845, 1990.
49. Lothman, E. W., Williamson, J. M., and Van Landingham, K. E., Intraperitoneal phenytoin suppresses kindled responses: effects on motor and electrographic seizures, *Epilepsy Res.*, 9, 11, 1991.

50. Rundfeldt, C. and Löscher, W., Anticonvulsant efficacy and adverse effects of phenytoin during chronic treatment in amygdala-kindled rats, *J. Pharmacol. Exp. Ther.*, 266, 216, 1993.

51. Löscher, W. and Honack, D., Differences in anticonvulsant potency and adverse effects between dextromethorphan and dextrorphan in amygdala-kindled and non-kindled rats, *Eur. J. Pharmacol.*, 238, 191, 1993.

52. McIntyre, D. C. and Kelly, M. E., Are differences in dorsal hippocampal kindling related to amygdala-piriform area excitability?, *Epilepsy Res.*, 14, 49, 1993.

53. McNamara, J. O., Kindling: an animal model of complex partial epilepsy, *Ann. Neurol.*, 16, 72, 1984.

54. Sato, M., Racine, R. J., and McIntyre, D. C., Kindling: basic mechanisms and clinical validity, *Electroenceph. Clin. Neurophysiol.*, 76, 459, 1990.

55. Löscher, W., Wahnschaffe, U., Honack, D., and Rundfeldt, C., Does prolonged implantation of depth electrodes predispose the brain to kindling?, *Brain Res.*, 697, 197, 1995.

56. Burchfiel, J. L. and Applegate, C. D., Stepwise progression of kindling: perspectives from the kindling antagonism model, *Biobehav. Rev.*, 13, 289, 1989.

57. Cain, D. P., Seizure development following repeated electrical stimulation of central olfactory structures, *Annals NY Acad. Sci.*, 200, 1977.

58. Cain, D. P., Corcoran, M. E., Desborough, K. A., and McKitrick, D. J., Is the deep pre-pyriform cortex a crucial forebrain site for kindling?, *Soc. Neurosci. Abstr.*, 14, 1149, 1988.

59. McIntyre, D. C. and Kelly, M. E., Is the pyriform cortex important for limbic kindling?, in *Kindling 4*, Wada. J. A., Ed., Raven Press, New York, 1990, 21.

60. McIntyre, D. C. and Racine, R. J., Kindling mechanisms: current progress on an experimental epilepsy model, *Progr. Neurobiol.*, 27, 1, 1986.

61. McIntyre, D. C., Kindling and the pyriform cortex, in *Kindling 3*, Wada, J. A., Ed., Raven Press, New York, 1986, 249.

62. Racine, R. J., Paxinos, G., Mosher, J. M., and Kairiss, E. W., The effects of various lesions and knife-cuts on septal and amygdala kindling in the rat, *Brain Res.*, 454, 264, 1988.

63. Burchfiel, J. L., Applegate, C. D., and Samoriski, G. M., Evidence for piriform cortex as a critical substrate for the stepwise progression of kindling, *Epilepsia*, 31, 632, 1990.

64. Löscher, W. and Ebert, U., The role of the piriform cortex in kindling, *Progr. Neurobiol.*, 50, 427, 1996.

65. Kelly, M. E. and McIntyre, D. C., Perirhinal cortex involvement in limbic kindled seizures, *Epilepsy Res.*, 26, 233, 1996.

66. McIntyre, D. C. and Goddard, G. V., Transfer, interference and spontaneous recovery of convulsions kindled from the rat amygdala, *Electroenceph. Clin. Neurophysiol.*, 35, 533, 1973.

67. Burnham, W. M., Primary and 'transfer' seizure development in the kindled rat, *Can. J. Neurol. Sci.*, 2, 417, 1975.

68. McIntyre, D.C. and Edson, N., Facilitation of secondary site kindling in the dorsal hippocampus following forebrain bisection, *Exp. Neurol.*, 96, 569, 1987.

69. McCaughran, J. A., Corcoran, M. E., and Wada, J. A., Role of the forebrain commissures in amygdaloid kindling in rats, *Epilepsia*, 19, 19, 1978.

70. Racine, R. J., Kindling: the first decade, *Neurosurgery*, 3(2), 234, 1978.

71. McNamara, J. O., Constant Byrne, M., Dasheiff, R. M., and Fitz, J. G., The kindling model of epilepsy: a review, *Progr. Neurobiol.*, 15, 139, 1980.

72. Corcoran, M. E., Characteristics and mechanisms of kindling, in *Sensitization of the Nervous System*, Kalivas, P. and Barnes, C., Eds., Telford Press, New Jersey, 1988, 81.

73. Morrell, R. and Tsura, N., Kindling in the frog: development of spontaneous epileptiform activity, *Electroenceph. Clin. Neurophysiol.*, 40, 1, 1976.

74. Rial, R. V. and Gonzalez, J., Kindling effect in the reptilian brain: motor and electrographic manifestations, *Epilepsia*, 19, 581, 1978.

75. Leech, C. K. and McIntyre, D. C., Kindling rates in inbred mice: an analog to learning?, *Behav. Biol.*, 16, 439, 1976.

76. Wada, J. A. and Sato, M., Generalized convulsive seizures induced by daily electrical amygdaloid stimulation in Senegalese baboons (*Papio papio*), *Neurology*, 40, 413, 1974.

77. Wada, J. A., Amygdaloid and frontal cortical kindling in subhuman primates, in *Limbic Epilepsy and the Dyscontrol Syndrome*, Girgis, M. and Kiloh, L. G., Eds., Elsevier/North-Holland Biomedical, Amsterdam, 1980, 133.

78. Watanabe, Y., Johnson, R. S., Butler, L. S., Binder, D. K., Spiegelman, B. M., Papaioannou, V. E., and McNamara, J., Null mutation of c-fos impairs structural and functional plasticities in the kindling model of epilepsy, *J. Neurosci.*, 16(12), 3827, 1996.

79. Campbell, I. L., Abraham, C. R., Masliah, E., Kemper, P., Inglis, J. D., Oldstone, M. B., and Mucke, L., Neurologic disease induced in transgenic mice by cerebral overexpression of interleukin 6, *Proc. Natl. Acad. Sci. U.S.A.*, 90(21), 10061, 1993.

80. Tecott, L. H., Sun, L. M., Akana, S. F., Strack, A. M., Lowenstein, D. H., Dallman, M. F., and Julius, D., Eating disorder and epilepsy in mice lacking 5-HT2c serotonin receptors, *Nature*, 374(6522), 542, 1995.

81. Löscher, W., Horstermann, D., Honack, D., Rundfeldt, C., and Wahnschaffe, U., Transmitter amino acid levels in rat brain regions after amygdala-kindling or chronic electrode implantation without kindling: evidence for a pro-kindling effect of prolonged electrode implantation, *Neurochem. Res.*, 18, 775, 1993.

82. Blackwood, D. H. R., Martin, M. J., and McQueen, J. K., Enhanced rate of kindling after prolonged electrode implantation into the amygdala of rats, *J. Neurosci. Meth.*, 5, 343, 1982.

83. Burnham, W. M., The GABA hypothesis of kindling, in *Kindling 4*, Wada, J. A. , Ed., Raven Press, New York, 1990, 127.

84. Goddard, G. V., The kindling model of epilepsy, *Trends Neurosci.*, 6(7), 275, 1983.

85. Peterson, S. L. and Albertson, T. E., Neurotransmitter and neuromodulator function in the kindled seizure and state, *Progr. Neurobiol.*, 19, 237, 1982.

86. Engel, J. J. and Cahan, L., Potential relevance of kindling to human partial epilepsy, in *Kindling 3*, Wada, J. A., Ed., Raven Press, New York, 1986, 37.

87. Ebert, U., Rundfeldt, C., and Löscher, W., Development and pharmacological suppression of secondary afterdischarges in the hippocampus of amygdala-kindled rats, *European J . Neruosci.*, 7, 732, 1995.

88. Clark, M., Post, R. M., Weiss, S. R. B., Cain, C. J., and Nakajima, T., Regional expression of *c-fos* mRNA in rat brain during the evolution of amygdala kindled seizures, *Mol. Brain Res.*, 11, 55, 1991.

89. Khurgel, M., Racine, R. J., and Ivy, G. O., Kindling causes changes in the composition of the astrocytic cytoskeleton, *Brain Res.,* 592, 338, 1992.

90. Kokaia, Z., Kelly, M. E., Elmer, E., Kokaia, M., McIntyre, D. C., and Lindvall, O., Seizure-induced differential expression of messenger RNAs for neurotrophins and their receptors in genetically FAST and SLOW kindling rats, *Neuroscience,* 75(1), 197, 1996.

91. McIntyre, D. C. and Stuckey, G. N., Dorsal hippocampal kindling and transfer in split-brain rats, *Exp. Neurol.,* 87, 86, 1985.

92. Corcoran, M. E. and Mason, S. T., Role of forebrain catecholamines in amygdaloid kindling, *Brain Res.,* 190, 473, 1980.

93. Albertson, T. E., Joy, R. M., and Stark, L. G., A pharmacological study in the kindling model of epilepsy, *Neuropharmacology,* 23(10), 1117, 1984.

94. McNamara, J. O., Russell, R. D., Rigsbee, L., and Bonhaus, D. W., Anticonvulsant and anti-epileptogenic actions of MK-801 in the kindling and electroshock models, *Neuropharmacology,* 27(6), 563, 1988.

95. Wada, J. A., Ed., *Kindling,* Plenum Press, New York, 1976.

96. Monroe, R. R., Limbic ictus and atypical psychoses, *J. Nerv. Ment. Dis.,* 170, 711, 1982.

97. Sramka, M., Sedlak, P., and Nadvornik, P., in *Neurosurgical Treatment in Psychiatry, Pain and Epilepsy,* Sweet, W. H., Ed., University Park Press, Baltimore, 1976, 651.

98. Goddard, G. V., Maru, E., and Kairiss, E. W., The kindling effect and epilepsy, *Neurosurgeons,* 5, 213, 1986.

Chapter **4**

Rapid Kindling: Behavioral and Electrographic

Janet L. Stringer

Contents

I.   Introduction

Repetitive electrical stimulation of discrete brain areas is well recognized as an experimental means to study chronic partial epilepsy. With appropriate intervals of repetitive stimulation, two conditions occur. The first is called kindling and refers to the progressive enhancement of the behavioral seizures with each successive stimulation. The second condition is called the kindled state. This is an enduring and possibly permanent ability to trigger a reproducible motor convulsion. Kindling and the kindled state can be produced by stimulation in a variety of brain regions, but the hippocampus and the amygdala are most commonly utilized. Early in kindling there are mild behavioral seizures that are thought to be models of complex partial seizures in humans and appear to represent seizure activity in limbic circuits. Later in kindling, more severe motor convulsions are seen that may model secondary generalization of the partial seizures.

Kindling is customarily studied with amygdala stimulation administered once daily. Using this protocol (as described in Chapter 3), kindling takes approximately 2 weeks and the establishment of a stable kindled state, where each behavioral seizure is the same, requires 3 to 5 weeks. In the early studies of kindling it was observed that the ability of a stimulus train to elicit a seizure soon after an initial seizure was dependent on the interval between the two seizures.[1,2] When traditional kindling stimuli (60 Hz, 1 s trains to the amygdala) are administered less than 20 min apart they failed to trigger afterdischarges.[1] Mucha and Pinel[3] determined that stimulus trains needed to be administered at least 60 to 90 min apart in order to reliably trigger a kindled motor seizure. As opposed to motor seizures, afterdischarges can be triggered as close together as every 30 min. Similar results have been obtained when stimulating in the hippocampus.[4] These results suggest that the traditional kindling stimulus parameters activate inhibitory processes that block subsequent motor seizures and afterdischarges when stimulus trains are administered close together.

Several years ago, while trying to find a way to elicit several seizures within a 45- to 60-min period, Lothman and colleagues[5] discovered that repeated administration of suprathreshold stimuli could elicit seizures several minutes apart. Short duration trains (1 s) were compared to longer duration trains (10 s) at different stimulus intensities. At each stimulus intensity the short duration trains induced a 60- to 90-min refractory period in which another seizure could not be initiated. The longer duration trains did not induce this refractory period and seizures could be elicited every few minutes. This discovery led to a stimulus protocol for kindling in which the stimulus trains are administered every 5 min. This protocol is now often referred to as rapid kindling and it results in kindled motor seizures and a lengthened afterdischarge. Some of the features of this rapid kindling protocol have also been utilized in anesthetized rats, in which no motor seizures occur. Repeated stimulation leads to a progressive lengthening of the afterdischarge in the anesthetized rat that mimics the lengthening of the afterdischarge in the awake animal during kindling acquisition. Since there are no motor seizures, this procedure can be referred to as electrographic kindling. This chapter details the methodology for rapid kindling in awake and rapid electrographic kindling in urethane-anesthetized rats.

II. Methodology

A. Rapid Kindling in Awake Rats

1. Surgery and Implantation of Electrodes

Details of intracranial electrode implantation are also presented in Chapter 3. Here, we focus on the differences between more traditional kindling and rapid kindling. All surgery is done using sterile gloves and instruments. The person doing the surgery also wears a mask. Almost all experiments have been done using adult male Sprague-Dawley rats (250 to 300 g). Rats are anesthetized with ketamine/xylazine (50/25 mg/kg, i.p.). The hair on the top of the head is shaved and then the rat is placed into a stereotaxic frame. After a scalp incision, the skull is carefully cleaned and dried. A small burr hole is drilled into the frontal sinus for the ground electrode and another burr hole is drilled into the skull for stimulating/recording electrode placement. A bipolar stimulating electrode is placed into the ventral hippocampus on either side (AP –3.6, ML 4.9, DV –5.0 to dura, incisor bar +5.0). Either purchased concentric electrodes (SNEX, Rhodes Biomedical Inst., Tujunga, CA) or homemade "twist" electrodes can be used. The twist electrodes are made by twisting coated 0.01" diameter stainless steel wire (several sources, including World Precision Inst., Sarasota, FL) and then scraping the coating off the distal 1 to 2 mm with a scalpel blade. Male pin connectors (one source is Wire Pro, Inc. "Relia-Tac," available through Allied Electronics, Ft. Worth, TX) are soldered or crimped to the other end. A ground electrode (coated 0.01" stainless steel wire) is placed in a small hole drilled into the frontal sinus. All electrodes are connected to male pin connectors which are fitted into a matching strip connector. The electrodes and connector strips are attached to the skull with dental acrylate. The stability of the headset is greatly enhanced by putting one jeweler's screw in the skull. Additional recording electrodes can be inserted at this time. The male Amphenol connections are protected by putting a "cap" on the connector strip. This cap consists of a length of strip connector that covers the exposed pins and is held in place by one screw in the same way that the recording cable is held in place. After the dental acrylate has dried, the animal is allowed to recover from the anesthesia.

2. Determination of Afterdischarge Threshold

One week after surgery, experiments are initiated. Animals are placed in separate plexiglass cages, connected to stimulating/recording equipment and allowed 30 min acclimation. Electrographic recordings should be amplified and are usually displayed on a chart recorder. A constant-current, isolated stimulator is used to deliver 1 msec biphasic square-wave pulses. An electronic switch allows stimulation and recording through the same electrode. During stimulation, leads from the animal are connected to the stimulator while the recording amplifier is grounded. After stimulation, the electrode leads are transiently connected to ground by a relay circuit before the recording mode is activated.

Afterdischarge thresholds are determined using an ascendancy method with stimuli given every minute until an afterdischarge is produced.[5] All stimulus trains are 10 s in duration and 50 Hz. The current is initially set at 40 µA and a stimulus train is administered. If no afterdischarge is elicited, the stimulus intensity is increased in 10 µA steps until 150 µA is reached and then the increases are changed to 20 µA. Stimulus trains for afterdischarge threshold determination are administered 1 min apart. The afterdischarge threshold for most animals is below 100 µA. Levels higher than this suggest less than optimal electrode placement. Animals should be excluded if afterdischarges are not elicited by 250 µA, as this indicates poor electrode placement. For elicitation of seizures, the suprathreshold current of 400 µA is used when the afterdischarge threshold is less than 150 µA and 500 µA when the threshold is between 150 and 250 µA. The supramaximal stimulus intensity is necessary to overcome adaptation and relative refractoriness.

3.   Measurement of Behavioral Seizures and Afterdischarge Durations

Behavioral seizures and electrographic afterdischarges are assessed in response to each stimulus train. Behavioral seizures are scored on the standard 5 point scale.[6] No motor seizure activity is class 0. Class 1 seizures consist of wet dog shakes, facial twitches, and chewing. Class 2 seizures include head bobbing, with or without head jerking. Class 1 and class 2 seizures are commonly considered mild limbic seizures. Class 3 seizures are intermediate in severity and consist of fore-limb clonus. We found that this class designation is rarely used. Class 4 and class 5 seizures are more severe motor seizures. Class 4 seizures consist of forelimb clonus with rearing and class 5 seizures include loss of balance. For consistency of the rating, it is best if all of the motor seizures in a set of experiments are rated by the same person.

The duration of the afterdischarge in the ventral hippocampus is measured from the end of the stimulus train to the end of the primary discharge (Figure 4.1A). The afterdischarge is defined as high amplitude spikes or polyspike epileptiform activity having at least twice the amplitude of the background EEG activity and present for at least 3 s after the end of the stimulus train. In general, the afterdischarge duration is simply measured directly from the chart recorder and is defined as the time from the end of the stimulus train to the end of the afterdischarge. Measurements of the afterdischarge duration are relatively simple unless a secondary afterdischarge appears (Figure 4.1B). This is an afterdischarge that appears after clear termination of the primary afterdischarge. The secondary afterdischarge is not generally included in the afterdischarge duration measurement, because it appears only after the animal has had many seizures.

A Measurement of Afterdischarge Duration

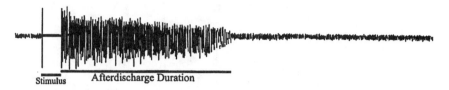

Stimulus   Afterdischarge Duration

B Appearance of Secondary Afterdischarge

Stimulus

FIGURE 4.1

Afterdischarge in awake rats. The measurement of the afterdischarge duration is illustrated in (A) on a stylized afterdischarge. The stimulus duration is 10 s. (B) The appearance of a secondary afterdischarge is illustrated.

4.   Stable Kindled Responses and Protocol to Measure Drug Effects

Animals are kindled using 20 or 50 Hz stimulus trains, 10 s in duration, using the suprathreshold stimulus intensity (usually 400 μA) every 5 min for 6 h, for a total of 72 stimulus trains.[5,7] During this initial kindling period there is not a progressive increase in afterdischarge duration and worsening of the behavioral seizures (Figure 4.2). Most of the behavioral seizures are mild limbic seizures (score 1) with occasional seizures with a score of 2. About halfway through the day a full motor seizure (score 5) will appear. Several additional class 5 seizures will appear through the rest of the day at irregular intervals. The afterdischarge duration follows a similar pattern. Periodically there is a longer afterdischarge that correlates with the worsening behavioral seizures.

Two to 3 d after the initial 1 d treatment with stimulus trains administered every 5 min for 6 h, the animals are put on an alternate day schedule (for example, Monday, Wednesday, Friday) with stimulus trains administered every 30 min. Twelve stimulus trains are given over a 6 h period. On each day of testing, the afterdischarge threshold is determined in the morning and again at the end of the day. Rats often need a "priming" stimulus at the beginning of each day to express a fully kindled motor seizure, therefore 5 min after the final determination of the afterdischarge threshold, a 400 μA stimulus train is administered. The response to this priming stimulus is not analyzed.

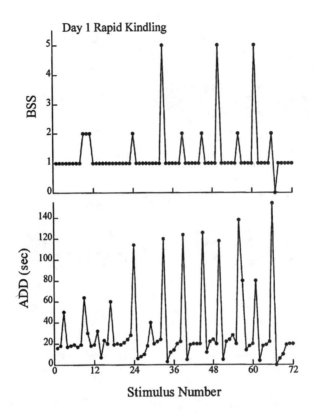

FIGURE 4.2

The behavioral seizure score (BSS) and afterdischarge duration (ADD) in response to each stimulus train on the day of kindling are graphed. Stimulus trains (50 Hz, 10 s, 400 μA, 1 msec biphasic) are administered every 5 min for 6 h (total 72 stimulations). (Adapted from Lothman, E. W. et al., *Brain Res.,* 360, 83, 1988. With permission.)

Animals are stimulated with this alternate-day protocol until a stable kindled state is reached. Here, a stable kindled state is defined as having been achieved when the animals respond to all stimulus trains with class 4 and 5 seizures and the afterdischarge durations are within 15% of each other.[7] Counting the initial kindling day, it takes about 2 weeks for all of the rats in a set to reach a stable kindled state. Animals are entered into drug trials after at least three consecutive testing days (Monday, Wednesday, and Friday) of stable kindled behavioral seizures.

On the day on which the drug is to be given the protocol is modified slightly.[7] After determination of the afterdischarge threshold, four suprathreshold stimulus trains are administered 30 min apart to serve as controls. The drug is then administered and the stimulus trains are administered every 30 min for an additional 6 h for a total of 16 stimulus trains (4 predrug and 12 postdrug). Figure 4.3 shows the behavioral seizure score for one animal in the drug testing protocol. Different doses of a drug can be tested in one animal or replicability can be tested by administering the same dose of a drug to an animal on different days. After drug administration the animals still receive stimulus trains on an alternate-day basis until the responses

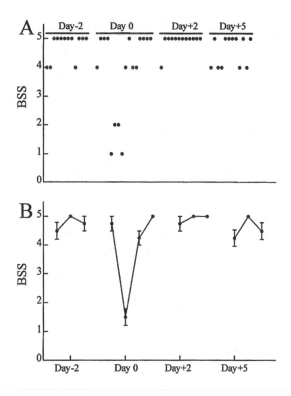

FIGURE 4.3

Drug protocol using rapidly recurring hippocampal seizures. This figure illustrates the results from a single animal which had been kindled using the rapid kindling protocol and achieved stable kindled responses. The animal is now on an alternate-day testing protocol of stimulation every 30 min for 6 h (total 12 stimulations per day). An example of one of these days is day –2. Day 0 is the drug testing day. The drug in this case was valproic acid (300 mg/kg) and it was administered intraperitoneally after the fourth stimulus train on day 0. Day 0 is extended 2 h (four stimulations) in order to have four stimulations prior to drug administration and 6 h of data after drug administration. After day 0, the animal is continued on a Monday, Wednesday, Friday schedule of testing. Day +2 and day +5 are Friday and Monday, respectively. (A) The behavioral seizure scores (BSS) in response to each individual stimulation. (B) The averages of four behavioral seizure scores in response to consecutive stimulations (±SEM) are presented as response blocks. (Adapted from Lothman, E. W. et al., *Epilepsy Res.*, 2, 367, 1988. With permission.)

have returned to control values. An additional 2 d are then given for recovery before the animal receives another dose of drug or a different drug. A summary of the effects of seven commonly used antiepileptic drugs is presented in Table 4.1.

It is possible to use this protocol for longer periods of time.[7] Criteria for kindled behavioral seizures are met for up to 18 h, but after this the motor responsiveness decreases. In addition, after prolonged testing there is suppression of normal responses to stimulation for up to a week after the prolonged test day. If the test procedure is carried out every day instead of the alternate-day schedule, there is a gradual decrease in responsiveness on days 4 and 5. Both the behavioral seizure score decreases and the afterdischarge duration shortens. As with the prolonged

TABLE 4.1
Effect of Antiepileptic Drugs Against Rapidly Recurring Hippocampal Seizures in Rats

Drug	Dose (mg/kg i.p.)	Suppression of kindled motor seizures[a]	Suppression of limbic seizures[b]	Shortening of afterdischarge[c]
Carbamazepine	30	+	+	++
	50	+	+	++
Phenytoin	80	+	+	NE
Phenobarbital	30	+	+	+
	60	+	+	++
Primidone	150	+	+	+
Valproic acid	200	+	+	+
	300	+	+	++
Diazepam	5	+	+	+
	10	+	+	+
Ethosuximide	300	NE	NE	NE

[a] For motor seizures, + indicates the drug reduced seizures to $\leq$ class 3 seizure.

[b] For limbic seizures, + indicates the drug reduced seizures to $\leq$ class 1 seizure.

[c] For afterdischarge duration, + indicates the drug reduced afterdischarge duration at least 15%; ++ indicates the drug reduced afterdischarge duration at least 50%.

NE is no effect.

Data adapted from Reference 8.

testing on one day, it takes about a week to recover normal responses after a week of every-day testing.

For data presentation and analysis, it is convenient to average the responses (both behavioral seizure scores and afterdischarge durations) to four consecutive stimulus trains (Figure 4.3B). Thus, results are presented as response blocks. This presentation has proven quite useful for studying the effects of antiepileptic drugs.[8] Values are presented as means ± standard errors of the means and statistical analysis of the animals within one group can be done, as well as comparisons between groups of animals.

Toxicity of drugs can also be determined with some simple observations. Sedation can be tested by observing the animal's reaction to sudden noises or to touch. Motor dysfunction can be tested by observation of the gait and testing of the righting reflex and muscle tone. More complicated toxicity tests (as described in Chapter 8) can be performed within the limitations of having the animal's head attached to a recording cable and the time limits (30 min) between stimulations.

The protocol described above can be used to test the effects of drugs on kindled seizures. The above protocol cannot be used to examine drug effects on the kindling process. However, a protocol of stimulating every 30 min can be utilized to examine the kindling process (Figure 4.4).[9] After implantation of the electrodes and determination of the afterdischarge threshold, suprathreshold stimulus trains of 20 Hz for

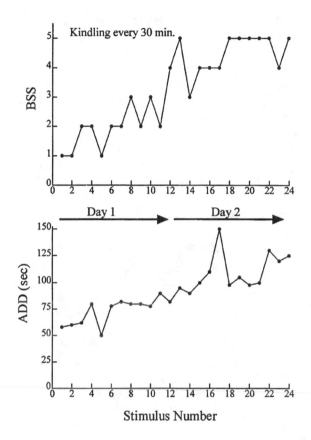

FIGURE 4.4

Protocol to test kindling acquisition. These graphs present the behavioral seizure score (BSS) and afterdischarge duration (ADD) for an animal that received 20 Hz stimulation (10 s, 400 μA) every 30 min over the course of two consecutive days (6 h each day). Notice the relatively steady increase in seizure score and afterdischarge duration. (Adapted from Lothman, E. W. and Williamson, J. M., *Epilepsy Res.*, 14, 209, 1993. With permission.)

2 or 10 s are administered every 30 min on consecutive days, for up to 4 d. This procedure results in a gradual and steady increase in the behavioral seizure score, until the rat is consistently having class 5 seizures in response to each stimulus train. The afterdischarge duration gradually increases over the first 2 d and then plateaus at an increased level.

B.   Electrographic Kindling in Urethane-Anesthetized Rats

1.   Seizure Definition

Several years ago, to explore mechanisms of epileptogenesis and changes in extracellular ions in the hippocampus, responses to stimulus trains administered to either the CA3 region of the dorsal hippocampus or to the angular bundle were characterized in the intact rat.[10,11] Patterns of activation in response to trains of electrical stimulation were described in both CA1 and the dentate gyrus. Maximal activation of the dentate gyrus is characterized by the presence of bursts of large-amplitude population spikes associated with a secondary rise in the extracellular potassium concentration and an abrupt negative shift of the extracellular DC potential. In the normal animal, bilateral maximal dentate activation is necessary for an afterdischarge to occur.[12] Maximal dentate activation in the dentate gyrus in intact rats is always associated with epileptiform activity in CA1, CA3, and the entorhinal cortex, suggesting that maximal dentate activation is an indicator of reverberatory activity throughout the hippocampal-parahippocampal circuit (Figure 4.5).[13] In other words, when maximal dentate activation is present there is always seizure activity in the entorhinal cortex and hippocampus proper. Since the onset and presence of maximal dentate activation are quite readily detectable, it can be used as a marker for the onset and duration of hippocampal seizure activity.

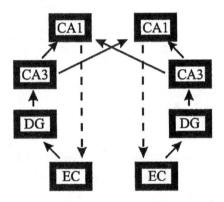

FIGURE 4.5
The main excitatory connections within the hippocampal-parahippocampal circuit are shown. On each side the entorhinal cortex (EC) sends afferents to the dentate gyrus (DG), which sends excitatory afferents to the CA3 region of the hippocampus proper. CA3 then projects to the CA1 region on both sides of the brain. One of the outputs of the CA1 region is back to the entorhinal cortex through the subiculum (dashed line). There are also some connections between the two sides of the brain that are not shown.

2.   Surgery and Implantation of Electrodes

To date, adult rats (male and female) from Wistar, Sprague-Dawley, and Long Evans strains have been used. In addition, rats as young as 10 d old have been successfully used.[14] The surgery does not have to be sterile, but should be clean. The rats are anesthetized with urethane (1.2 g/kg i.p.) and placed in a stereotaxic frame. Urethane is safe for acute usage and is relatively long-lasting (no supplementation needed), but too much urethane will inhibit seizure onset.[15] The animal should have sufficient urethane to reach surgical anesthesia, but no more. Other anesthetics can be used, but with caution. Ketamine/xylazine/acepromazine (25/5/0.8 mg/kg i.p.) has a relatively short duration of action and the seizure durations will be quite variable as the anesthetic wears off. This is the anesthetic that we use if we want the rats to recover rapidly from the anesthetic and we are not testing the effect of drugs on the seizure parameters. Inhalational anesthetics can also be used if the equipment is available. Barbiturates, used as anesthetics, will block seizure onset.

The scalp is split and the skull carefully cleaned and dried. Small burr holes are drilled in the skull for stimulating and recording electrode placement. These burr holes should be large enough to move the electrodes if necessary. A stimulating electrode (we use a concentric bipolar, SNEX 100, Rhodes Biomedical Inst., Tujunga, CA) is placed in the CA3 region of the left dorsal hippocampus at an angle of 5 to 10° (AP −3 mm from bregma, 4 mm lateral, Figure 4.6). The placement of the stimulating electrode is critical for the initiation of the seizure discharge. It is most important that the electrode be placed in the middle of the hippocampus, too high or too low will not work well. To achieve the best placement, put a recording electrode in the contralateral CA1 cell layer (AP −3 mm from bregma, 2 mm lateral, depth 2 to 3 mm) and adjust the position of the stimulating electrode to give the largest possible evoked response in CA1 (Figure 4.7). The stimulation will often still work if the stimulating electrode is more medial. The AP dimension is not very critical as long as the electrode is in the hippocampus. Recording electrode(s) are generally placed on the opposite side of the brain from the stimulating electrode, because there is more space and because onset of seizure activity on the contralateral side to the stimulating electrode indicates the presence of reverberatory seizure activity.[13]

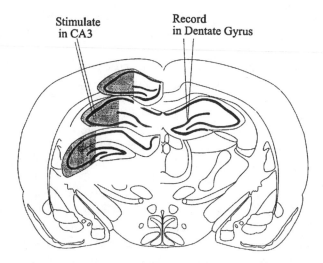

Stimulate
in CA3

Record
in Dentate Gyrus

FIGURE 4.6

Placement of stimulating and recording electrodes for electrographic kindling. This scheme presents the optimal placement of the stimulating electrode for the elicitation of repeated episodes of maximal dentate activation. The tip of the stimulating electrode should be within the gray shaded area of one hippocampus. Three hippocampal sections (one anterior and one posterior to the main section) are presented on the left to indicate the extent of the hippocampus that can be successfully stimulated. Recording is most commonly done in the dentate gyrus on the opposite side from the stimulating electrode (as shown), but recording can be done in other areas also.

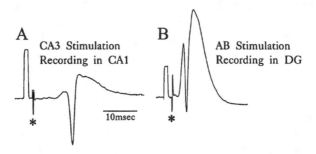

FIGURE 4.7
Typical evoked responses in CA1 and DG. (A) A stimulus was administered to the left CA3 region (stimulus artifact is marked with an *) and the extracellular field potential response recorded in the right CA1 is shown. (B) A stimulus was administered to the right angular bundle (AB, the fiber tract from the entorhinal cortex to the dentate gyrus) while recording in the right dentate gyrus (DG). Each stimulus artifact is preceded by a calibration pulse of 10 mV.

The onset and termination of reverberatory seizure activity is best determined with the recording electrode in the dentate gyrus and recording maximal dentate activation. To record from the dentate gyrus, the recording electrode is placed 2 mm lateral in the same anteroposterior plane as the stimulating electrode. The depth of the recording electrode in the dentate gyrus is determined by stimulating through an electrode in the angular bundle or entorhinal cortex on the ipsilateral side (AP –8 mm, lateral 4.4 mm, depth 3 mm, Figure 4.7). Many types of recording electrodes can theoretically be used during these experiments (ion-sensitive, extracellular field, single unit, whole cell). The most common recording is of the extracellular field potentials with DC recording. The onset of maximal dentate activation is most distinct with DC recording (Figure 4.8). Extracellular recording electrodes can be made from glass or metal (AC recording). Most commonly capillary glass is used and pulled to a tip with an electrode puller (almost any model will do). The electrode is filled with NaCl (2 M) with 1% Fast Green to give an impedance of 0.5 to 10 MΩ. Generally, the lower the impedance the lower the signal-to-noise ratio.

At the end of every experiment, electrode positions should be marked for confirmation of correct position by histology. Fast Green can be iontophoresed from the recording electrode (–20 nA for 20 min) and current passed through the stimulating electrode (1 mA for 10 s). The animal is then perfused through the heart with 1% potassium ferrocyanide in 10% buffered formalin (100 ml). The Fast Green leaves a green spot. The current passed through the stimulating electrodes leave either a lesion or will plate some metal into the tissue which reacts with the ferrocyanide to leave a blue spot. The brain is subsequently equilibrated in sucrose-formalin, cut on a sliding microtome, and stained with Cresyl Violet. Routine examination of the electrode positions prevents a gradual shift in electrode placement between experiments and may help determine why a particular experiment did not work as well as another.

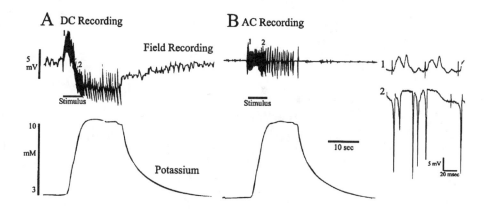

FIGURE 4.8

Chart recordings of maximal dentate activation. The responses to two consecutive stimulus trains are presented. In this experiment a double barrel electrode was used — one side records the extracellular field potential and the other side contains an ion-exchange resin sensitive to potassium. The extracellular potassium concentration can be determined from a differential between the two barrels of the electrode. In each panel the top is the field recording from the electrode placed in the dentate gyrus. The bottom record is the extracellular potassium level. Each stimulus train was 20 Hz at 400 μA for 6 s. Responses to stimuli at two different points during the stimulation are shown at a faster time scale on the right side of the figure (**1** and **2**). (A) The extracellular recording was DC-coupled and the onset of maximal dentate activation is defined by the negative shift of the extracellular DC potential, along with the appearance of the large-amplitude population spikes and the secondary rise in extracellular potassium (which in this case is obscured by the rapid time to onset of maximal dentate activation). (B) The extracellular recording was AC-coupled, which results in loss of one marker for the onset of maximal dentate activation — the DC shift. In this example the onset of maximal dentate activation can only be determined by the appearance of the large-amplitude population spikes. Calibrations are indicated on the chart records.

3. Stimulation Protocol

Twenty hertz stimulus trains (1 msec biphasic pulses) are delivered through the CA3 stimulating electrode for a maximum of 30 s. Initially, trains are administered every 2 to 3 min with increasing stimulus intensities (in 40- to 100-μA steps, beginning with 400 μA) to determine the threshold for maximal dentate activation. Maximal dentate activation is defined by the appearance of large-amplitude population spikes in the extracellular recording (10 to 40 mV) associated with a rapid decrease in the DC potential (Figure 4.8A). The stimulus threshold for elicitation of maximal dentate activation is usually 500 to 800 μA. If the animal requires higher stimulus intensities, then either the stimulating electrode is not positioned correctly or the animal has received too much urethane. A stimulus train sufficient to produce maximal dentate activation is then administered every 5 min until an afterdischarge appears and then every 10 min. An afterdischarge is defined as at least two bursts of population spikes after the end of a stimulus train. For every stimulus train, the stimulus is stopped 2 to 3 s after maximal dentate activation begins (Figure 4.9). This causes most of the paroxysmal discharges of the granule cells to be in the form of an afterdischarge.

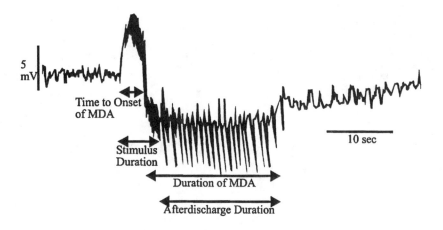

FIGURE 4.9

Measurements of the parameters of maximal dentate activation. One chart recording of the extracellular DC potential during and after the stimulus train to the CA3 region is shown. The stimulus duration is indicated. The time to onset of maximal dentate activation (MDA) is defined as the time from the beginning of the stimulus train to the midpoint of the positive-to-negative DC shift. The duration of maximal dentate activation is defined from the midpoint of the positive-to-negative DC shift to the midpoint of the return of the DC shift to baseline. The afterdischarge duration is defined as the time from the end of the stimulus train to the return of the DC potential back to baseline. Normally the stimulus duration is stopped 2 to 3 s after the onset of maximal dentate activation, so the afterdischarge duration is 2 to 3 s shorter than the duration of maximal dentate activation.

This repeated elicitation of reverberatory seizures in the hippocampal circuit is thought to closely mimic the changes seen in afterdischarge duration during kindling and, thus, could be referred to as electrographic kindling.

4.   Measurement of Time to Onset and Duration

Two parameters of maximal dentate activation have been defined (Figure 4.9). The time to onset is the time from the beginning of the stimulus train to the midpoint of the positive-to-negative DC shift and appears to be a measure of the ease with which maximal dentate activation can be initiated. The duration of maximal dentate activation is measured from the midpoint of the positive-to-negative DC shift to the midpoint of the return of the DC potential to baseline. The duration of maximal dentate activation is a measure of the ability of the brain to terminate the epileptic discharge. Because the stimulus is always stopped manually 2 to 3 s after the appearance of maximal dentate activation, the afterdischarge duration is always 2 to 3 s less than the duration of maximal dentate activation (Figure 4.9). To get more consistent measurements, it is best if the same person measures the time to onset and durations for a set of experiments.

5.   How to Measure Drug Effects

Once the afterdischarge is lengthening with each subsequent stimulus train, then the drug-testing protocol can begin.[16,17] If the stimulus trains are repeated every 10 min in the absence of drug, the time to onset will decrease and the duration of maximal dentate activation will increase (Figure 4.10). Drugs are administered after three to five stimulus trains that are followed by lengthening afterdischarges. Drugs should always be compared to vehicle injections.

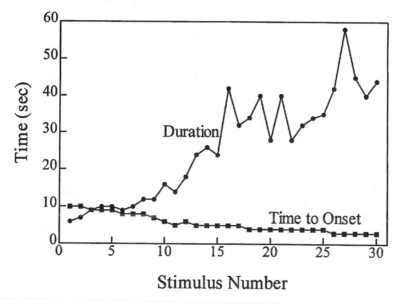

FIGURE 4.10

Changes in duration and time to onset of maximal dentate activation with repeated elicitation. One control experiment is graphed. The time to onset and duration of maximal dentate activation were measured as shown in Figure 4.9. These measurements were then graphed for each stimulus. Notice that there is a gradual decrease in the time to onset of maximal dentate activation (filled squares). There is a gradual increase in the duration of maximal dentate activation (filled circles) until stimulus 15–20. Often spreading depression appears in response to a stimulus train after 15–20 stimulus trains have been administered. This pattern of changes in time to onset and duration is consistent across animals.

To make comparisons across animals, the data can be "normalized," by subtracting the duration of maximal dentate activation in response to the first stimulus from the duration of maximal dentate activation measured after each subsequent stimulus train. Thus, for each stimulus train after the first, a change in duration of maximal dentate activation is calculated. Data from separate animals can be averaged and comparisons made across groups of animals (Figure 4.11). Comparisons of time to onset can be made in the same way. This "normalization" is necessary because the first afterdischarge that begins the lengthening process varies considerably across animals, but the resulting increase in duration with each stimulus train is quite reproducible, with minimal variability.

The effect of a number of drugs on these two parameters has been investigated.[18-25] The results are summarized in Table 4.2. The effect of the antiepileptic agents on the duration of maximal dentate activation for the most part parallels the effectiveness of the same agents in partial complex epilepsy. The time to onset of

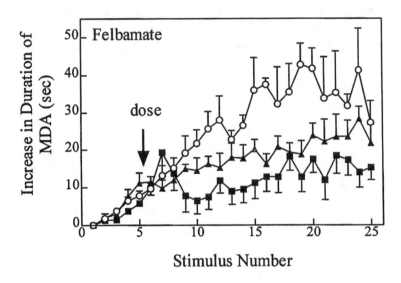

FIGURE 4.11

Effect of a drug on the time to onset and duration of maximal dentate activation. In order to determine the effects of drugs on the time to onset and duration of maximal dentate activation, measurements are made as described in Figures 4.9 and 4.10. The data are normalized as described in the text, so that changes in the parameters are averaged for each dose of drug and for the vehicle controls. This graph summarizes the effect of 300 mg/kg (filled triangles) and 450 mg/kg (filled squares) felbamate on the increase in the duration of maximal dentate activation. The drug (or vehicle control) was administered just after stimulation number 5. (From Xiong, Z.-Q. and Stringer, J. L., *Epilepsy Res.*, 27, 187, 1997. With permission.)

maximal dentate activation is altered by drugs that are usually classified as neuro-modulators in the hippocampus (cholinergic and adrenergic). Reduction of GABAergic inhibition shortens the time to onset, and augmentation of GABAergic inhibition lengthens the time to onset. Adenosine has been postulated to be an endogenous antiepileptic agent and the findings summarized in Table 4.2 are consistent with that hypothesis. Adenosine agonists shorten the duration of maximal dentate activation and antagonists lengthen the duration.

III. Interpretation

A. Epileptogenesis in Rapid Kindling

The rapid kindling protocol rapidly increases epileptogenesis, but the animals do not meet the criteria for a stable kindled state after 1 d of stimulation. Over the course of that first day (with the interstimulus period of 5 min) there appears to be increased epileptogenesis, but the expression of this increased excitability requires at least overnight to consolidate. This suggests that during this time there is an additional process occurring, which requires time to complete. This observation has

TABLE 4.2

Effect of Pharmacological Agents on Parameters of Maximal Dentate Activation (MDA) in Urethane-Anesthetized Animals

	Drug	Dose (mg/kg)	Duration of MDA	Time to onset
AEDs	Phenytoin	80	NE	NE
	Carbamazepine	50	↓	NE
	Phenobarbital	60	⇓	NE
	Ethosuximide	300	NE	⇑
	Valproic acid	300	NE	NE
GABA agents	Diazepam	3	↓	⇑
	Bicuculline	0.3	↑	↓
	Picrotoxin	5	↑	NE
	Muscimol	3	↓	↑
	Baclofen	10	↓	NE
Cholinergic	Pilocarpine	50	↑	⇑
	Atropine	50	↓	↑
Adrenergic	Propranolol	10	↓	↑
	Clonidine	0.5	↓	⇑
EAA	Ketamine	30	↓	⇑
	MK-801	2	↓	NE
Adenosine	2-chloroadenosine	10	⇓	NE
	Cyclopentyl adenosine	3	↓	NE
	DPCPX	0.05	↑	NE
NO	L-NAME	100	NE	NE
	L-arginine	500	↓	NE
	8-br-cGMP	icv	⇓	⇑
	Methylene blue	icv	NE	NE

Abbreviations: AEDs, antiepileptic drugs; GABA, gamma-aminobutyric acid; EAA, excitatory amino acids; NO, nitric oxide; icv, intracerebroventricular; L-NAME, 1-N^G-nitro-l-arginine methyl ester; 8-br-cGMP, 8 bromoguanosine-3′-5′-cyclic monophosphate; DPCPX, 1,3-dipropyl-8-cyclo-phenylxanthine.

↑ or ↓ indicates blocking the change (either the increase or decrease) of the duration or time to onset; ⇑ or ⇓ indicates lengthening or shortening the duration or time to onset beyond that recorded before drug administration.

NE is no effect.

Data from References 17–24.

been supported by several other studies. One examined the permanence of the enhanced epileptogenesis after different patterns and timing of the stimulation.[26] It was quite clear that time for consolidation of the enhanced epileptogenicity is necessary. The second study[27] examined permanence of the kindling electrophysiologically and measured sprouting, and several biochemical markers of plasticity. The results also indicate that enhanced excitability is induced by the 1 d of stimulation, but suggest that the permanent kindled state requires several weeks of time. This time lag has implications for the interpretation of any pharmacological experiments that are carried out using this technique. For instance, if one wanted to test the ability of a drug to block the heightened epileptogenesis that appears during the first day of every 5 min stimulation, one would administer the drug at the beginning of that day. The difficulty would be in determining when to test the drug effects. Alternatively, one could test the ability of a drug to block the consolidation process by administering the drug after the first day of stimulation.

B. Circuits Involved in the Seizure Discharges

The kindled seizures in the rapid kindling model appear to involve the same circuits as traditional kindling.[9] This is based on observations of the behavioral seizures. The seizures are forebrain seizures, which begin in the limbic system and then spread to other cortical regions. Therefore, if a drug has an effect in the rapid kindling model, the locus of the drug action cannot be determined. In the model of electrographic kindling under urethane anesthesia, all evidence suggests that the seizures are limited to the hippocampal-parahippocampal circuits only. In this case, if a drug blocks the seizure discharge, then the effect of the drug can be localized at least to this circuit.

C. Evaluation of Drug Activity

The use of the kindling model for the study of anticonvulsants has been analyzed by Burnham.[28] In essence, the effect of drugs can be studied on either the development of seizures (kindling acquisition) or on the occurrence of seizures (stable kindled state). The development of kindling could be used to identify prophylactic treatments that will prevent or attenuate the progression in the severity of epilepsy. Alternatively, stable kindled seizures may be used to identify drugs useful in a chronic, stable epileptic state. Despite this potential, studies of antiepileptic drugs with kindled seizures have been limited, in large part because of technical issues. The traditional approach has been to use 60 Hz, 1 s stimulus trains applied daily. Using once a day stimulation, it is difficult to determine the time course for the effects of antiepileptic drugs against kindled seizures.

The protocol of rapid kindling of behavioral seizures described in this chapter has definite advantages for drug testing. Once the animal has achieved a stable kindled state, multiple kindled responses can be triggered within a 6- to 12-h period, making it possible to determine the time course of action of a drug. The effects of drugs on both afterdischarge duration and behavioral seizures can be measured, as well as obvious behavioral changes. In addition, it is often possible to separate effects of drugs on limbic (class 1 and class 2) seizures and generalized motor (class 4 and class 5) seizures. However, this method is limited to testing the effects of drugs on stable kindled seizures and does not examine the effects of drugs on kindling acquisition. As mentioned earlier, the fact that a drug can alter the seizures in this model does not give any indication of where that effect is taking place or by what mechanism the drug is exerting its antiepileptic effect. The basic stimulation protocol can be altered to examine drug effects on kindling development,[9] but this requires a different set of seizure naïve rats.

Using the protocol of electrographic kindling described in this chapter will elucidate different aspects of drug effects. As with the rapid kindling in awake rats, this method will determine a time course of action of a drug, especially if the time course is less than 4 to 5 h. In contrast to rapid kindling in awake rats, electrographic kindling in anesthetized rats is limited to the circuits involved in the seizure discharge and will provide no information about behavioral seizures. But this can also be an advantage of this methodology. For example, if a drug blocks generalized motor seizures (class 4 and class 5) in the awake model, but has no effect on limbic seizures (class 1 and class 2), then one might predict that the drug will have no effect in the anesthetized rat, since the seizures are limited to the hippocampal-parahippocampal circuits. This type of hypothesis is readily tested with this model. This model starts to determine the region that is affected by drug actions. The electrographic kindling method allows study of the hippocampal circuit in the intact animal, as opposed to using hippocampal slices or combined entorhinal cortex/hippocampal slices. Also, as more is learned about the mechanisms behind initiation (time to onset) and termination (duration of maximal dentate activation) of the seizures within these circuits, additional clues about mechanisms of action of new (and old) drugs can be determined. One serious limitation of this technique is the possibility of drug interactions with the urethane anesthetic.

References

1. Goddard, G. V., McIntyre, D. C., and Leech, C. K., A permanent change in brain function resulting from daily electrical stimulation, *Exp. Neurol.*, 25, 295-330, 1969.
2. Racine, R. J., Burnham, W. M., Gartner, J. G., and Levitan, D., Rates of motor seizure development in rats subjected to electrical brain stimulation: strain and interstimulation interval effects, *Clin. Neurophysiol.*, 35, 553-560, 1973.
3. Mucha, R. F. and Pinel, J. P. J., Postseizure inhibition of kindled seizures, *Exp. Neurol.*, 54, 266-282, 1977.
4. Sainsbury, R. S., Bland, B. H., and Buchan, D. H., Electrically induced seizure activity in the hippocampus: time course for postseizure inhibition of subsequent kindled seizures, *Behav. Biol.*, 22, 479-488, 1978.
5. Lothman, E. W., Hatlelid, J. M., Zorumski, C. F., Conry, J. A., Moon, P. F., and Perlin, J. B., Kindling with rapidly recurring hippocampal seizures, *Brain Res.*, 360, 83-91, 1985.
6. Racine, R. J., Modification of seizure activity by electrical stimulation: II Motor seizure, *Electroencephalogr. Clin. Neurophysiol.*, 32, 281-294, 1972.
7. Lothman, E. W., Perlin, J. B., and Salerno, R. A., Response properties of rapidly recurring hippocampal seizures in rats, *Epilepsy Res.*, 2, 356-366, 1988.
8. Lothman, E. W., Salerno, R. A., Perlin, J. B., and Kaiser, D. L., Screening and characterization of antiepileptic drugs with rapidly recurring hippocampal seizures in rats, *Epilepsy Res*, 2, 367-379, 1988.

9. Lothman, E. W. and Williamson, J. M., Rapid kindling with recurrent hippocampal seizures, *Epilepsy Res.*, 14, 209-220, 1993.

10. Stringer, J. L., Williamson, J. M., and Lothman, E. W., Induction of paroxysmal discharges in the dentate gyrus: frequency dependence and relationship to afterdischarge production, *J. Neurophysiol.*, 62, 126-135, 1989.

11. Stringer, J. L. and Lothman, E. W., Maximal dentate activation: characteristics and alterations after repeated seizures, *J. Neurophysiol.*, 62, 136-143, 1989.

12. Stringer, J. L. and Lothman, E. W., Bilateral maximal dentate activation is critical for the appearance of an afterdischarge in the dentate gyrus, *Neuroscience*, 46, 309-314, 1992.

13. Stringer, J. L. and Lothman, E. W., Reverberatory seizure discharges in hippocampal-parahippocampal circuits, *Exp. Neurol.*, 116, 198-203, 1992.

14. Stringer, J. L. and Lothman, E. W., Ontogeny of hippocampal afterdischarges in the urethane-anesthetized rat, *Dev. Brain Res.*, 70, 223-229, 1992.

15. Cain, D. P., Raithby, A., and Corcoran, M. E., Urethane anesthesia blocks the development and expression of kindled seizures, *Life Sci.*, 44, 1201-1206, 1989.

16. Stringer, J. L. and Lothman, E. W., Maximal dentate activation: a tool to screen compounds for activity against limbic seizures, *Epilepsy Res.*, 5, 169-176, 1990.

17. Stringer, J. L. and Lothman, E. W., Use of maximal dentate activation to study the effect of drugs on kindling and kindled responses, *Epilepsy Res.*, 6, 180-186, 1990.

18. Stringer, J. L. and Lothman, E. W., NMDA receptor dependent paroxysmal discharges in the dentate gyrus, *Neurosci. Lett.*, 92, 69-75, 1988.

19. Stringer, J. L. and Lothman, E. W., Pharmacological evidence indicating a role of GABAergic systems in termination of limbic seizures, *Epilepsy Res.*, 7, 197-204, 1990.

20. Stringer, J. L. and Lothman, E. W., A1 adenosinergic modulation alters the duration of maximal dentate activation, *Neurosci. Lett.*, 118, 231-234, 1990.

21. Stringer, J. L. and Lothman, E. W., Cholinergic and adrenergic agents modify the initiation and termination of epileptic discharges in the dentate gyrus, *Neuropharmacology*, 30, 59-65, 1991.

22. Stringer, J. L. and Higgins, M. G., Interaction of phenobarbital and phenytoin in an experimental model of seizures in the rat, *Epilepsia*, 35, 216-220, 1994.

23. Stringer, J. L., Valproic acid and ethosuximide slow the onset of maximal dentate activation in the rat hippocampus, *Epilepsy Res.*, 19, 229-235, 1994.

24. Stringer, J. L. and Erden, F., In the hippocampus *in vivo*, nitric oxide does not appear to function as an endogenous antiepileptic agent, *Exp. Brain Res.*, 105, 391-401, 1995.

25. Xiong, Z.-Q. and Stringer, J. L., Effects of felbamate, gabapentin and lamotrigine on seizure parameters and excitability in the rat hippocampus, *Epilepsy Res.*, 27, 187-194, 1997.

26. Lothman, E. W. and Williamson, J. M., Closely spaced recurrent hippocampal seizures elicit two types of heightened epileptogenesis: a rapidly developing, transient kindling and a slowly developing, enduring kindling, *Brain Res.*, 649, 71-84, 1994.

27. Elmer, E., Kokaia, M., Kokaia, Z., Ferencz, I., and Lindvall, O., Delayed kindling development after rapidly recurring seizures: relation to mossy fiber sprouting and neurotrophin, GAP-43 and dynorphin gene expression, *Brain Res.*, 712, 19-34, 1996.

28. Burnham, M., Anticonvulsants and the kindling model: a critical analysis, in *Kindling and Synaptic Plasticity*, Morrell, F., Ed., Birkhauser, Boston, 1991, chap. 16.

Chapter **5**

Experimental Models of
Status Epilepticus

Jeffrey H. Goodman

Contents

0-8493-3362-8/98/$0.00+$.50
© 1998 by CRC Press LLC

I.   Introduction

Status epilepticus (SE) has been defined clinically as continuous seizure activity lasting more than 30 min or multiple seizures without regaining consciousness lasting more than 30 min.[1] In humans, SE is a medical emergency that, if untreated, can result in brain damage and/or death.[2,3] It has been estimated that in the U.S. the number of cases ranges from 60,000 to 250,000 per year,[4] with a mortality rate of 10% to 12%.[5] Given the evidence that SE in childhood can contribute to epilepsy in the adult[6,7] it is clear that SE is a significant clinical problem. Clinically, there are three different types of SE: (1) generalized convulsive status epilepticus (GCSE); (2) nonconvulsive status epilepticus; and (3) continuous focal activity without loss of consciousness. GCSE is the most common form of status.[8]

A single seizure is usually self-limiting, with a short duration. In SE, seizure activity is not limited. The end of one seizure becomes blurred by the start of the next seizure. During SE in humans, a series of predictable and distinct progressive changes occur in the electroencephalogram (EEG).[9] These electrographic changes, as described by Treiman and colleagues are (1) a series of discrete discharges; (2) the discrete seizures begin to merge, generating a waxing and waning pattern; (3) the electrographic seizure activity becomes continuous; (4) the continuous discharge pattern starts to be interrupted by periods of EEG flattening; and (5) the EEG has a flat background with periodic epileptiform discharges (PEDs).[9]

Several experimental models of SE have been developed that approximate specific aspects of the clinical event and the subsequent alterations that occur in neuronal structure and function. However, an experimental model is only an approximation of the clinical syndrome. Few experimental models duplicate all aspects of the human disorder. The model of experimental SE chosen by the investigator should be determined by the experimental question and the type of data that will be collected. Experimental models of SE that are used to study the mechanisms underlying the transition of a single seizure into SE or to test therapeutic agents that block SE must induce the same sequential electroencephalographic changes observed by Treiman and colleagues[9] during human SE. The requirements for experimental models of SE that are used as a tool to study epileptogenesis or a specific aspect of seizure-induced changes in brain structure and function are less stringent. Currently, models of experimental SE are being used to study the transition of a single episode of SE into chronic epilepsy; general mechanisms of neuronal injury and selective neuronal vulnerability; hippocampal sclerosis; synaptic reorganization (sprouting); seizure-induced changes in gene expression and growth factors; and the development of new therapeutic anticonvulsants and neuroprotectants.

Experimental SE can be induced with pharmacologic agents or by electrical stimulation. Pharmacologic models of SE include kainic acid,[10,11] pilocarpine,[12,13] cobalt/homocysteine thiolactone,[14] and flurothyl.[15,16] Models of SE that induce seizure activity with electrical stimulation include perforant path stimulation[17-19] and self-sustained limbic status epilepticus (SSLSE).[20] Experimental SE also has been studied in immature animals[21-23] and recently several *in vitro* models of SE have been developed.[24-26]

This chapter discusses each model of SE from the perspective of how to induce SE (animal preparation, route of administration), pathophysiology and neuropathology, postseizure care and behavior, the efficacy of standard anticonvulsant drugs, as well as the advantages and limitations of each model. While several of these models have been tested in a number of different species, the rat is by far the most commonly used animal in experimental models of SE. For this reason the focus of this chapter is on models that use the rat.

II. Pharmacologic Models of Status Epilepticus

In several of the following models chronic electrodes are stereotaxically implanted into the brain so the investigator can record electrographic seizure activity. Although electrodes may not be necessary to use a given model, electrodes may be necessary to address specific questions. A detailed description of the methods used to implant intracranial electrodes appears elsewhere, in Chapter 3. In this chapter, the discussion of electrode implantation will be limited to the type of electrodes used and the stereotaxic coordinates for their placement. The cobalt/homocysteine thiolactone model also employs the use of focal epilepsy techniques. An in-depth description of focal models can be found in Chapter 7.

A. Kainic Acid

The kainic acid model of SE is one of the most extensively studied seizure models. It is regularly used to induce SE and shares many of the features of human temporal lobe epilepsy (TLE).[11] Kainic acid, an extract of the seaweed *Digenea simplex*,[27] is a rigid analog of glutamate that binds to a subset of glutamate receptors. Kainic acid was originally used as a lesioning agent because it kills cell bodies of neurons but spares glia and axons passing through the injection site.[28-30] When the brains from kainic acid-injected animals are examined, additional damage is found in brain regions distant to the injection site.[31] This suggests there are two mechanisms by which kainic acid induces neuronal damage, a direct excitotoxic effect and seizure-induced damage at a distance from the injection site. The distant damage is likely due to the synaptic release of glutamate secondary to kainic acid-induced seizure activity.[31,32]

1.   Routes of Administration

Kainic acid infused into the lateral ventricle at doses of 0.4 to 1.5 µg,[10,33-35] or directly into the brain parenchyma at doses of 1 to 2 µg,[31,36] induces limbic seizures accompanied by direct damage at the site of the injection, as well as distant damage due to seizure activity. A stereotaxically placed microinfusion guide cannula can be used to administer kainic acid directly into brain tissue or intracerebroventricularly (i.c.v.). The cannula can be temporary, left in place only long enough to infuse the toxin, or it can be cemented in place (see Chapter 3). The rat is anesthetized, placed in a stereotaxic frame, and the skull surface is exposed. For hippocampal injection a burr hole is drilled and the guide cannula is placed at the following coordinates: AP +4.6 mm from the intra-auricular line, ML ±2.0 mm from the sagittal suture, DV –3.4 mm from the dural surface.[36] The coordinates for an intraventricular cannula are AP +0.4 mm, ML +1.3 mm, DV –3.5 mm measured from the bregma suture and skull surface.[37,38] For chronic implantation the animal is allowed to recover a minimum of 1 week. Kainic acid (Sigma, St. Louis, MO) is infused through the cannula over a 1- to 10-min period. If the cannula is temporary it is removed at the end of the infusion and the wound is closed. The coordinates for guide cannula placement for direct infusion of kainic acid into other brain sites are dependent on the specific brain structure chosen by the investigator. In general, infusion into tissue requires infusion of a smaller volume of drug over a longer period of time, usually 10 min. The infusion needle should be left in place a minimum of 10 min after the drug is infused to make sure there is no reflux of the drug through the cannula when the needle is removed. An alternative approach is to use a 1 µl Hamilton syringe attached to a stereotaxic carrier. The needle of the Hamilton syringe is placed at the appropriate coordinates and the drug is infused as described above.[39]

Systemic administration of kainic acid also induces SE and is an easier way to deliver the drug.[34] Kainic acid injected s.c.,[40,41] i.p.,[42,43] or IV[44,45] at a dose of 8 to 12 mg/kg consistently induces SE. However, kainic acid appears to induce a steep dose-response effect since systemic administration of kainic acid at doses less than 8 mg/kg does not induce SE and does not cause brain damage in a majority of the rats tested.[37] Sperk and colleagues[46,47] have successfully induced SE by injecting kainic acid (10 mg/kg, s.c. or i.p.) into male Sprague-Dawley rats weighing 260 to 350 g. In their series of experiments they were able to induce SE in 60% to 80% of the animals with a 90% survival rate. Doses of kainic acid greater than 10 mg/kg result in a 50% mortality rate.[37] A selective summary of studies that have administered kainic acid to induce SE, via different routes and doses, to address a variety of questions associated with experimental epilepsy, can be found in Table 5.1.

TABLE 5.1
Doses of Kainic Acid that Induce SE by Different Routes of Administration

Route	Dose	Ref.
Intracerebroventricular (i.c.v.)	0.4–0.8 µg	10,33,35,115,116
Intracerebral	1–2 µg	31,36,51,53,54
Subcutaneous (s.c.)	8–12 mg/kg	40,41,46,47,60
	18 mg/kg	120
Intraperitoneal (i.p.)	8–12 mg/kg	42,43,117–119
	5–7 mg/kg + 3 nmol ouabain	37
Intravenous (IV)	8–12 mg/kg	44,45

2. Pathophysiology and Neuropathology

Whether administered systemically or directly into the brain, kainic acid will induce seizure activity. SE induced with kainic acid exhibits the same progressive, sequential electroencephalographic changes[48] observed by Treiman and colleagues[9] during human SE. The behavioral seizure activity associated with kainic acid-induced SE is limbic motor seizures similar to those observed during a kindled seizure.[49] However, unlike a kindled seizure, which lasts 60 to 90 s, kainic acid-induced SE is characterized by repetitive seizure activity lasting up to 6 h after injection.

Kainic acid-induced behavioral seizure activity is rated by a scale devised by Racine[50] to score kindled seizures. Wet dog shakes (WDS), head nodding and facial clonus are given a seizure score of stage 1 to 2, forelimb clonus is stage 3, rearing is stage 4, and continued rearing and falling is stage 5. Initial behavioral changes during the first hour after kainic acid injection include staring episodes followed by head bobbing and numerous WDS. These behavioral changes are followed by isolated limbic motor seizures which increase in frequency, eventually leading to SE.

A proper assessment of seizure duration and severity requires the implantation of intracranial electrodes. Without electrodes the assessment of seizure severity and duration is limited to the observation of behavioral seizure activity. Nonconvulsive seizure activity that does not involve motor structures will go undetected. To implant electrodes each rat is anesthetized and placed in a stereotaxic frame, the skull surface exposed and burr holes are drilled at the appropriate sites. See Chapter 3 for a more detailed description of electrode implantation.

Kainic acid-induced alterations in electrographic activity correspond with changes in behavior. EEG changes first appear in the hippocampus during staring episodes and WDS. During isolated limbic motor seizures paroxysmal discharges occur in the hippocampus, amygdala and other limbic structures and during SE the neocortex, striatum, and thalamus become involved. The hippocampus appears to be extremely sensitive to kainic acid and plays a central role in the initiation of SE.

Intracerebral or intraventricular injection of kainic acid results in a more restricted and a more easily duplicated pattern of neuronal damage than when kainic acid is administered systemically. Intracerebroventricular injection consistently results in hippocampal damage localized to CA3 pyramidal neurons.[10]

Systemically induced SE results in extensive damage in several brain regions. These include the hippocampus, amygdala, pyriform cortex, entorhinal cortex, septum, and medial thalamus. Although kainic acid-induced SE after systemic administration results in damage throughout the hippocampus, the ventral hippocampus appears to be particularly vulnerable.[44] Damage in the hippocampus includes pyramidal neurons in CA3 and CA1 and hilar neurons in the dentate gyrus. The CA2 pyramidal cells and dentate granule cells appear to be resistant to damage induced by kainic acid SE. The pattern of brain damage after kainic acid-induced SE is symmetrical in that bilateral structures exhibit the same degree of damage. However the pattern of damage is often variable among rats receiving the same treatment.[34,37,47]

One of the long-term consequences of kainic acid-induced SE is the occurrence of a decrease in seizure threshold[51,52] and a chronic epileptic state characterized by spontaneous limbic seizures.[53,54] When the brains of these animals are examined histologically a process of synaptic reorganization (sprouting) occurs in the mossy fiber pathway of the dentate gyrus.[43,55] The sprouting is similar to what has been observed in human hippocampal tissue surgically removed from cryptogenic epileptics.[56-58] New axon fibers from the granule cells grow into the inner molecular layer of the dentate. The function of these fibers is controversial. There is evidence that the new pathway is functional and potentially proconvulsant, synapsing on the dendrites of granule cells, thereby creating a recurrent excitatory pathway.[43,55] However, there is a report that suggests the sprouted pathway is anticonvulsant with some of the new fibers synapsing on the dendrites of inhibitory interneurons whose cell bodies are located in the granule cell layer.[41] Recently it has been demonstrated that blockade of synaptic reorganization with cycloheximide in kainic acid treated rats does not prevent recurrent spontaneous seizures.[59]

One of the major drawbacks of the kainic acid model is the variable sensitivity of rats of different strains, sex, age, and weight to kainic acid.[34,37,60] Wistar rats are more sensitive to kainic acid than Long-Evans or Sprague-Dawley rats.[60] Older, heavier rats exhibit SE at lower doses with a greater amount of brain damage.[61,62] It has been hypothesized that the extreme variability of seizure response to kainic acid is due to the poor transport of kainic acid across the blood-brain barrier.[63] It is possible that if one could open the blood-brain barrier in conjunction with kainic acid administration a more consistent and uniform result could be obtained. Several investigators have tried different modifications of the kainic acid model in an attempt to decrease the variable response to kainic acid treatment (see below).

3. Ouabain

Brines and colleagues[37] reported that pretreatment with ouabain, a cardiac glycoside, can enhance kainic acid-induced seizure activity and cause a subsequent increase in excitotoxic cell death in the rat hippocampus. It has been suggested that excitotoxic cell death is mediated by two mechanisms: an early phase that results from osmotic injury mediated by changes in fluxes of Na^+ and Cl^- across the cell membrane, and a delayed cell death mediated through an increase in intracellular Ca^{2+}.[64] It has been hypothesized that since Na^+,K^+-ATPase is an important regulator of ion gradients across the cell membrane, manipulation of this enzyme could influence the relative vulnerability and seizure threshold of hippocampal neurons. Ouabain inhibits sodium pump function and at high concentrations is toxic to neurons and glia. However, at low concentrations ouabain can decrease pump activity without blocking it entirely and without causing neuron or glial cell death.[37]

The use of ouabain in combination with kainic acid requires the implantation of an intraventricular cannula. Male Sprague-Dawley rats (200 to 250 g) are anesthetized, placed in a stereotaxic frame where a microinjection guide cannula is implanted into the right lateral ventricle. If the investigator determines that intracranial electrodes are needed they should be implanted at the same time. Each animal is allowed to recover a minimum of 1 week after surgery at which time kainic acid is injected (5 to 7 mg/kg, i.p.) followed 30 min later by the infusion of ouabain (3 nmol, i.c.v.) through the guide cannula into the lateral ventricle. The combination of subconvulsant doses of kainic acid with ouabain results in severe limbic seizure activity within 2 h of the kainic acid injection with essentially no mortality. The reduction of sodium pump activity by ouabain appears to enhance the ability of kainic acid to induce SE.

4. Multiple Kainic Acid Injections

This unique method of inducing SE with kainic acid was developed to produce a consistent lesion accompanied by synaptic reorganization and spontaneous seizures.[65] Adult rats are injected with kainic acid (5 mg/kg, dissolved in 150 mM NaCl, i.p.) once per hour for up to 10 h. Repeated stages 4 and 5 limbic seizures occur over a 4- to 6-h period. Some of the hourly kainic acid injections can be skipped in animals that exhibit continuous stage 4 and stage 5 seizures. All rats must have a minimum of 6 h of seizure activity to ensure consistent damage to the central nervous system. The survival rate using this method is approximately 80%.

5.   Postseizure Care and Animal Behavior

For the first 2 to 3 d after kainic acid-induced SE animals will require help eating and drinking. This can be accomplished by injections of lactated Ringers (2 ml/d, s.c.) or by hand feeding.[65] Sperk and colleagues[47] observed that rats become hyperthermic during kainic acid-induced SE so they house the animals at 18°C once the seizures have stopped. One month after kainic acid-induced SE the animals exhibit increased aggressive behavior and spontaneous recurrent limbic motor seizures.[66]

6.   Anticonvulsant Efficacy

An important factor that must be taken into account when comparing the effectiveness of potential anticonvulsant treatments on SE is that the longer SE persists the more difficult it is to control.[14,67,68] A detailed review of the effect of potential anticonvulsant drugs on kainic acid-induced SE has been compiled by Sperk.[34] Phenytoin (50 to 210 mg/kg)[69-72] and carbamazepine (80 mg/kg)[71,72] are ineffective against kainic acid-induced SE. Valproate (100 to 250 mg/kg) has been reported to have both anticonvulsant[72] and proconvulsant[71] effects in this model. Both diazepam (3 to 5 mg/kg)[71-74] and phenobarbital (50 mg/kg)[69,71,75] are effective anticonvulsants in this model. These results indicate that not all of the standard anticonvulsants used to treat clinical SE are effective in kainic acid-induced SE.

7.   Advantages and Limitations

Kainic acid, regardless of the route of administration, induces SE that electrographically resembles SE in man.[48] The pattern of seizure-induced neuronal damage also resembles what has been found in human epileptogenic tissue.[11] The kainic acid model has been used to study the relationship between synaptic reorganization and chronic epilepsy as well as seizure-induced alterations in growth factors and neuronal gene expression.[35,118] Although kainic acid is a good model of SE there are two complications associated with the model. The first is the direct excitotoxic action of kainic acid which can make it difficult to separate direct neuronal damage from seizure-induced neuronal damage. The second drawback to the kainic acid model is the extreme variability in the sensitivity to the toxin between rats of different strains, sex, weight, and age. Animals within the same litter can exhibit the same amount and severity of behavioral seizure activity but not exhibit the same pattern of neuronal damage.[47] Kainic acid-induced SE may not be a good model for testing new anticonvulsant agents for the treatment of SE, since not all of the currently effective drugs are effective in this model.

B.   Pilocarpine

1.   Routes of Administration

Systemic administration of pilocarpine, a cholinergic agonist, has also been used to induce SE.[12,76,77] The pilocarpine model of SE also shares many of the characteristics of human TLE.[77] While initiation of the seizure is due to activation of the cholinergic system, the histopathology, neuronal loss, and spontaneous seizure activity occur secondary to seizure-induced glutamate release.[13,80]

As with kainic acid, the investigator must determine whether it is necessary to implant intracranial electrodes before initiation of pilocarpine-induced SE. Sprague-Dawley rats (170 to 400 g) are pretreated with atropine methylbromide (1 mg/kg, Sigma, St. Louis, MO) or scopolamine methylnitrate (1 mg/kg, Sigma, St. Louis, MO) by i.p. or s.c. injection to prevent the peripheral effects of pilo-carpine. These particular muscarinic antagonists are used because they do not cross the blood-brain barrier, thereby blocking only the peripheral actions of pilocarpine. Thirty minutes later pilocarpine hydrochloride (320 to 380 mg/kg, Sigma, St. Louis, MO) dissolved in physiologic saline is injected i.p. or s.c. Limbic motor seizures are usually triggered by pilocarpine within 30 min of the injection. Gibbs et al.[78] administer a second dose of pilocarpine (175 mg/kg, i.p.) to animals that do not exhibit a stage 3 behavioral seizure within 1 h of the initial pilocarpine injection. The latency from the time of the pilocarpine injection until the onset of behavioral seizures and SE appears to be dose dependent.[79] With higher doses more animals exhibit SE; however, this is accompanied by an increase in mortality. Pilocarpine induces SE through a cholinergic mechanism since pretreatment with atropine sulfate, which crosses the blood-brain barrier, prevents the initiation of SE. Atropine sulfate has no effect on established SE, leading several investigators to hypothesize that pilocarpine initiates SE but continuation of the seizure activity is likely through a glutaminergic mechanism.[13,80]

2. Pathophysiology and Neuropathology

Pilocarpine-induced SE has been extensively characterized by Turski and col-leagues.[12] The ability of pilocarpine to induce SE is dose and time dependent. While pilocarpine at a dose of 100 to 200 mg/kg induces electrographic and behavioral changes it is not sufficient to induce SE.[12,13] Pilocarpine at a dose of 400 mg/kg induces SE in a majority of animals tested.

Initial behavioral changes include staring spells, mouth movements, head bob-bing, chewing, salivation, and eye blinking which usually last no longer than 45 min from time of injection.[12] These initial behavioral changes are followed by isolated limbic motor seizures, which are accompanied by salivation, clonus, rearing and falling similar to stage 5 kindled seizures.[50] Motor seizures may occur every 5 to 15 min, with a maximum frequency of 13 per hour.[12] By 1 to 2 h after the pilocarpine injection the isolated motor seizures progress into SE, which may last 5 to 6 h. Unlike kainic acid, WDS are seldom observed in this model, except at the end of a motor seizure.

Turski et al.[12] reported that pilocarpine induces electrographic seizure activity similar to what is observed after kainic acid. Changes in electrographic activity after pilocarpine first appear in the hippocampus followed by the amygdala and neocortex. However, in a later study that recorded EEG activity from more brain areas, initial alterations in EEG activity after pilocarpine were detected in the ventral forebrain.[13] This could explain the lack of WDS at the beginning of pilocarpine-induced SE.

Initial hippocampal electrographic changes that occur 15 to 20 min after injection are characterized by high-voltage, fast activity superimposed over theta activity with isolated high-voltage spikes but no electrographic change in the amygdala or neocortex.[12] This activity eventually spreads to the amygdala and neocortex and corresponds to the episodes of staring and facial automatisms. Typical electrographic seizures that correspond to isolated motor seizures appear 40 to 45 min after the pilocarpine injection. This ictal activity corresponds to the motor seizures that occur every 5 to 15 min and is followed by periods of EEG depression. The isolated ictal activity progresses into SE, which parallels the behavioral seizure activity. Within 24 h of pilocarpine injection the EEG returns to normal, although there is a decrease in hippocampal theta activity under conditions where it normally would be present.[12]

Pilocarpine-induced SE results in extensive brain damage similar to what has been observed after kainic acid.[12,13,81] When brains are examined 24 to 27 h after the pilocarpine injection, damage is found in the olfactory cortex, the amygdaloid complex, thalamus, neocortex, hippocampus, and substantia nigra.[12,81] Extreme damage, characterized by shrunken neuronal cell bodies with swollen edematous neuropil, is present in the anterior olfactory, pyriform, and entorhinal cortex. The basal amygdala and ventral hippocampus are particularly sensitive. In the dorsal hippocampus the majority of the damage occurs in CA3 and the dentate hilus while in the ventral hippocampus most of the damage occurs in CA3 and CA1. Neocortical cell loss occurs mostly in layer 2 and layer 3, with some cell loss in layer 5. The pars reticulata of the substantia nigra is also extensively damaged.[13]

The results from the study by Turski et al.[12] indicate that SE induced by pilocarpine is similar to SE induced by kainic acid. Clifford et al.[13] demonstrated that the two models differ in the site of initial electrographic changes and although the pattern of neuronal damage is the same for both models, pilocarpine induces greater neocortical damage while kainic acid is more likely to damage the hippocampus.

Similar to kainic acid, a long-term consequence of pilocarpine-induced SE is the development of spontaneous limbic motor seizures and synaptic reorganization of the mossy fiber pathway in the hippocampal dentate gyrus.[86] Since synaptic reorganization is a common feature of human epileptogenic tissue,[56-58] the pilocarpine model is often used to examine the relationship between synaptic reorganization and spontaneous limbic motor seizures.[86]

3.   Lithium Chloride

Pilocarpine at a dose of 400 mg/kg (i.p. or s.c.) does not always induce SE. In an attempt to enhance the action of pilocarpine several investigators pretreat animals with lithium chloride. Lithium chloride (Sigma, St. Louis, MO) is injected (3 mEq/kg or 3 mM/kg, i.p.) 19 to 24 h prior to the administration of a significantly lower dose of pilocarpine (25 to 30 mg/kg).[13,80-84] Pretreatment with lithium chloride appears to potentiate the effect of pilocarpine, since lithium in combination with a 30 mg/kg dose of pilocarpine consistently induced SE.[13,80-82] Behavioral and electrographic seizure activity and accompanying neuropathology after the combination of lithium and low-dose pilocarpine is the same as that observed after high-dose pilocarpine alone.[13] There is less variability in the time of onset of behavioral seizures and an increase in the number of animals that go into SE with the combination of lithium and low-dose pilocarpine. Atropine sulfate pretreatment of lithium-treated rats blocks pilocarpine-induced seizures, suggesting cholinergic activation is still necessary for SE to occur. Neither lithium at 3 mM/kg or pilocarpine at a dose of 30 mg/kg when administered by themselves induced seizure activity.[13,80]

4. Postseizure Care and Animal Behavior

Several investigators administer diazepam to pilocarpine-treated rats in an attempt to limit the time the rat spends in SE, thereby increasing the likelihood the rats will survive.[13,78] One approach is to give a single injection (5 to 10 mg/kg) while an alternative approach is to inject diazepam (4 to 5 mg/kg, i.p.) 1 h after the beginning of SE followed by additional injections at 2 and 3 h after initiation of SE if needed.[78]

Different approaches are currently being used to care for rats with pilocarpine-induced SE. Each approach is an attempt to increase the survival rate of the rats. Obenaus et al.[85] administer lactated Ringer's (2 ml/d, s.c.) and feed the rat moist rat chow for up to 1 week after SE. Other investigators offer each rat sliced apples or peaches (personal observation), oral sports drink mixed with sucrose,[86] or an oral mixture of powdered milk and sucrose for several days after SE.[78] These procedures are labor intensive and the individual investigator has to find a balance between the amount of postseizure care and the improvement in the survival rate of pilocarpine-treated rats. Similar to kainic acid-treated rats, pilocarpine-treated rats exhibit an increase in aggressive behavior after recovery from SE.

5. Anticonvulsant Efficacy

Anticonvulsants can be tested in the two pilocarpine models of SE in two ways. The first is to administer the anticonvulsant before the pilocarpine injection, thereby testing whether the drug can prevent initiation of SE. The second approach is to administer the drug after SE has become established. Although the lithium plus pilocarpine model appears to be the same as the high-dose pilocarpine model electrographically and behaviorally, the two models differ in their sensitivity to anticonvulsant drugs.

The lithium pilocarpine model is more sensitive to pretreatment with several anticonvulsants. However, in both models initiation of SE can be blocked with a number of anticonvulsants. Pretreatment with atropine (20 mg/kg, s.c.), scopolamine (20 mg/kg, s.c.), diazepam (10 to 20 mg/kg, i.p.), phenobarbital (25 to 30 mg/kg, i.p.) and valproate (100 to 300 mg/kg, i.p.) 30 min before the pilocarpine injection will effectively prevent initiation of SE in either model.[12,80-83,122,123] There are reports that MK-801 and felbamate are effective in preventing initiation of SE in the lithium pilocarpine model but not the high-dose pilocarpine model.[124,125] Carbamazepine and phenytoin, two drugs effective clinically in the treatment of SE, are ineffective in preventing initiation of SE in either model.[123] There is a single report that carbamazepine has some efficacy in the lithium pilocarpine model.[125] Pretreatment with ethosuximide and acetozolamine have been found to exacerbate pilocarpine-induced SE.[123]

As is the case in human SE these drugs become less effective once SE is established.[83] The effectiveness of a given drug can almost be predicted based upon the discharge pattern present in the EEG and how long the animal has been in SE. Once a waxing and waning discharge pattern appears in the EEG the effectiveness of diazepam (20 mg/kg, i.p.) at completely stopping the seizures is decreased by 50%.[83] Mello and Covolan[121] injected rats with thionembutal (25 mg/kg, i.p.) up to 2 h after the onset of SE to decrease the rate of mortality. It is not clear whether SE was blocked or just attenuated. Once SE is established neither model is responsive to conventional anticonvulsants.[122,123]

6.   Advantages and Limitations
As with kainic acid, pilocarpine-induced SE results in pathophysiologic and neuro-pathologic changes similar to what is observed after human SE. The animals exhibit spontaneous recurrent seizures as well as synaptic reorganization.[86] One of the drawbacks of this model is the high mortality rate (20% to 40%).[12,76,77,86,107] In one report pilocarpine at a dose of 400 mg/kg resulted in a mortality rate of 70%.[77] Pretreatment with lithium chloride allows the dose of pilocarpine to be decreased which results in a lower mortality rate and a higher percentage of animals exhibiting SE. In a series of experiments by Clifford et al.,[13] 81% of the animals went into SE after high-dose pilocarpine. The SE response improved to 97% with the lithium and low-dose pilocarpine treatment. Despite these observations both high-dose pilo-carpine and the lithium plus pilocarpine combination are popular experimental models of SE.

C.   Cobalt/Homocysteine Thiolactone

As recently as 1983 there were no good animal models of generalized convulsive status epilepticus that could be used to test new therapeutic agents for the treatment of SE.[14,87] One of the technical difficulties involved inducing SE in a way that did not rapidly kill the animal. Walton and Treiman[14] developed a model that mimics human SE and is well suited for testing anticonvulsant agents. They induce SE by injecting homocysteine thiolactone (HCTL) into rats that have an active epileptic focus previously induced with cobalt.

1.   Methods
Male Sprague-Dawley rats (200 to 250 g) are anesthetized and placed in a stereotaxic frame. The skull surface is exposed and four burr holes are drilled at the following coordinates relative to bregma: anterior +2.0 mm, ML ±3.0 mm, AP +4.0 mm, ML ±3.0 mm.[38] Cobalt powder (25 mg, Sigma, St. Louis, MO) is placed in the left anterior hole. Epidural electrodes made from stainless steel screws (No. 0-80 × 1/8 in., Small Parts, Inc., Miami Lakes, FL) are then placed in all four holes. Further details concerning preparations for focal epilepsy and chronic electrode implantation can be found in Chapters 7 and 3, respectively.

Animals are allowed to recover a minimum of 4 d, at which time EEG activity is monitored daily. Detailed methods for EEG monitoring can be found in Chapters 3, 4, and 7. Once behavioral and electrographic seizure activity are observed, each animal is injected with D,L-homocysteine thiolactone (5.5 mmol/kg, i.p., Sigma, St. Louis, MO). The HCTL is administered in a volume of 4 ml/kg of physiologic saline and should be mixed immediately prior to use. The average time from placement of cobalt until injection of homocysteine thiolactone is 9 ± 2.5 d.

2.   Pathophysiology

The pattern of electrographic seizure activity exhibited the same sequence of changes observed in human SE.[48] Initial electrographic changes occur 10 to 15 min after the HCTL injection. The epileptiform activity increases and spreads over the next 10 to 15 min until the first generalized tonic-clonic seizure (GTCS).[14] The first GTCS usually occurs within 30 min of the HCTL injection. The time between the first and second seizure averages 8 to 9 min. The EEG pattern then progresses through the sequence of changes reported to occur in human SE.[14] Approximately 45 min after the HCTL injection continuous spiking activity is present in the EEG. This continuous spiking can last as long as an hour.[14] Approximately 50% of the animals develop periodic epileptiform discharges (PEDs) similar to those that occur in late-stage human SE. PEDs in humans are characterized by a flat EEG background interrupted by periodic discharges. These discharges are considered ictal in origin and their presence suggests that the episode of SE has not ended.[8] Once PEDs appear in the EEG of humans or experimental animals SE becomes extremely resistant to anticonvulsant therapy.[8,14] Thirty percent of the animals treated with HCTL die before exhibiting PEDs.[14] However, in those animals that exhibit PEDs during HCTL-induced SE the abnormal EEG activity may still be present the following day.[14] It takes surviving animals 3 to 5 d for the EEG to return to preseizure activity.

3.   Anticonvulsant Efficacy

Walton and Treiman[14] tested anticonvulsants that are effective in the treatment of clinical SE in this model and found them all to be effective. Phenytoin (100 to 150 mg/kg) inhibits all types of seizure activity but only when serum concentrations reach 29.5 μg/ml. There is a poor correlation between the i.p. dose and serum concentration. Phenobarbital (60 mg/kg) effectively blocks GTCS in all animals tested, although some animals still exhibit brief tonic seizures after this treatment. Diazepam at a 2.5 mg/kg dose is essentially ineffective. Diazepam at a dose of 5 mg/kg blocks GTCS in all rats although some seizure activity remains. The ED_{50} of diazepam necessary to stop all behavioral and electrographic seizure activity is 5 mg/kg.[14] The ED_{50} for phenytoin and phenobarbital have not been determined in this model.

4.   Advantages and Limitations

This model appears to be well suited for anticonvulsant testing since drugs that are clinically effective against SE effectively inhibit cobalt/HCTL SE. One disadvantage is that the neuropathology associated with these seizures has not been characterized. This information is necessary so that relevance as a clinical model can be established. It appears that a significant number of animals die during SE. In their study Walton and Treiman[14] reported that 30% of the animals died before exhibiting PEDs. Because an active cobalt focus must be created and chronic electrodes implanted before SE can be induced the investigator must consider the labor and time it takes to prepare these animals.

D. Flurothyl

Flurothyl, hexaflurodiethyl ether (Flura Corp., Newport, TN), is a convulsant gas that when inhaled induces seizure activity.[15,16,88] It has been hypothesized that flurothyl induces seizure activity by opening sodium channels.[15] Flurothyl has been used to induce SE but not as often as the previously discussed models. The advantages of flurothyl are that it induces SE that is accompanied by a predictable pattern of irreversible neuronal damage; physiologic parameters during SE can be controlled and monitored; and the duration of the seizures can be determined by the investigator.[16] In order to use flurothyl, the laboratory must be equipped with a closed gas inhalation delivery system for rodents so the investigator is not exposed to the flurothyl, otherwise all experiments must be done in a laboratory fume hood. These animals also require significant instrumentation in order to monitor and control the physiologic changes that occur during SE.

1.   Methods
The animal is anesthetized with ether, intubated, and paralyzed with a nondepolarizing muscle blocker like curare (3 mg/kg, i.p., Sigma, St. Louis, MO). The animal is connected to a ventilator and a combination of 70% N_2O, 27% O_2, 3% halothane is delivered. Catheters are implanted into the femoral artery to allow for physiologic monitoring and the femoral vein to provide a route for drug infusion. Intracranial electrodes must be implanted so electrographic seizure activity can be detected. Because flurothyl induces a profound hypertension at the beginning of seizure activity the animal must be pretreated with phentolamine (0.1 mg/kg, IV, Sigma, St. Louis, MO) and blood must be removed through the venous catheter so that baseline blood pressure is no greater than 100/70.[16] Flurothyl is delivered to the paralyzed and ventilated animal by bubbling air through a 4:1 (v/v) mixture of water:flurothyl and then to the intake line of the ventilator. The inhalation of flurothyl is adjusted to induce electrographic SE on the EEG for a duration determined by the investigator. Seizure activity will stop within minutes once ventilation with flurothyl is stopped.[16]

2.   Pathophysiology and Neuropathology
Electrographic changes during flurothyl-induced SE have been poorly characterized. Lowenstein et al.[16] reported that EEG changes occur within minutes after initiation

of flurothyl ventilation. There is a correlation between animals that exhibit prolonged periods of high-frequency, high-amplitude, repetitive discharges and neuronal stress measured by heat shock protein expression and neuronal damage.[16]

Two hours of SE induced with flurothyl results in widespread neuronal damage. Areas particularly vulnerable include the cerebral cortex, hippocampus, amygdala, basal ganglia, thalamus, and midbrain.[15] It is interesting that in the hippocampus CA1 pyramidal neurons and dentate hilar neurons are vulnerable but, unlike other models of SE, the CA3 pyramidal neurons are resistant.[15]

3.   Advantages and Limitations

This model is well suited for examination of questions related to the characterization of SE-induced neuronal damage. The initiation and termination of seizure activity can be finely controlled by the investigator. However, this model is extremely labor intensive and requires mastery of a variety of surgical skills and experience with methods of inhalation anesthesia.

III. Stimulation-Induced Models of Status Epilepticus

A.   Perforant Path Stimulation

This seizure model was developed by Sloviter to demonstrate that by stimulating a pathway that contains axons of glutaminergic neurons one can duplicate much of the hippocampal damage observed in human TLE and other pharmacologic seizure models.[17,18] The advantage is that hippocampal seizures can be induced without the interpretive complication of direct excitotoxic damage that results from the chemical models. In this model anesthetized rats are intermittently stimulated unilaterally in the perforant path for 24 h.

1.   Methods

Sprague-Dawley rats are anesthetized with urethane (Sigma, St. Louis, MO) dissolved in physiologic saline (1.25 g/kg s.c.). Urethane is injected in four to six sites subcutaneously to increase the rate of absorption. Once anesthetized the rat is placed in a stereotaxic apparatus with the incisor bar set at −5.0. The animal is placed on a heating pad (Harvard Apparatus, Holliston, MA) connected to a rectal probe to maintain temperature at 37°C. This is survival surgery so aseptic procedures must be followed. The skull surface is exposed and two burr holes are drilled on the same side of the skull. The first hole is drilled ±4.5 mm lateral from the sagittal suture directly rostral to the lambda suture. A bipolar stimulating electrode (NE-200, Rhodes Medical Instruments, Tujunga, CA) is placed at this site and lowered approximately 2 mm below the skull surface. The second burr hole is drilled 3 mm rostral to lambda, ±2 mm lateral from the sagittal suture.[38] The dura is removed and a glass micropipette (0.5 to 3.5 Megohm, filled with 0.9% NaCl) is placed at the brain surface of the second hole. The stimulating electrode is connected through a stimulus isolation unit to a stimulator. The recording electrode is connected to a differential amplifier with output to an oscilloscope.

Proper placement of the recording electrode in the upper blade of the granule cell layer is accomplished by simultaneously delivering a paired-pulse stimulus (20 to 30 V, 0.1 msec pulse duration, 40 msec pulse separation, 0.2 Hz) while the recording electrode is slowly lowered approximately 3.5 mm below the skull surface. The recording electrode is lowered in 100 μm steps while evoked potentials are recorded until the signature evoked potential of the dorsal granule cell layer is observed.[89] The size of the evoked potential is then maximized by adjusting the position of the stimulating and recording electrodes. The stimulus voltage is then adjusted to the minimum voltage necessary to obtain a maximum response. Once optimal evoked potentials are obtained the paired pulse stimulus paradigm is initiated using a 40 msec interpulse interval delivered at 2 Hz. For 10 s of every minute of stimulation the rat receives 20 Hz of single pulse stimulation with all other stimulation parameters unchanged. Each rat is stimulated continuously for 24 h. Anesthetic supplements are given as needed. However, the investigator must check the level of anesthesia during the night and early morning. At the end of 24 h of stimulation the electrodes are removed, the wound is cleansed with a disinfectant and closed with sutures or wound clips (Stoelting, Wood Dale, IL).

2.   Pathophysiology and Neuropathology

The pathophysiology associated with this model is characterized by a loss of paired-pulse inhibition and occasionally the presence of multiple population spikes in the dentate gyrus in response to a single paired-pulse stimulation of the perforant path.[17,18] This alteration in hippocampal physiology appears to be permanent, since it is still present in animals tested more than a year after induction of SE.[19] Since the animals are anesthetized during the entire stimulation period there are no behavioral manifestations associated with this model.

The perforant path stimulation model results in cell death, primarily in the ipsilateral dentate hilus.[19] While vulnerable cell populations include hilar mossy cells, and somatostatin- and neuropeptide Y (NPY)-containing neurons, GABA-containing neurons in the hilus and granule cell layer and granule cells survive.[19] The lesion induced by perforant path stimulation is similar to the lesion found in sclerotic hippocampi removed from cryptogenic epileptic patients.[90,113] The survival of GABAergic neurons led to the formation of the dormant basket cell hypothesis to explain the seizure-induced loss of inhibition.[91] The dormant basket cell hypothesis states that inhibition in the dentate gyrus depends on the tonic activation of inhibitory interneurons by hilar mossy cells. A loss of mossy cells results in a functional loss of inhibition not because the inhibitory neurons are dead but because the inhibitory neurons are not receiving sufficient activation to respond to remaining inputs.

Initial studies using this model suggested that the granule cells in the stimulated and contralateral hippocampus are relatively resistant to the seizures such that the contralateral hippocampus could be used as an unstimulated control. However, a recent study revealed that some granule cells in both hippocampi undergo apoptotic cell death as a result of the seizure activity.[92]

3. Postseizure Care and Animal Behavior

Each animal is kept warm under a radiant heat lamp for 1 to 2 d or until the rat is able to right itself and move out from under the lamp. Since the rat is unable to eat or drink after the prolonged exposure to urethane, each rat is injected with lactated Ringer's (2 ml, s.c.) on the first day after stimulation followed by oral administration on the second day. Usually by the third day after stimulation the rat is able to eat and drink on its own. Once recovered these animals do not exhibit an increase in aggressive behavior commonly observed in other models of SE.

4. Anticonvulsant Efficacy

There has been limited anticonvulsant testing using this model of SE as described above. Several investigators have tested anticonvulsants in an awake version of the perforant path model.[93,94] Bilateral, chronic recording electrodes are implanted in the dentate gyrus of the dorsal hippocampus and bilateral stimulating electrodes are implanted in the angular bundle. The stereotaxic coordinates for the angular bundle electrodes are AP −7.0 mm, ML ±4.5 mm, DV −4.1 mm relative to bregma and skull surface.[38,94] The animals are allowed to recover a minimum of 1 week, the location of the electrodes is tested and the side with the better response to a 1.5 mA stimulus is chosen for the experiment. SE is induced by stimulating for 60 min (2 mA, 20 Hz, 0.1 msec pulse duration). Behavioral seizures induced by the stimulation resemble kindled seizures and are scored using the scale by Racine.[50] Anticonvulsant drugs are administered for 1 week before induction of SE and continued for 2 weeks after stimulation. Two anticonvulsants have been tested using this method: vigabatrin (125 mg/kg twice a day) and carbamazepine (10 mg/kg twice a day), both administered at 12-h intervals.[94] On the day of the experiment the anticonvulsants are administered 4 h before stimulation to ensure maximal blood levels. Both drugs effectively decrease the number and duration of generalized clonic seizures. The drugs do not interfere with granule cell spiking in response to the stimulus.[94]

5. Advantages and Limitations

Perforant path stimulation results in a limited lesion in the dorsal hippocampus. The model is well suited for studies that examine mechanisms of seizure-induced cell death, seizure-induced changes in hippocampal circuitry, and changes in gene expression. Some limitations of this model are that these seizures do not lead to significant synaptic reorganization or spontaneous seizures which are common in human epilepsy. The model is labor intensive in that the level of anesthesia requires long-term monitoring. However, it takes less than an hour to anesthetize the animal, perform the surgery, and begin the stimulation. The model as originally developed is not well suited for anticonvulsant testing due to the continued presence of the anesthetic urethane. It is not clear whether 1 h of perforant path stimulation in an awake animal used to test the efficacy of anticonvulsants is equivalent to 24 h of stimulation in a urethane-anesthetized animal.

B.   Self-Sustained Limbic Status Epilepticus (SSLSE)

This model, as developed by Lothman et al.,[20] shares some similarities with the perforant path stimulation model and advances ventral hippocampal stimulation models developed by McIntyre et al.[95] in hippocampal kindled animals and by Vicedomini and Nadler[96] in naïve animals. Unlike the model used by McIntyre et al.,[95] SSLSE can be induced in previously kindled or naïve rats and while SSLSE uses a similar stimulation paradigm to the one used by Vicedomini and Nadler,[96] in the SSLSE model a standardized amount of stimulation is delivered to each rat. All three models provide another way to induce SE without the confounding variable of the direct excitotoxic effect of a toxin or drug.

1.   Methods

Male rats (225 to 275 g) are anesthetized and placed in a stereotaxic frame. Twisted, bipolar electrodes made from Teflon-coated, stainless steel wire (AM Systems, Everett, WA) with a tip diameter of 0.4 mm and a tip separation of 0.5 mm are implanted bilaterally in CA3 of the ventral hippocampus at the following coordinates: incisor bar –3.3 mm, AP –5.3 mm from bregma, ML ± 4.9 mm, DV –5.0 mm.[97] A ground electrode is placed over the frontal sinus and all electrodes are connected to Amphenol pin connectors and attached to the skull. Details of chronic electrode implantation can be found in Chapter 3.

Each animal is allowed to recover for 1 week, at which time afterdischarge (AD) thresholds are determined with a standard stimulus of a 10 s train (biphasic 50 Hz, 1 msec pulse width, square waves). Only animals with an AD threshold less than 250 µA are used. At this time the continuous hippocampal stimulation (CHS) protocol is initiated. The animals are unanesthetized and freely moving during CHS. CHS consists of continuous electrical stimulation delivered to the hippocampus (biphasic 50 Hz, 1 msec pulse width, 400 µA peak to peak square waves). An individual stimulus epoch lasts 10 min consisting of 10 s on and 1 s off for 9 min. The stimulation is stopped for the tenth minute. The total duration of CHS is 90 min. Two patterns of responses occur at the end of CHS. Some animals exhibit electrographic SE that persists for 6 to 12 h. These animals are classified as exhibiting self-sustained limbic status epilepticus (SSLSE). The other response is a lack of SE and these animals are classified as non-SSLSE. Electrographic SE must be present a minimum of 30 min for an animal to be designated SSLSE.

2.   Pathophysiology and Neuropathology

Electrographic activity during CHS can be scored and used as a predictor as to which animals will exhibit SSLSE. The following scoring scale is used: 1, no afterdischarge; 2, stimulation-dependent afterdischarges; 3, afterdischarges during the stimulus off period that slow during the period; 4, autonomous ictus, no decrease in electrographic seizure activity during the stimulus off period. Successful SSLSE rats exhibit an EEG score of 3 to 4 by the fifth stimulation epoch. The EEG during SSLSE exhibits the same progressive sequential changes observed by Treiman and colleagues during human SE.[9]

Behavioral seizures during CHS have also been used to quantify seizure severity. Mild limbic seizures are equivalent to kindled seizure stages 1 and 2, while severe seizures are equivalent to kindled seizure stages 3 to 5.[50] Discrete ictal EEG activity is not always accompanied by behavioral seizure activity, but behavioral seizure activity is always accompanied by electrographic activity. This is an important observation because the investigator cannot assume an animal is not in SE due to a lack of behavioral seizure activity. Other observations that can be used to predict which rats will develop SSLSE are (1) synchronized ictal activity between the two hippocampi; (2) all stimulations resulting in seizure scores of 3 to 4; and (3) ictal activity in stimulus off periods during CHS.

In animals examined 1 month after SSLSE there is hippocampal pyramidal cell loss similar to what is observed in human TLE.[98,99] Bilateral cell loss is consistently found in CA1 accompanied by a shortening of the dentate granule cell layer. There is also evidence of dentate hilar cell loss, synaptic reorganization, and two types of spontaneous recurrent seizures.[98,99] The first type of spontaneous seizures appear to be similar to complex partial seizures observed in humans as they are characterized by staring accompanied by facial automatisms. The second type of spontaneous seizures are similar to kindled limbic motor seizures.[99,100]

3.   Postseizure Care and Animal Behavior

Eighty percent to 90% of the animals that exhibit SSLSE survive and do not require specialized postseizure care. However, these animals become extremely aggressive post-SSLSE. The investigator should use caution and wear protective gloves when handling these animals.[114]

4.   Anticonvulsant Efficacy

Bertram and Lothman[68] tested the efficacy of standard anticonvulsants in the SSLSE model. Each anticonvulsant was administered intraperitoneally 2 h after the end of 90 min of CHS. At this time point the EEG pattern is indicative of late stage SE when anticonvulsants are least likely to work. The effectiveness of each drug was rated on its ability to alter behavioral and electrographic seizure activity 1 h after injection. If the initial injection was ineffective an additional injection of the drug was given. This was continued up to a maximum of three injections. The effectiveness of diazepam was examined 30 min after each injection. Phenobarbital (40 mg/kg) and diazepam (5 to 8 mg/kg) suppress behavioral seizure activity after the first injection. A second injection of both drugs is occasionally necessary to block electrographic seizure activity. Phenytoin (50 mg/kg) is ineffective against behavioral and electrographic seizure activity after two injections. All drugs that are effective also induce sedation.

5.   Advantages and Limitations

SSLSE induces SE by prolonged stimulation of the ventral hippocampus without the interpretive complication associated with a chemical convulsant. Two types of SE develop, one resembles GCSE and the other is similar to complex partial seizures in humans. During SSLSE, there is a progressive sequence of electrographic changes analogous to those observed in human SE. SSLSE induces neuronal cell loss, synaptic reorganization, and spontaneous recurrent seizures, which are features of human TLE. This model does require chronic implantation of intracranial electrodes. Not all anticonvulsants currently effective in the treatment of human SE are effective in this model, so it may not be the best model for new anticonvulsant testing.

IV. Experimental Status Epilepticus in Immature Animals

The results from several retrospective, clinical studies suggest that childhood SE can lead to chronic epilepsy and neuronal loss in the adult.[6,7] It is unclear whether SE in the developing brain induces cell death or cell death precedes the SE that leads to epilepsy in later life. Experimental SE has been induced in immature animals to examine the relationship between early childhood SE and adult epilepsy. The immature rat is more likely to undergo SE and exhibit more severe seizures than an adult rat.[101-103] At issue is whether experimental SE in the immature brain causes neuronal damage with functional consequences in the adult animal.[104] One of the difficulties in interpreting data from immature animals is that it is not clear at what age the developing nervous system of a rat pup is equivalent to the nervous system of a human child. Experimental models that induce SE in adult animals will also induce SE in immature animals. Some of the models are adapted for use in immature animals by adjusting the dose of the convulsant agent.[104]

A.   Kainic Acid

Kainic acid induces SE accompanied by extensive neuronal damage, synaptic reorganization, and spontaneous recurrent seizures in adult rats. This is not the case in immature rats. Kainic acid SE-induced damage is age-dependent.[22,102,105,106] Sperber et al.[106] injected 5-, 15-, and 30-d-old rats with kainic acid. The dose of kainic acid was decreased to 5 mg/kg. While all of the immature rats had more severe seizures, animals 15 d old and younger were resistant to hippocampal damage. When paired pulse activation of the perforant path was examined in 15-d-old pups after kainic acid SE there was no evidence of an alteration in physiologic function.[106] Therefore, the lack of a physiologic change after kainic acid SE is consistent with the lack of a morphological change after kainic acid SE in immature rats. The results in this model suggest that there is a dissociation between SE and neuronal damage in immature rats.

B. Pilocarpine

Pilocarpine also induces SE in immature animals. Similar to the effect of kainic acid, pilocarpine induces severe seizures that do not result in neuronal damage, further supporting the observation that SE is dissociated from neuronal damage in the immature rat.[79,107]

C. Perforant Path Stimulation

Perforant path stimulation in 14- to 16-d-old rat pups results in a loss of paired pulse inhibition accompanied by hippocampal neuronal cell loss.[23] However, the pattern of cell loss differs from what has been observed after perforant path stimulation in adult animals. In the immature rat there is a selective loss of hilar neurons and cells at the base of the granule cell layer.[23] This pattern of neuronal vulnerability is consistent with what occurs after ischemia in immature rats.[108] Therefore, the results obtained after SE induced by perforant path stimulation differs from those obtained after kainic acid- and pilocarpine-induced SE in immature animals. The mechanisms responsible for the differences between the different models of SE in immature animals remain to be determined.

V. *In Vitro* Models of Status Epilepticus

A detailed description of *in vitro* slice preparation can be found in Chapter 10. This section focuses on the type of slice, electrode placement, stimulation parameters, and the resultant epileptiform activity induced by the models described below.

A. Low Magnesium in Hippocampal-Parahippocampal Slices

Numerous studies have reported that decreasing the extracellular concentration of magnesium in hippocampal slice preparations results in the generation of epileptiform activity.[109-112] Dreier and Heinemann[24] use a modified slice which is thinner (400 μm) and includes neocortical temporal area Te3, perirhinal cortex, the entorhinal cortex, and the hippocampal formation. Decreasing the extracellular concentration of Mg^{2+} in this expanded slice induces all aspects of epileptogenesis observed *in vivo*, including interictal spikes and tonic-clonic discharges.[24] Synchronous electrographic activity can be recorded in the entorhinal cortex, Te3, and the subiculum similar to that observed during SE in humans and animals[24] and suggests SE can be studied *in vitro*. These observations contributed to the development of the *in vitro* stimulation model discussed below.

B.   Stimulation of Hippocampal-Parahippocampal Slices

1.   Methods

This model was developed by Rafiq and colleagues.[25,26] Combined hippocampal-parahippocampal slices are made from male 21- to 30-d-old Sprague-Dawley rats. The rat is anesthetized with halothane and then decapitated. The brain is removed and placed in cold (4°C), oxygenated artificial cerebrospinal fluid (ACSF) composed of the following: 200 mM sucrose, 3 mM KCl, 1.25 mM Na$_2$PO$_4$, 10 mM glucose, 0.5 to 0.9 mM MgCl$_2$, and 2 mM CaCl$_2$ for 90 min.

2.   Slice Preparation

The two hemispheres are separated by a midsagittal cut and placed in ACSF. The individual hemisphere is blocked, and the dorsal surface glued at a 12° incline in the transverse plane with the rostral end up. The brain tissue is then placed in a vibratome filled with cold ACSF, where two or three 450 μm thick slices are cut. The slices are trimmed so that only the parahippocampus and hippocampus proper remained. This slice maintains the connections between the hippocampus, the parahippocampus, and the adjacent cortical areas.

   The slices are transferred to a holding chamber containing warmed (32°C) oxygenated ACSF for 1 to 2 h. The modified ACSF in the holding chamber is composed of: 130 mM NaCl, 3 mM KCl, 1.25 mM Na$_2$PO$_4$, 26 mM NaHCO$_3$, 10 mM glucose, 0.5 to 0.9 mM MgCl$_2$, and 2 mM CaCl$_2$. For stimulation and recording purposes each slice is transferred to an interface-type recording chamber (Medical Systems, Greenvale, New York) and perfused with the modified ACSF warmed to 35°C at a rate of 1 to 1.5 ml/min.

3.   Stimulation and Recording

Insulated tungsten stimulating electrodes (AM Systems, Everett, WA) are placed in the Schaffer collaterals in stratum radiatum of CA1. A stimulus of 8 to 10 V (2 s, 60 Hz, pulse width 100 μsec) is delivered every 10 to 30 min to generate hippocampal afterdischarges. The stimulus frequency is adjusted for each individual slice to minimize the postictal refractory period. Insulated tungsten electrodes are placed in the cell body layer of CA1 and CA3, the granule cell layer of the dentate gyrus, and the entorhinal cortex to record extracellular field potentials. Intracellular and patch recordings can also be made in this preparation.

4.   Generation of *In Vitro* Status Epilepticus

Initial stimulations elicit a primary AD that progresses in duration and complexity with each successive stimulus. The AD can be recorded throughout the hippocampus, dentate gyrus, and entorhinal cortex. By four to six stimulations, the primary AD has reached its maximum duration and complexity. Continued stimulation results in a secondary AD which increases in duration, some lasting longer than 30 min.

   A majority of rostral slices exhibit repeated ictal-like discharges with a duration of 3 to 5 min. The interval between these ictal-like events is less than 15 min. These discharges can last for hours. A second type of secondary ictal discharge occurs in 5% to 10% of the slices. The second type of abnormal discharge is characterized by a continuous discharge with a duration of 30 to 120 min. The first type of secondary AD resembles ictal discharges recorded during complex partial SE in humans, while the second type of AD resembles the ictal activity recorded during GCSE. Both types of secondary AD can be completely blocked with diazepam (100 nM to 1 μM) and enhanced with the NMDA antagonist 2-amino-5-phosphonovaleric acid (APV, 50 μM).

5.   Advantages and Limitations

An important advantage of being able to induce SE in a slice preparation is that it provides an easy way to test new anticonvulsant drugs. This model is particularly intriguing since the types of epileptiform activity observed are consistent with the definition of clinical SE, that is, continuous seizure activity lasting more than 30 min or intermittent seizure lasting more than 30 min without regaining consciousness. Obviously one cannot discuss consciousness in a slice preparation but the electrographic activity recorded in this model resembles what has been reported to occur clinically.

VI. Interpretation

At the beginning of this chapter a series of experimental questions were listed in which experimental models of SE have been used to study specific aspects of SE or epilepsy. Having characterized the different models of SE this section discusses which models are best suited to examine specific epilepsy-related questions.

The primary requirement of an experimental model of SE to be used in the testing of new anticonvulsants is that it exhibit the same sequential electroencephalographic changes that have been shown to occur in human SE. This requirement is met by HCTL, kainic acid, both pilocarpine models, SSLSE, and the *in vitro* models. Kainic acid has the complications of variability in response, direct excitotoxicity, and it is not responsive to anticonvulsants currently effective in human SE. Pilocarpine-induced SE has the complications of high mortality and also does not respond to anticonvulsants effective in human SE. The *in vitro* models are relatively new and need further characterization. This leaves HCTL and SSLSE, which are both labor intensive but only HCTL is responsive to anticonvulsants effective in human SE.

Kainic acid, pilocarpine, perforant path stimulation, SSLSE, and flurothyl all result in seizure-induced damage. It would be hard to say that one of these models is superior to the others in examining questions related to seizure-induced damage. Since all of these models induce damage they would all be candidates for use in the development of new neuroprotectants. All of these models have limitations, so the individual investigator has to decide which model is best suited for the specific question and available facilities. The flurothyl model is probably the most complicated and requires the most resources and expertise. Perforant path stimulation generates a limited lesion that resembles hippocampal sclerosis and is similar to what has been observed in tissue removed from epileptic patients.[113] Synaptic reorganization and spontaneous seizures occur after SE induced by kainic acid, pilocarpine, and SSLSE. However, given the recent report by Longo and Mello,[59] the significance of synaptic reorganization is unclear. Although SSLSE may be more labor intensive than the other models it has fewer complications. Spontaneous recurrent seizures in the pilocarpine model are effectively blocked by anticonvulsants that are currently effective against complex partial seizures in humans.[126]

In conclusion, there are a number of different experimental models of SE. None perfectly duplicate the human condition but, as with all models, they duplicate some aspect of human SE. The ultimate decision of which model is best falls upon the individual investigator. There is always room for the development of new models of SE.

References

1. Delgado-Escueta, A. V., Wasterlain, C. G., Treiman, D. M., and Porter, R. J., Status epilepticus: summary, in *Advances in Neurology,* Vol. 34, *Status Epilepticus: Mechanisms of Brain Damage and Treatment,* Delgado-Escueta, A. V., Wasterlain, C. G., Treiman, D. M., and Porter, R. J., Eds., Raven Press, New York, 1983, 539.
2. Lothman, E. W., The biochemical basis and pathophysiology of status epilepticus, *Neurology,* 40(S2), 13, 1990.
3. DeLorenzo, R. J., Towne, A. R., Pellock, J. M., and Ko, D., Status epilepticus in children, adults and the elderly, Epilepsia, 33(S4), 14, 1992.
4. Sperling, M. R., Introduction: status epilepticus, *Epilepsia,* 34(S1), S1, 1993.
5. Hauser, W. A., Status epilepticus: frequency, etiology, and neurological sequelae, in *Advances in Neurology,* Vol. 34, *Status Epilepticus: Mechanisms of Brain Damage and Treatment,* Delgado-Escueta, A. V., Wasterlain, C. G., Treiman, D. M., and Porter, R. J., Eds., Raven Press, New York, 1983, 3.
6. Aicardi, J. and Chevrie, J. J., Consequences of status epilepticus in infants and children, *Adv. Neurol.,* 34, 115, 1983.
7. Sagar, H. J. and Oxbury, J. M., Hippocampal neuron loss in temporal lobe epilepsy: correlation with early childhood convulsions, *Ann. Neurol.,* 22, 334, 1987.
8. Treiman, D. M., Generalized convulsive status epilepticus in the adult, *Epilepsia,* 34(S1), S2, 1993.
9. Treiman, D. M., Walton, N. Y., and Kendrick, C., A progressive sequence of electrographic changes during generalized convulsive status epilepticus, *Epilepsy Res.,* 5, 49, 1990.
10. Nadler, J. V., Perry, B. W., and Cotman, C. W., Intraventricular kainic acid preferentially destroys hippocampal pyramidal cells, *Nature,* 271, 676, 1978.
11. Ben-Ari, Y., Limbic seizure and brain damage produced by kainic acid: mechanisms and relevance to human temporal lobe epilepsy, *Neuroscience,* 14, 375, 1985.
12. Turski, W. A., Cavalheiro, E. A., Schwarz, M., Czuczwar, S. J., Kleinrok, Z., and Turski, L., Limbic seizures produced by pilocarpine in rats: a behavioural, electroencephalographic neuropathological study, *Behav. Brain Res.,* 9, 315, 1983.

13. Clifford, D. B., Olney, J. W., Maniotis, A., Collins, R. C., and Zorumski, C. F., The functional anatomy and physiology of lithium-pilocarpine, and high-dose pilocarpine seizures, *Neuroscience,* 23, 953, 1987.

14. Walton, N. Y. and Treiman, D. M., Experimental secondarily generalized convulsive status epilepticus induced by D,L-homocysteine thiolactone, *Epilepsy Res.,* 2, 79, 1988.

15. Nevander, G., Invar, M., Auer, R., and Siesjö, B. K., Status epilepticus in well-oxy-genated rats causes neuronal necrosis, *Ann. Neurol.,* 18, 281, 1985.

16. Lowenstein, D. H., Simon, R. P., and Sharp, F. R., The pattern of 72-kDa heat shock protein-like immunoreactivity in the rat brain following flurothyl-induced status epilepticus, *Brain Res.,* 531, 173, 1990.

17. Sloviter, R. S., Epileptic brain damage in rats induced by sustained electrical stimulation of the perforant path. I. Acute electrophysiological and light microscopic studies, *Brain Res. Bull.,* 10, 675, 1983.

18. Sloviter, R. S., Decreased hippocampal inhibition and a selective loss of interneurons in experimental epilepsy, *Science,* 235, 73, 1987.

19. Sloviter, R. S., Feedforward and feedback inhibition of hippocampal principal cell activity evoked by perforant path stimulation: GABA-mediated mechanisms that regulate excitability *in vivo, Hippocampus,* 1, 31, 1991.

20. Lothman, E. W., Bertram, E. H., Bekenstein, J. W., and Perlin, J. B., Self-sustaining limbic status epilepticus induced by "continuous" hippocampal stimulation; electrographic and behavioral characteristics, *Epilepsy Res.,* 3, 107, 1989.

21. Wasterlain, C. G., Effects of neonatal status epilepticus on rat brain development, *Neurology,* 26, 975, 1986.

22. Sperber, E. S., Developmental profile of seizure-induced hippocampal damage, *Epilepsia,* 33(S3), 44, 1992.

23. Thompson, K. W., Holm, A. M., and Wasterlain, C. G., Loss of hippocampal interneurons in a model of neonatal status epilepticus, *Soc. Neurosci. Abstr.,* 20, 1668, 1994.

24. Dreier, J. P. and Heinemann, U., Regional and time dependent variations of low Mg^{2+} induced epileptiform activity in rat temporal cortex slices, *Exp. Brain Res.,* 87, 581, 1991.

25. Rafiq, A., DeLorenzo, R. J., and Coulter, D. A., Generation and propagation of epileptiform discharges in a combined entorhinal cortex/hippocampal slice, *J. Neurophysiol.,* 70, 1962, 1993.

26. Rafiq, A., Zhang, Y.-F., Delorenzo, R. J., and Coulter, D. A., Long-duration self-sustained epileptiform activity in the hippocampal-parahippocampal slice: a model of status epilepticus, *J. Neurophysiol.,* 74, 2028, 1995.

27. Takemoto, T., Isolation and identification of naturally occurring excitatory amino acids, in Kainic Acid as a Tool in Neurobiology, McGeer, E. G., Olney, J. W., and McGeer, P. L., Eds., Raven Press, New York, 1978, 1.

28. Olney, J. W., Rhee, V., and DeGubareff, T., Kainic acid: a powerful neurotoxic analogue of glutamate, *Brain Res.,* 77, 507, 1974.

29. Coyle, J. T. and Schwarcz, R., Models of Huntington's chorea: lesion of striatal neurons with kainic acid, *Nature,* 263, 244, 1976.

30. Coyle, J. T., Molliver, M. R., and Kuhar, M. J., In situ injection of kainic acid: a new method for selectively lesioning neuronal cell bodies while sparing axons of passage, *J. Comp. Neurol.,* 180, 301, 1978.

31. Ben-Ari, Y., Tremblay, E., Ottersen, O. P., and Naquet, R., Evidence suggesting secondary epileptogenic lesions after kainic acid: pre-treatment with diazepam reduces distant but not local damage, *Brain Res.,* 165, 362, 1979.

32. Collins, R. C. and Olney, J. W., Focal cortical seizures cause distant thalamic lesions, *Science,* 218, 177, 1982.

33. Sundstrom, L. E., Mitchell, J., and Wheal, H. V., Bilateral reorganization of mossy fibres in the rat hippocampus after unilateral intracerebroventricular kainic acid injection, *Brain Res.,* 609, 321, 1993.

34. Sperk, G., Kainic acid seizures in the rat, *Prog. Neurobiol.,* 42, 1, 1994.

35. Gall, C., Murray, K., and Isackson, P. J., Kainic acid-induced seizures stimulate increased expression of nerve growth factor mRNA in rat hippocampus, *Mol. Brain Res.,* 9, 113, 1991.

36. Schwarcz, R., Zaczek, R., and Coyle, J. T., Microinfusion of kainic acid into the rat hippocampus, *Eur. J. Pharmacol.,* 50, 209, 1978.

37. Brines, M. L., Dare, A. O., and de Lanerolle, N. C., The cardiac glycoside ouabain potentiates excitotoxic injury of adult neurons in rat hippocampus, *Neurosci. Lett.,* 191, 145, 1995.

38. Paxinos, G. and Watson, C., *The Rat Brain in Stereotaxic Coordinates,* 2nd ed., Academic Press, San Diego, 1986.

39. Goodman, J. H. and Sloviter, R. S., Evidence for commissurally projecting parvalbumin-immunoreactive basket cells in the dentate gyrus of the rat, *Hippocampus,* 2, 13, 1992.

40. Milgram, N.W., Yearwood, T., Khurgel, M., Ivy, G. O., and Racine, R., Changes in inhibitory processes in the hippocampus following recurrent seizures induced by systemic administration of kainic acid, *Brain Res.,* 551, 236, 1991.

41. Sloviter, R. S., Possible functional consequences of synaptic reorganization in the dentate gyrus of kainate-treated rats, *Neurosci. Lett.,* 137, 91, 1992.

42. Ben-Ari, Y., Tremblay, E., Riche, D., Ghilini, G., and Naquet, R., Electrographic, clinical, and pathological alterations following systemic administration of kainic acid, bicuculline, and pentylenetetrazole: metabolic mapping using the deoxyglucose method with special reference to the pathology of epilepsy, *Neuroscience,* 6, 1361, 1981.

43. Tauck, D. L. and Nadler, J. V., Evidence of functional mossy fiber sprouting in the hippocampal formation of kainic acid-treated rats, *J. Neurosci.,* 5, 1016, 1985.

44. Lothman, E. W. and Collins, R. C., Kainic acid-induced limbic motor seizures: metabolic, behavioral, electroencephalographic and neuropathological correlates, *Brain Res.,* 218, 299, 1981.

45. Zucker, D. K., Wooten, G. F., Lothman, E. M., and Olney, J. W., Neuropathological changes associated with intravenous kainic acid, *J. Neuropathol. Exp. Neurol.,* 40, 324, 1981.

46. Sperk, G., Lassman, H., Baran, H., Kish, S. J., Seitelberger, F., and Hornykiewicz, O., Kainic acid-induced seizures: neurochemical and histopathological changes, *Neuroscience,* 10, 1301, 1983.

47. Sperk, G., Lassman, H., Baran, H., Seitelberger, F., and Hornykiewicz, O., Kainic acid-induced seizures: dose relationship of behavioural, neurochemical and histopathological changes, *Brain Res.,* 338, 289, 1985.

48. Treiman, D. M., Walton, N. Y., Wickboldt, C., and DeGiorgio, C. M., Predictable sequence of EEG changes during generalized convulsive status epilepticus in man and three experimental models of status epilepticus in the rat, *Neurology,* 37(S1), 244, 1987.

49. Goddard, G. V., McIntyre, D. C., and Leetch, C. K., A permanent change in brain function resulting from daily electrical brain stimulation, *Exp. Neurol.,* 25, 295, 1969.

50. Racine, R. J., Modification of seizure activity by electrical stimulation. II. Motor seizure, *Electroencephalogr. Clin. Neurophysiol.,* 32, 281, 1972.

51. Feldblum, S. and Ackermann, R. F., Increased susceptibility to hippocampal and amygdala kindling following intrahippocampal kainic acid, *Exp. Neurol.,* 97, 255, 1987.

52. Marksteiner, J., Lassmann, H., Saria, A., Humpel, C., Meyer, D. K., and Sperk, G., Neuropeptide levels after pentylenetetrazol kindling in the rat, *Eur. J. Neurosci.,* 2, 98, 1990.

53. Pisa, M., Sanberg, P. R., Corcoran, M. E., and Fibiger, H. C., Spontaneous recurrent seizures after intracerebral injections of kainic acid in rat: a possible model of human temporal lobe epilepsy, *Brain Res.,* 200, 481, 1980.

54. Cavalheiro, E., Riche, D., and Le Gal La Salle, G., Long-term effects of intrahippocampal kainic acid in rats: a method for inducing spontaneous recurrent seizures, *Electroencephalogr. Clin. Neurophysiol.,* 53, 581, 1982.

55. Cronin, J., Obenaus, A., Houser, C. R., and Dudek, F. E., Electrophysiology of dentate granule cells after kainate-induced synaptic reorganization of the mossy fiber, *Brain Res.,* 573, 305, 1992.

56. Sutula, T., Casino, G., Cavazos, J., Parada, I., and Ramirez, L., Mossy fiber synaptic reorganization in the epileptic human temporal lobe, *Ann. Neurol.,* 26, 321, 1989.

57. Houser, C. R., Miyashiro, J. E., Swartz, B. E., Walsh, G. O., Rich, J. R., and Delgado-Escueta, A. V., Altered patterns of dynorphin immunoreactivity suggest mossy fiber reorganization in human hippocampal epilepsy, *J. Neurosci.,* 10, 267, 1990.

58. Babb, T. L., Pretorius, J. K., Crandall, P. H., and Livesque, M. F., Synaptic reorganization by mossy fibers in human epileptic fascia dentata, *Neuroscience,* 42, 351, 1991.

59. Longo, B. M. and Mello, L. E. A. M., Blockade of pilocarpine- or kainate-induced mossy fiber sprouting by cycloheximide does not prevent subsequent epileptogenesis in rats, *Neurosci. Lett.,* 226, 163, 1997.

60. Golden, G. T., Smith, G. G., Ferraro, T. N., Reyes, P. F., Kulp, J. K., and Fariello, R. G., Strain differences in convulsive response to excitotoxin kainic acid, *Neuroreport,* 2, 141, 1991.

61. Wozniak, D. F., Stewart, G. R., Miller, J. P., and Olney, J. W., Age-related sensitivity to kainate neurotoxicity, *Exp. Neurol.,* 114, 250, 1991.

62. Dawson, R. and Wallace, D. R., Kainic acid-induced seizures in aged rats: neurochemical correlates, *Brain Res. Bull.,* 29, 459, 1992.

63. Berger, M. L., Lefauconnier, J. M., Tremblay, E., and Ben-Ari, Y., Limbic seizures induced by systemically applied kainic acid: how much kainic acid reaches the brain?, in *Excitatory Amino Acids and Epilepsy,* Schwarcz, R. and Ben-Ari, Y., Eds., Plenum Press, New York, 1986, 199.

64. Rothman, S. M. and Olney, J. W., Excitotoxicity and the NMDA receptor, *Trends Neurosci.,* 10, 299, 1987.

65. Wuarin, J.-P. and Dudek, F. E., Electrographic seizures and new recurrent excitatory circuits in the dentate gyrus of hippocampal slices from kainate-treated epileptic rats, *J. Neurosci.,* 16, 4438, 1996.

66. Baran, H., Lassman, H., and Hornykiewicz, O., Behavior and neurochemical changes 6 months after kainic acid, *Soc. Neurosci. Abstr.,* 14, 472, 1988.

67. Lowenstein, D. H., Aminoff, M. J., and Simon, R. P., Barbiturate anesthesia in the treatment of status epilepticus: clinical experience with 14 patients, *Neurology,* 38, 395, 1988.

68. Bertram, E. H. and Lothman, E. W., NMDA antagonists and limbic status epilepticus: a comparison with standard anticonvulsants, *Epilepsy Res.,* 5, 177, 1990.

69. Stone, W. E. and Javid, M. J., Effects of anticonvulsants and glutamate antagonists on the convulsive action of kainic acid, *Arch. Intern. Pharmacodyn. Ther.,* 243, 56, 1980.

70. Fuller, T. A. and Olney, J. W., Only certain anticonvulsants protect against kainate neurotoxicity, *Neurobehav. Toxic. Teratol.,* 3, 355, 1981.

71. Braun, D. E. and Freed, W. J., Effect of nifidipine and anticonvulsants on kainic acid-induced seizures in mice, *Brain Res.,* 533, 157, 1990.

72. Turski, W. A., Niemann, W., and Stephens, D. N., Differential effects of antiepileptic drugs and β-carbolines on seizures induced by excitatory amino acids, *Neuroscience,* 39, 799, 1990.

73. De Bonnel, G. and De Montigny, C., Benzodiazepines selectively antagonize kainate-induced activation in the rat hippocampus, *Eur. J. Pharmacol.,* 93, 45, 1983.

74. Baran, H., Sperk, G., Hörtnagl, H., Sapetschnig, G., and Hornykiewicz, O., Alpha2-adrenoreceptors modulate kainic acid-induced limbic seizures, *Eur. J. Pharmacol.,* 113, 263, 1985.

75. Ault, B., Gruenthal, M., Armstrong, D. R., and Nadler, J. V., Efficacy of baclofen and phenobarbital against kainic acid limbic seizure brain damage syndrome, *J. Pharmacol. Exp. Ther.,* 239, 612, 1986.

76. Cavalheiro, E. A., Leite, J. P., Bortolotto, Z. A., Turski, W. A., Ikonomidou, C., and Turski, L., Long-term effects of pilocarpine in rats: structural damage of the brain triggers kindling and spontaneous recurrent seizures, *Epilepsia,* 32, 778, 1991.

77. Liu, Z., Nagao, T., Desjardins, G. D., Gloor, P., and Avoli, M., Quantitative evaluation of neuronal loss in the dorsal hippocampus in rats with long-term pilocarpine seizures, *Epilepsy Res.,* 17, 237, 1994.

78. Gibbs, J. W., III, Shumate, M. D., and Coulter, D. A., Differential epilepsy-associated alterations in postsynaptic GABA-A receptor function in dentate granule and CA1 neurons, *J. Neurophysiol.,* 77(4), 1924, 1997.

79. Cavalheiro, E. A., Silva, D. F., Turski, W. A., Calderazzo-Filho, L. S., Bortolotto, Z. A., and Turski, L., The susceptibility of rats to pilocarpine-induced seizures is age-dependent, *Dev. Brain Res.,* 37, 43, 1987.

80. Jope, R. S., Morrisett, R. A., and Snead, O. C., Characterization of lithium potentiation of pilocarpine-induced status epilepticus in rats, *Exp. Neurol.,* 91, 471, 1986.

81. Honchar, M. P., Olney, J. W., and Sherman, W. R., Systemic cholinergic agents induce seizures and brain damage in lithium-treated rats, *Science,* 220, 323, 1983.

82. Morrisett, R. A., Status epilepticus produced by cholinergic agonists, *Exp. Neurol.,* 98, 594, 1987.

83. Walton, N. Y. and Treiman, D. M., Response of status epilepticus induced by lithium and pilocarpine to treatment with diazepam, *Exp. Neurol.,* 191, 267, 1988.

84. Handforth, A. and Treiman, D. M., Functional mapping of the early stages of status epilepticus: a 14C-2-deoxyglucose study in the lithium-pilocarpine model in rat, *Neuroscience,* 64, 1057, 1995.

85. Obenaus, A., Esclapez, M., and Houser, C. R., Loss of glutamate decarboxylase MRNA-containing neurons in the rat dentate gyrus following pilocarpine-induced seizures, *J. Neurosci.,* 13, 4470, 1993.

86. Mello, L. E. A. M., Cavalheiro, E. A., Tan, A. M., Kupfer, W. R., Pretorius, J. K., Babb, T. L., and Finch, D. M., Circuit mechanisms of seizures in the pilocarpine model of chronic epilepsy: cell loss and mossy fiber sprouting, *Epilepsia,* 34, 985, 1993.

87. Woodbury, D. M., Experimental models of status epilepticus and mechanisms of drug action, in *Advances in Neurology,* Vol. 34, *Status Epilepticus: Mechanisms of Brain Damage and Treatment,* Delgado-Escueta, A. V., Wasterlain, C. G., Treiman, D. M., and Porter, R. J., Eds., Raven Press, New York, 1983, 441.

88. Ingvar, M., Morgan, P. F., and Auer, R. N., The nature and timing of excitotoxic neuronal necrosis in the cerebral cortex, hippocampus and thalamus due to flurothyl-induced status epilepticus, *Acta Neuropathol. (Berlin),* 75, 362, 1988.

89. Andersen, P., Holmqvist, B., and Voorhoeve, P. E., Entorhinal activation of the dentate granule cells, *Acta Physiol. Scand.,* 66, 448, 1966.

90. Babb, T. L. and Pretorius, J. K., Pathological substrates of epilepsy, in *The Treatment of Epilepsy: Principles and Practice,* Wylie, E., Eds., Lea and Febiger, Philadelphia, 1993, 55.

91. Sloviter, R. S., Permanently altered hippocampal structure, excitability and inhibition after status epilepticus in the rat: the "dormant basket cell" hypothesis and its possible relevance to temporal lobe epilepsy, *Hippocampus,* 1, 41, 1991.

92. Sloviter, R. S., Dean, E., Sollas, A. L., and Goodman, J. H., Apoptosis and necrosis induced in different hippocampal neuron populations by repetitive perforant path stimulation in the rat, *J. Comp. Neurol.,* 366, 516, 1996.

93. Ylinen, A., Valjakka, A., Lathtinen, H., Miettinen, R., Freund, T. F., and Riekkinen, P., Vigabatrin pre-treatment prevents hilar somatostatin cell loss and the development of interictal spiking activity following sustained stimulation of the perforant path, *Neuropeptides,* 142, 393, 1991.

94. Pitkänen, A., Tuunanen, J., and Halonen, T., Vigabatrin and carbamazepine have different efficacies in the prevention of status epilepticus induced neuronal damage in the hippocampus and amygdala, *Epilepsy Res.,* 24, 29, 1996.

95. McIntyre, D. C., Stokes, K. A., and Edson, N., Status epilepticus following stimulation of a kindled hippocampus in intact and commissurotomized rats, *Exp. Neurol.,* 94, 554, 1986.

96. Vicedomini, J. P. and Nadler, J. V., A model of status epilepticus based on electrical stimulation of hippocampal afferent pathways, *Exp. Neurol.,* 96, 681, 1987.

97. Pellegrino, L. S., Pellegrino, A. S., and Cushman, A. J., *A Stereotaxic Atlas of the Rat Brain,* Plenum Press, New York, 1979.

98. Bertram, E. H., Lothman, E. W., and Lenn, N. J., The hippocampus in experimental chronic epilepsy: a morphometric analysis, *Ann. Neurol.,* 27, 43, 1990.

99. Lothman, E. W. and Bertram, E. H., Epileptogenic effects of status epilepticus, *Epilepsia,* 34(S1), S59, 1993.

100. Lothman, E. W., Bertram, E. H., Kapur, J., and Stringer, J. L., Recurrent spontaneous hippocampal seizures in the rat as a chronic sequela to limbic status epilepticus, *Epilepsy Res.,* 6, 110, 1990.

101. Moshé, S. L., Albala, B. J., Ackermann, R. F., and Engel, J., Increased seizure susceptibility of the immature brain, *Dev. Brain Res.,* 7, 81, 1983.

102. Albala, B. J., Moshé, S. L., and Okada, R., Kainic acid-induced seizures: a developmental study, *Dev. Brain Res.,* 13, 139, 1984.

103. Okada, R., Moshé, S. L., and Albala, B. J., Infantile status epilepticus and future seizure susceptibility in the rat, *Dev. Brain Res.,* 15, 177, 1984.

104. Sankar, R., Wasterlain, C. G., and Sperber, E. S., Seizure-induced changes in the immature brain, in *Brain Development and Epilepsy,* Schwartzkroin, P. A., Moshé, S. L., Noebels, J. L., and Swann, J. W., Eds., Oxford University Press, New York, 1995, 268.

105. Holmes, G. L. and Thompson, J. L., Effects of kainic acid on seizure susceptibility in the developing brain, *Dev. Brain Res.,* 39, 51, 1988.

106. Sperber, E. F., Haas, K. Z., Stanton, P. K., and Moshé, S. L., Resistance of the immature hippocampus to seizure-induced synaptic reorganization, *Dev. Brain Res.,* 60, 89, 1991.

107. Priel, M. R., Santos, N. F. D., and Cavalheiro, E. A., Developmental aspects of the pilocarpine model of epilepsy, *Epilepsy Res.,* 26, 115, 1996.

108. Goodman, J. H., Wasterlain, C. G., Massarweh, W. F., Dean, E., Sollas, A. L., and Sloviter, R. S., Calbindin-D28K immunoreactivity and selective vulnerability to ischemia in the dentate gyrus of the developing rat, *Brain Res.,* 606, 309, 1993.

109. Anderson, W. W., Lewis, D. V., Swartzwelder, H. S., and Wilson, W. A., Magnesium-free medium activates seizure-like events in the rat hippocampal slice, *Brain Res.,* 398, 215, 1986.

110. Jones, R. S. G. and Heinemann, U., Synaptic and intrinsic responses of medial entorhinal cortical cells in normal and magnesium-free medium *in vitro, J. Neurophysiol.,* 59, 1476, 1988.

111. Mody, I., Lambert, J. D. C., and Heinemann, U., Low extracellular magnesium induces epileptiform activity and spreading depression in rat hippocampal slices, *J. Neurophysiol.,* 57, 869, 1987.

112. Tancredi, V., Hwa, G. G. C., Zona, C., Brancati, A., and Avoli, M., Low magnesium epileptogenesis in the rat hippocampal slice: electrophysiological and pharmacological features, *Brain Res.,* 511, 280, 1990.

113. de Lanerolle, N. C., Kim, J. H., Robbins, J., and Spencer, D.D., Hippocampal neuron loss and plasticity in human temporal lobe epilepsy, *Brain Res.,* 495, 387, 1989.

114. Bertram, E. H., personal communication.

115. Lancaster, B. and Wheal, H. V., A comparative histological and electrophysiological study of some neurotoxins in the rat hippocampus, *J. Comp. Neurol.,* 211, 105, 1982.

116. Anderson, W. R., Franck, J. E., Stahl, W. L., and Maki, A. A., Na,K-ATPase is decreased in hippocampus of kainate-lesioned rats, *Epilepsy Res.,* 17, 221, 1994.

117. Schwob, J. E., Fuller, T., Price, J. L., and Olney, J.W., Widespread patterns of neuronal damage following systemic or intracerebral injections of kainic acid: a histological study, *Neuroscience,* 5, 991, 1980.

118. Sperk, G., Marksteiner, J., Saria, A., and Humpel, C., Differential changes in tachykinins after kainic acid-induced seizures in the rat, *Neuroscience,* 34, 219, 1990.
119. Pennypacker, K. R., Thai, L., Hong, J.-S., and McMillian, M. K., Prolonged expression of AP-1 transcription factors in the rat hippocampus after systemic kainate treatment, *J. Neurosci.,* 14, 3998, 1994.
120. Meier, C. L., Obenaus, A., and Dudek, F. E., Persistent hyperexcitability in isolated hippocampal CA1 of kainate-lesioned rats, *J. Neurophysiol.,* 68, 2120, 1992.
121. Mello, L. E. A. M. and Covolan, L., Spontaneous seizures preferentially injure interneurons in the pilocarpine model of chronic spontaneous seizures, *Epilepsy Res.,* 26, 123, 1996.
122. Morrisett, R. A., Jope, R. S., and Snead, O. C., Effects of drugs on the initiation and maintenance of status epilepticus induced by administration of pilocarpine to lithium-pretreated rats, *Exp. Neurol.,* 97, 103, 1987.
123. Turski, W. A., Cavalheiro, E. A., Coinmbra, C., da Penha Berzaghi, M., Ikonomidou-Turski, C., and Turski, L., Only certain anticonvulsant drugs prevent seizures induced by pilocarpine, *Brain Res.,* 434, 281, 1987.
124. Ormandy, G. C., Jope, R. S., and Snead, O. C., Anticonvulsant actions of MK-801 on the lithium-pilocarpine model of status epilepticus in rats, *Exp. Neurol.,* 106, 172, 1989.
125. Sofia, R. D., Gordon, R., Gels, M., and Diamantis, W., Effects of felbamate and other anticonvulsant drugs in two models of status epilepticus in the rat, *Res. Commun. Chem. Pathol. Pharmacol.,* 79, 335, 1993.
126. Leite, J. P. and Cavalheiro, E. A., Effects of conventional antiepileptic drugs in a model of spontaneous recurrent seizures in rats, *Epilepsy Res.,* 20, 93, 1995.

Chapter

Audiogenic Seizures in Mice and Rats

Charles E. Reigel

Contents

I. Introduction

Audiogenic or sound-induced seizures occur in some mice and rats in response to exposure to intense acoustic stimulation. Susceptible animals exhibit a wild running response which terminates in either a violent generalized clonic or tonic convulsion. Nonsusceptible animals demonstrate no convulsive behavior in response to the same auditory stimulation. Audiogenic seizure susceptibility can be genetically determined or can be induced in previously nonsusceptible animals through audiogenic priming, a process that involves prior exposure to the intense acoustic stimulus. The study of audiogenic seizures has provided considerable insight into the genetic contributions to epilepsy, the pathophysiology of the epileptic brain, and the responsiveness of innately epileptic animals to antiepileptic drugs.

II. Methodology

A. Elicitation of Audiogenic Seizures

The following is a description of the audiogenic seizure testing technique. Detailed descriptions of the testing apparatus and seizure scoring scales are described in later sections. Mice or rats to be tested for audiogenic seizure susceptibility are placed in the audiogenic seizure chamber (Figure 6.1) and allowed to habituate for 15 s. At this point the investigator initiates the acoustic stimulus and simultaneously starts a timer located on top of the chamber. The investigator carefully observes the subject for the appearance of audiogenic seizure responses. At the onset of the running phase, the investigator quickly glances up at the timer to see the latency and imme-diately returns to observation of the rat. It is critical to continue the audiogenic stimulus until the actual convulsive response begins. If the audiogenic stimulus is stopped during the running phase, this will result in termination of the running episode without elicitation of a clonic or tonic seizure. When the actual convulsion begins, the acoustic stimulus and timer are simultaneously stopped. The investigator does not look up at the timer, but rather continues to observe the convulsive pattern of the subject. The terminal convulsive response is rated according to one of the audiogenic seizure rating scales described below. At this time the investigator records the latency to running phase observed at a glance, records the latency to convulsion

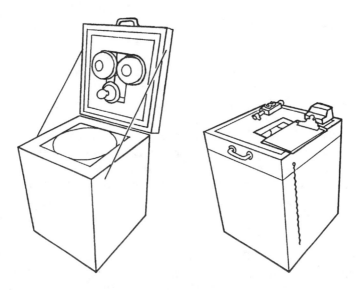

FIGURE 6.1
Sound attenuated chamber utilized to elicit audiogenic seizures in rats. The chamber is drawn with the hinged lid open and closed. The open chamber reveals the inner cylindrical metal chamber 40 cm in diameter by 50 cm in height in which the rat is placed. Two electric fire bells, a light, and an observation window are located on the inside of the lid. The observation window, timer, and switches for the light and firebells can be seen on the closed chamber lid.

on the stopped timer, and records the seizure severity score. If the subject fails to exhibit any audiogenic seizure response within 90 s, the audiogenic stimulus is discontinued and the animal is considered to be nonsusceptible.

The most reliable end point for the onset of convulsion is the brief tonic forelimb flexion described below that immediately precedes the generalized clonic or tonic convulsion. When training new technicians, it is better to allow them to continue the stimulus presentation beyond this end point to the clear onset of generalized clonus or tonus until they are comfortable in their ability to recognize the onset of the brief tonic forelimb flexion. Technicians will soon recognize that as forelimb flexion begins, the running phase slows as the animal plows forward, driven only by its hindlimbs. However, even an experienced investigator can be fooled when testing anticonvulsant drugs, as some can slow the running phase or can actually produce a pause in the running phase which may reflexively cause the investigator to terminate the stimulus. When training technicians, it is best not to allow technicians to train technicians. Rather, the investigator should train all new personnel to prevent erosion of the audiogenic screening method. This is particularly true in the case of the rat audiogenic seizure rating scale described below, which makes distinctions between the degree of tonic extension. Following this audiogenic seizure screening protocol results in extremely reproducible evaluation of audiogenic seizures between laboratory personnel and even different laboratories.

B. Convulsive Response

Patterns of audiogenic seizures are very similar in mice and rats.[1] The basic convulsive response in mice and rats consists of a wild running episode which terminates in either a generalized clonic convulsion or some degree of generalized tonic convulsion. Each of these behaviors is considered to be of brainstem origin.[2] One difference between the genetically susceptible mouse and rat strains involves the selective breeding for specific audiogenic seizure traits in rats. Within a mouse strain, the audiogenic seizure response can be a wild running episode, a generalized clonic convulsion, or a generalized tonic convulsion.[3] In rats, selective breeding for specific audiogenic seizure traits allows the investigator to utilize strains of susceptible rats that will reliably exhibit either a generalized clonic convulsion or a full tonic extensor convulsion as the terminal convulsive response.[4]

1. Convulsive Response in Mice

Following the onset of the audiogenic stimulus, mice typically exhibit a latent period of a few seconds.[1,3] The initial response in mice is a wild running phase consisting of an explosive burst of poorly coordinated locomotor behavior.[3,5] This wild running phase terminates in a generalized clonic convulsion with the animal lying on its side.[3,5,6] The clonic convulsion may progress to tonic flexion and then extension of head, trunk, forelimbs, and hindlimbs.[1,3,5] Full tonic extension or a full tonic extensor convulsion are phrases utilized in the audiogenic seizure literature to refer to tonic extension that progresses to hindlimb extension. The reader should be cautioned that different authors may utilize hindlimb extension or full tonic extension to refer to the same seizure response. Full tonic extension resembles that seen in maximal electroshock (see Chapter 1) and may last for 10 to 20 s.[1] Respiratory arrest, indicated by relaxation of the pinnae, can occur during the tonic extension phase.[1,3,6] If the mouse is not resuscitated, it usually will die.[3,6] Resuscitation can be easily accomplished by forcing air through a piece of soft tubing placed over the mouse's snout. Air is forced by squeezing a wash bottle attached to the other end of tubing. Following the tonic extension phase, mice often exhibit some terminal clonus followed by a cataleptic stupor.[1] Not all mice exhibit the complete pattern of audiogenic seizure.[1,3] Some mice may exhibit only a wild running episode. Other mice may exhibit multiple running episodes separated by periods of grooming prior to their terminal convulsion. Some mice may exhibit a wild running episode that terminates in only a generalized clonic convulsion.

2. Convulsive Response in Rats

Like mice, rats demonstrate a latent period following the initiation of the audiogenic stimulus. The initial response in rats is an explosive wild running episode which is terminated by brief (approximately 1 s) tonic flexion of the forelimbs and dorsiflexion of the back. Due to the tonic flexion of the forelimbs, the rat plows to a stop. This brief tonic forelimb flexion represents a branch point from which each of the two genetically susceptible strains exhibits its phenotypic audiogenic convulsive response.

a. Generalized Clonus

Following the brief tonic forelimb flexion, rats of one genetically susceptible strain will begin a violent generalized clonic convulsion that involves both the forelimbs and hindlimbs. The back remains arched during the generalized clonic convulsion. These animals may or may not roll over on their sides during the clonic convulsion. A characteristic clonic audiogenic seizure is depicted in Figure 6.2. Following the clonic convulsion, rats exhibit a rage reaction which may include vocalization. Rats should not be handled at this time as they are extremely likely to bite. This rage reaction lasts 2 to 3 min and is followed by a mild postictal depression.

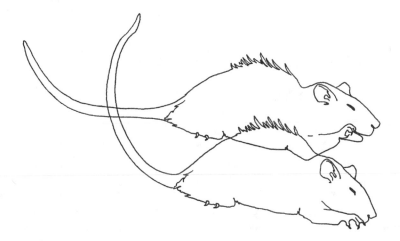

FIGURE 6.2
Generalized clonic convulsion in a rat elicited by intense auditory stimulation. The superimposed drawing illustrates the range of motion seen in a generalized clonic convulsion.

b. Tonic Extension

In the other genetically susceptible strain, the brief tonic flexion of the forelimbs and dorsiflexion of the back progresses to ventriflexion of the trunk and neck and tonic extension of the forelimbs. At this point the animal is usually on its side. The seizure typically progresses to full tonic extension of the hindlimbs. Thus, the full tonic extensor convulsion resembles that of maximal electroshock in rats and similarly proceeds in a rostral to caudal fashion. Like the mice, the tonic extension phase typically lasts between 10 and 20 s. A characteristic full tonic extensor convulsion is depicted in Figure 6.3. Following the tonic extension phase, the rat may then exhibit some mild clonic movements of the forelimbs and hindlimbs. Respiration, which arrests during the tonic phase, returns spontaneously with a gasp early in this mild clonic period. Unlike the mice, rats do not need to be resuscitated after the tonic convulsion. Finally, the rat exhibits a period of profound postictal depression, characterized by catatonia.

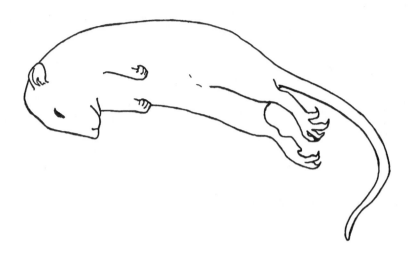

FIGURE 6.3
Full tonic extensor convulsion in a rat elicited by intense auditory stimulation.

c. Status Epilepticus

Although rats do not die as a result of respiratory arrest immediately following an audiogenic seizure, a small percentage do exhibit status epilepticus following full tonic extension. Of these, approximately 50% will die. Following full tonic extension, respiration spontaneously reappears and these animals immediately right themselves. There is no postictal depression. Rather, these animals vocalize and begin exhibiting a continuous running episode. This episode is not as well organized as the initial audiogenic wild running episode, as the running is interspersed with clonic jerks of the hindlimbs. The animals appear fatigued and wobble from side to side as they continue to ambulate. Some animals also exhibit vertical jumping due to hindlimb clonus. All of the animals continue to vocalize throughout the status episode. At intermittent intervals, tonic extensor convulsions occur, terminating the running episode. Following tonic extension, the running episode resumes. A given animal typically exhibits three to five tonic extensor convulsions during the period of status epilepticus, which can last from 20 to 30 min.

Status epilepticus, when it occurs, tends to run in litters. Unexplained deaths also occur in littermates of rats exhibiting status epilepticus after an audiogenic seizure. It is possible that these littermates experience spontaneous status epilepticus and die. Group housing of susceptible rats would facilitate status epilepticus. The noise of one rat in status running into the cage wall appears to induce audiogenic seizures in its siblings. By the time these litters have undergone the three audiogenic screenings required for selection of breeders and research subjects (described below), few remain. Attempts to breed these rats have failed due to unexplained deaths of the remaining breeders.

Spontaneous seizures have been reported by Dailey et al.[7] in audiogenic susceptible rats bred for full tonic extensor convulsions. Reported incidence was 1 to 2 episodes per month in a colony of 600 animals. However, the actual incidence may be higher as these reported incidences were limited to chance observations by vivarium staff and technicians. Spontaneous seizures in some of these rats progress to a pattern of status epilepticus that resembles the status epilepticus occurring after a sound-induced tonic extensor convulsion.

C. Evaluation of Seizure Severity

Seizure rating scales developed to quantitate audiogenic seizure severity in mice and rats all possess a number of similarities. All audiogenic seizure rating scales are ordinal, following the same progression: no response, wild running only, generalized clonus, and finally tonic extension. There are differences in these scales based upon the characteristics of the species and preferences of the investigator.

1. Audiogenic Seizure Rating Scales in Mice

Audiogenic seizures in mice are commonly rated in severity from 0 to 3.[6] A seizure severity score of 0 indicates that no audiogenic seizure occurred. A wild running episode receives a score of 1. A wild running episode that terminates in a clonic convulsion receives a score of 2. Finally, a wild running episode that terminates in a tonic convulsion receives a score of 3. It should be noted that a score of 3 does not distinguish between a seizure that consists of only tonic forelimb extension from a seizure that also includes hindlimb extension (full tonic extension). Other rating scales have added a score of 4 for death.[8] However, death has been reported following the less severe clonic seizures in mice.[9] Thus, it has been argued that death occurs independent of seizure severity and should not be part of a linear scale describing seizure severity. A final variant of the mouse rating scale ranges from 0 to 2.[3,10] In this classification scheme, wild running episodes that terminate in either a clonic or tonic seizure receive a score of 2. Such a seizure rating scale obscures actual incidences of clonic and tonic convulsions and may not be as desirable as the 0 to 3 scale.

2. Audiogenic Seizure Rating Scales in Rats

The most widely used audiogenic seizure rating scale in rats is the audiogenic response score (ARS) system of Jobe et al.,[11] depicted in Figure 6.4. This system reflects an expansion of the "no response to tonic convulsion continuum" utilized in mice to include distinct differences in seizure patterns that can be observed in rats. Most notably, distinctions are made that reflect the degree of tonic extension and/or multiple running episodes prior to convulsion. The ARS system extends from 0 (no response) to 9 (a single wild running phase that terminates in full tonic hindlimb extension). A wild running episode that terminates in a generalized clonic convulsion receives an ARS of 3. A wild running episode that terminates in tonic ventriflexion of the trunk and neck, tonic extension of the forelimbs, and tonic flexion of the hindlimbs receives an ARS of 5. An ARS of 7 reflects a seizure pattern that also includes partial tonic extension of the hindlimbs (all but the hind feet). The occurrence of two or more running episodes separated by pauses prior to the terminal

ARS Score	RESPONSE TO SOUND STIMULATION	Characteristic Convulsive Posture
0	No response	no convulsion
1	Running only	
2	two running phases; } -clonic convulsion	
3	one running phase;	
4	two running phases; } tonus of neck, trunk and	
5	one running phase; } forelimb; hindlimb clonus	
6	two running phases; } nearly complete tonic	
7	one running phase; } extension except hindfeet	
8	two running phases; } -complete tonic extension	
9	one running phase;	

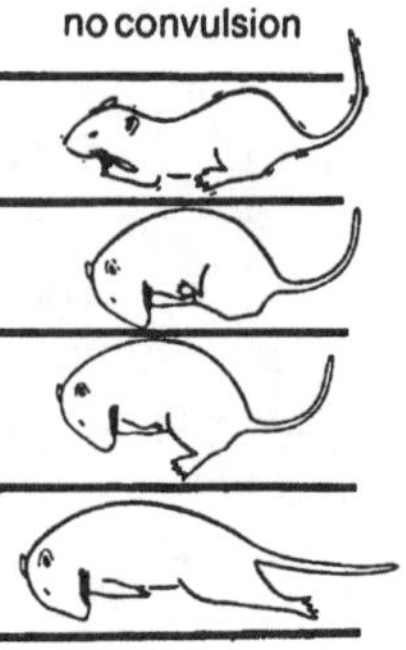

FIGURE 6.4

The audiogenic response score (ARS) system of Jobe et al.[11] utilized to rate convulsion severity in rats. (From Dailey, J.W. and Jobe, P.C., *Fed. Proc.*, 44, 2640, 1985. With permission.)

convulsive response results in the reduction of the ARS by 1 point. In this situation, the initial running episode does not generalize into the terminal convulsive response. This seizure cannot be rated as severe as a seizure in which the initial running episode does generalize. In this rating scale, a generalized clonic convulsion preceded by two running episodes receives an ARS of 2 rather than 3. Full tonic extension preceded by two running episodes receives an ARS of 8 rather than 9.

Although greatly expanded from the mouse scale, the ARS system of Jobe et al.[11] is clearly an ordinal scale reflecting increasing seizure severity. As described above, a clonic audiogenic seizure preceded by two running phases (ARS of 2) is clearly not as severe as an audiogenic seizure preceded by a single running phase (ARS of 3). Likewise, tonic ventriflexion of the neck and trunk and extension of the forelimbs following two running episodes (ARS of 4) is more severe than a generalized clonic convulsion preceded by a single running episode (ARS of 3). Tonic seizures are invariably considered to be more severe seizures than clonic seizures. The linear nature of the ARS system is further supported by the observation that anticonvulsant drugs produce dose-dependent reductions in seizure severity that correspond to linear decreases in ARS.[12] Finally, the pattern of increasing severity in the ARS system resembles the current dependent increases in convulsion severity described by Browning[13] for electroshock seizures (also described in Chapter 1).

D. Genetically Susceptible Strains

1. Mice: DBA/2 and Other Mouse Strains

The DBA/2 is currently the most widely utilized mouse strain genetically susceptible to audiogenic seizures. This is largely due to its commercial availability and applicability to genetic studies. Studies of recombinant inbred strains derived from C57BL/6 (seizure resistant) and DBA/2 (seizure sensitive) progenitor strains have revealed insights into the mode(s) of audiogenic seizure inheritance in the DBA/2 mouse.[3,9,10,14] DBA/2 mice can be obtained from The Jackson Laboratory, Bar Harbor, ME, Charles Rivers Laboratories, Wilmington, MA, or Harlan Sprague Dawley, Inc., Indianapolis, IN.

Audiogenic seizure susceptibility in the DBA/2 occurs during a limited period of development.[5] A low incidence of the initial audiogenic seizure response, wild running, occurs at 13 d of age.[15] Audiogenic seizure incidence and severity rapidly increase with age, reaching 100% incidence by 17 d of age, with a 90% incidence of tonic seizures. It should be noted that tonic seizures in this report included both tonic hindlimb flexion and hindlimb extension and thus do not represent a 90% incidence of a maximal audiogenic seizure response (hindlimb extension).[15] In another report, peak incidence of audiogenic seizures occurs at 21 d of age in which over 91% of the mice exhibited either clonic or tonic seizures.[16] Again the data presentation does not reveal the actual incidence of hindlimb extension. Still another report indicates an 87% incidence of the "most severe form of the seizure, i.e., the full clonic-tonic seizure," occurs at 21 d of age.[17] Presumably this description is referring to hindlimb extension as the most severe form of audiogenic seizure. These three reports are illustrative of the confusion that exists in the audiogenic seizure literature due to inconsistent or incomplete descriptions of seizure responses. However, there is general consensus that audiogenic seizures in DBA/2 mice peak in incidence and severity between 16 and 21 d of age.[5]

Beyond 21 d of age, incidence and severity of audiogenic seizure begin to dissipate.[18] At 28 d of age, incidence of tonic seizures and clonic seizures are reduced to 13% and 40%, respectively, although wild running incidence remains high.[17] By 42 d of age, no DBA/2 mice exhibit tonic seizures and incidence of clonic seizures and wild running is down to 7% and 27%, respectively.[17] Complete loss of audiogenic seizure susceptibility occurs by 80 d of age in DBA/2 mice.[18] In other DBA/2 mice, complete loss of audiogenic susceptibility occurs as early as 42 d of age.[8] The reader should be cautioned that the actual audiogenic seizure responsiveness of DBA/2 mice may vary over time or in mice from different vendors or laboratories.

Numerous mouse strains have been reported in the literature that are genetically susceptible to audiogenic seizures. All share the pattern of seizure described above, but may vary in their developmental profile of audiogenic seizure susceptibility. The strains described below are limited to O'Grady, Frings, and Rb mice. These strains are now most likely extinct unless maintained in individual laboratories. Their inclusion here is important due to their prevalence in the audiogenic literature, particularly in comparison to DBA/2 mice.

O'Grady mice were derived from Swiss albino mice by 30 to 40 generations of selective breeding for audiogenic seizure susceptibility.[19,20] When an audiogenic stimulus in the standard range is utilized, 100% of O'Grady mice exhibit full tonic extensor convulsions.[20] Nonsusceptible Swiss mice typically serve as controls.[19] Developmentally, the initial audiogenic seizure response is wild running, appearing in low incidence at 10 d of age.[21] By 15 d of age, 100% of the O'Grady mice are audiogenic seizure susceptible; however, the dominant response is wild running and clonus with only 18% exhibiting hindlimb flexion. Hindlimb extension appears first at 17 d of age, reaching 100% by 22 d of age. Although incidence of audiogenic seizure susceptibility decreases with age in the O'Grady strain, it does not do so as rapidly as in the DBA/2 mice.[21,22] Over 52% of the O'Grady mice are no longer susceptible to audiogenic seizures by 60 d of age.[22] However, approximately 29% do still exhibit hindlimb extension at this young adult age.

Frings mice represent another strain of albino mice that exhibit a high degree of audiogenic seizure susceptibility.[22] At 8 d of age, a small percentage of Frings mice exhibit wild running episodes or clonic seizures in response to sound. Wild running (only) reaches a peak incidence of approximately 15% by 13 d of age. Incidence of clonic seizures peaks at over 64% at 14 d of age. Peak audiogenic seizure incidence occurs at 20 d of age, when 100% of Frings mice exhibit hindlimb extension. Seizure incidence and severity do decrease with increasing age, but at a much slower rate than the DBA/2 or O'Grady mice. By 60 d of age seizure incidence remains at 100%, but incidence of hindlimb extension drops to approximately 80%.[22] At 180 d of age, 30% of Frings mice are no longer susceptible to audiogenic seizures and the incidence of hindlimb extension drops to 45%. By 1 year, 75% of these mice are no longer susceptible to audiogenic seizures and only 8% exhibit hindlimb extension.

Rb mice were derived from Swiss albino mice by selective breeding for audiogenic seizure susceptibility.[23] Rb mice begin to demonstrate initial audiogenic seizure susceptibility at 13 d of age.[15] Seizures at these ages are primarily wild running episodes, but rapidly mature in incidence and severity. By 18 d of age, all Rb mice are audiogenic seizure susceptible, with over 90% exhibiting full tonic extension. Lethality following tonus occurs only in Rb mice under 18 d of age. Unlike the DBA/2 mice, audiogenic seizure susceptibility persists into adulthood in Rb mice.

The differential developmental patterns of audiogenic seizure susceptibility in the various mouse strains has important implications toward developmental aspects of human epilepsy. In addition, it is clear that appropriate age-matched controls must be utilized in any pathophysiological studies of audiogenic seizure susceptibility in mice.

2.   Rats: GEPR and Other Rat Strains

The GEPR (genetically epilepsy-prone rat) is presently the most widely utilized rat strain genetically susceptible to audiogenic seizures. The GEPR was derived from Sprague-Dawley rats by selective breeding for audiogenic seizure susceptibility.[4] The GEPR currently exists as two separate strains, each bred for a specific pattern of audiogenic seizure. The GEPR-3 strain was bred to exhibit a generalized clonic seizure following a single wild running episode in response to sound (Figure 6.2).[4] Such a seizure would be rated at an ARS of 3 according to the audiogenic seizure rating scale of Jobe et al.[11] (Figure 6.4). The GEPR-9 strain was bred to exhibit a full tonic extensor convulsion following a single running phase in response to sound (Figure 6.3).[4] Such a seizure would be rated at an ARS of 9 (Figure 6.4). More detailed descriptions of the audiogenic seizures in GEPR-3s and GEPR-9s can be found above, in Section II.B.2.

Selective breeding for the current GEPR-3 colony was initiated by Phillip C. Jobe at Northeast Louisiana University in 1971.[4] The GEPR-3 colony maintained by Dr. Jobe at the University of Illinois College of Medicine in Peoria has reached 51 generations of brother-sister pairing for generalized clonic convulsions (personal communication, P.C. Jobe). As such, approximately 99.6% of the members of this colony exhibit an audiogenic seizure upon first exposure to the audiogenic stimulus.[2]

The current GEPR-9 colony maintained by Dr. Jobe at the University of Illinois College of Medicine in Peoria has reached 40 generations of brother-sister pairing for full tonic extensor convulsions (personal communication, P.C. Jobe). Initial exposure to the audiogenic stimulus results in a seizure incidence of 97.6% in members of the GEPR-9 colony.[2] The history of the current GEPR-9 colony is much more complicated than that of the GEPR-3 colony. The current GEPR-9s are descendants of an audiogenic seizure-susceptible strain developed by Albert Picchioni and Lincoln Chin at the University of Arizona in 1957.[4] This early strain was derived from Sprague-Dawley rats by breeding for audiogenic seizure susceptibility. Members of the early Arizona colony exhibited audiogenic seizures that ranged from wild running episodes to full tonic extension. In 1976, the Arizona colony was split into two colonies by selective breeding for specific seizure traits. The "minimal" colony was bred to exhibit clonic seizures following two individual wild running episodes. These seizures would receive an ARS of 2 according to the method of Jobe et al.[11] The "maximal" colony was bred to exhibit full tonic extension following a single wild running episode (ARS of 9).[4] In 1980, Dr. Jobe established the current GEPR-9 colony from "maximal" stock received from the University of Arizona. The 40 generations of brother-sister pairing for full tonic extension attributed to the current GEPR-9 colony above began when Dr. Jobe began breeding the Arizona animals. For a more detailed description of the origin of the GEPR and the evolution of its nomenclature, see Reigel et al.[4] or Dailey et al.[7]

Control subjects utilized in GEPR research are varied, depending upon the nature of the experiment. The most common control rats are derived from Sprague-Dawley rats by selective breeding for resistance to audiogenic seizure susceptibility.[4] In studies of seizure severity, GEPR-3s, lacking the tonic trait, become control subjects for GEPR-9s. Infrequently, progeny of GEPR-9 breeders are produced that are completely nonsusceptible to audiogenic seizures. These nonsusceptible progeny become valuable controls for the study of audiogenic seizure susceptibility in GEPR-9s. In fact, two types of nonsusceptible progeny of GEPR-9s have been identified. The first lacks the peripheral hearing impairment characteristic of the GEPR which is believed to be responsible for development of the audiogenic seizure focus.[24,25] The second possesses the hearing deficit but lacks the deficit in norepinephrine content in the midbrain characteristic of the GEPR-9.[26] Finally, because audiogenic seizure responsiveness is so consistent in GEPRs that have been screened for susceptibility, GEPRs are utilized as their own controls in the evaluation of anticonvulsant drugs.[12]

Unlike the DBA/2 mouse, once audiogenic seizure susceptibility develops in the GEPR, it persists through adulthood.[27] Initial susceptibility to audiogenic seizures begins at 15 and 16 d of age in GEPR-3s and GEPR-9s, respectively.[28] Seizures are minimal at these ages, consisting of wild running or some clonus. Seizure incidence increases rapidly, with 100% of the members of each colony exhibiting some form of audiogenic seizure by 21 d of age.

Seizure severity matures in a linear fashion in the GEPR-9.[28] Tonic seizures first occur in low incidence at 18 d of age. Incidence of tonic seizures increases with age, with incidence of tonus first exceeding incidence of clonus in 25-d-old GEPR-9s. By 45 d of age, 100% of the GEPR-9s exhibit tonic seizures.

Maturation to adult patterns of audiogenic seizures is more complicated in the GEPR-3.[28] Seizure severity rapidly increases with age in GEPR-3s. By 21 d of age, incidence of generalized clonus reaches 100%, resembling the adult pattern. However, tonic seizures characteristic of the adult GEPR-9 appear in GEPR-3s between 19 and 30 d of age. Peak incidence of tonic seizures in GEPR-3s is 70%, occurring at 23 d of age. By 45 d of age, 100% of the GEPR-3s exhibit their characteristic clonic convulsion. Thus, the development of seizure severity is biphasic in GEPR-3s.

The immature GEPR-3s also demonstrate a second pattern of seizure not seen in the adults. Between the ages of 15 and 30 d, some immature GEPR-3s that exhibit clonic seizures also exhibit a secondary rearing seizure 10 to 15 s after their audiogenic seizure.[28] These animals exhibit a series of two to five rearing seizures accompanied by clonus of the forelimbs. These seizures resemble limbic seizures described in Chapters 1 and 3. Peak incidence is 100% in GEPR-3s between 16 and 21 d of age. The secondary seizures are never observed following a tonic seizure in developing GEPR-3s.

Some rat strains susceptible to audiogenic seizures reported in the earlier literature are now extinct. However, a new strain of audiogenic seizure-susceptible rats has been developed recently from Wistar rats.[28,29] Audiogenic seizures in this new strain may be limited to one or two wild running episodes or may progress to what the authors refer to as a tonic seizure.[29] This tonic phase is described as consisting of dorsal hyperextension. These rats also exhibit an open mouth and slight tremor of the mouth and limbs during this phase of the audiogenic seizure. This description seems to better resemble the generalized clonic seizure characteristic of the GEPR-3 than the full tonic extensor convulsion characteristic of the GEPR-9.

E. Audiogenic Priming

Acoustic priming, audiogenic sensitization, or audiogenic priming refer to the production of audiogenic seizure susceptibility in mice or rats not previously susceptible to audiogenic seizures through exposure to intense acoustic stimulation at critical developmental periods. In a typical acoustic priming procedure, mice or rats are exposed to intense acoustic stimulation for some period of time. No audiogenic seizure is elicited by the initial acoustic exposure. However, a subsequent acoustic exposure days later results in an audiogenic seizure. Thus, the animal has been "primed" for susceptibility to audiogenic seizures. Audiogenic priming can occur without the initial acoustic priming stimulus. Exposure to ototoxic agents such as kanamycin at critical developmental periods can induce subsequent audiogenic seizure susceptibility in previously nonsusceptible mice and rats. For the purpose of this chapter, "audiogenic priming" will be utilized to refer to induced forms of audiogenic seizure susceptibility, regardless of methodology. Audiogenic priming is a valuable seizure model as it allows mechanistic examination of the induction of audiogenic seizure susceptibility. Mechanistically, many parallels exist between mice and rats genetically susceptible to audiogenic seizures and those in which susceptibility is acquired through audiogenic priming.

1. Mice

Audiogenic priming was first reported in C57BL/6J mice following a 30 s exposure to an electric bell (103 dB).[31] The optimal age of priming was 16 d of age for mice tested for audiogenic susceptibility at 21 d of age. Alternately, optimal age of priming was 19 d of age for mice tested for audiogenic susceptibility at 28 d of age. In a subsequent report in CF#1 mice, intense acoustic exposure at 20 d of age resulted in a maximal susceptibility to clonic-tonic seizures 3 d later.[32] Finally, a 20 s exposure to an electric bell (102 to 104 dB) at 21 d of age induced audiogenic seizure susceptibility 72 h later in SJL/J mice.[33] Once primed, SJL/J mice remained susceptible to audiogenic seizures for at least 21 weeks. The optimal age for audiogenic priming was from 3 to 4 weeks of age in this strain. From these three early studies, it is clear that audiogenic priming is an extremely complex phenomenon. In addition to age, subsequent studies have revealed genetic differences in the optimal interval between the priming stimulus and maximal audiogenic seizure susceptibility.[6] Genetic differences also exist in the duration of priming-induced audiogenic seizure susceptibility.[6] The intensity and duration of the acoustic stimulus have also been demonstrated to be critical determinants of acoustic priming in mice.[34]

Treatment with the ototoxic aminoglycoside, kanamycin, produced audiogenic priming in BALB/C mice that was dependent on the age at which kanamycin was administered and the age at which audiogenic seizure testing occurred.[35] Mice that received kanamycin (400 mg/kg, intraperitoneally) from day 5 through day 21 exhibited a high degree of audiogenic seizure susceptibility at 28 d of age. Mice receiving the same kanamycin treatment from days 17 through 27 did not exhibit any audiogenic seizures at 28 d of age.

2. Rats

Audiogenic priming in rats is a more recent observation than in mice. The first report of audiogenic priming utilized the ototoxic aminoglycoside kanamycin rather than noise and was performed in Wistar rats.[36] In that study, the optimal kanamycin dosage regimen was 100 mg/kg, given intraperitoneally on postnatal days 9 through 12. Audiogenic seizure incidence was 100% at postnatal day 28 or 32 in rats treated with this regimen. Seizure severity was also greatest for this regimen. Higher doses of kanamycin given at this optimal period and other periods resulted in a lower incidence and severity of audiogenic seizures.

Somewhat different results were obtained in a kanamycin audiogenic priming study utilizing Sprague-Dawley rats.[37] In that study, 100 mg/kg, given intraperitoneally on postnatal days 9 through 12, was the most effective dosage regimen in producing audiogenic seizure susceptibility at 30 d of age, but only when the pups were pretreated with RO4-1284 45 min prior to testing for audiogenic seizure susceptibility. RO4-1284 is a drug that acutely depletes monoamines. Kanamycin treatment alone or the RO4-1284 treatment alone failed to induce audiogenic seizure susceptibility. Lower and higher doses of kanamycin were similarly ineffective. Interestingly, the optimal kanamycin regimen produced only a transient hearing loss as measured by auditory brainstem potential thresholds and yet this treatment was capable of inducing an audiogenic seizure focus functional at postnatal day 30. Further, this audiogenic seizure focus appeared to be permanent as every subject susceptible to audiogenic seizures at postnatal day 30 following acute monoamine depletion remained susceptible when tested at 90 d of age, again following acute monoamine depletion.

Differences in the two studies can best be attributed to strain differences. The Sprague-Dawley rats appear to lack a mechanism that allows generalization of audiogenic seizures. The kanamycin treatment induces a focal audiogenic seizure mechanism, but monoamine depletion is required to provide the generalization mechanism necessary for audiogenic seizure susceptibility.

Audiogenic priming is also possible with noise exposure in Wistar rats. Pierson and Swann[38] demonstrated that an 8 min noise exposure at 125 dB on postnatal day 14 produced a 100% incidence of tonic audiogenic seizures at postnatal day 28 in Wistar rats. Initial noise exposure at other ages or of less duration resulted in a reduction of seizure incidence and severity. Postnatal day 28 was determined to be the optimal age of susceptibility as well.

F. Audiogenic Stimulus

The typical audiogenic stimulus is produced by electric bells that produce a broad frequency range (10 to 20 kHz) and an intensity ranging from 90 to 120 dB as measured by a standard sound level meter.[6] Such a stimulus should be considered a maximal audiogenic stimulus, much as are the currents utilized in maximal electroshock. A mouse or rat either exhibits an audiogenic seizure in response to this stimulus or does not. Genetically susceptible strains such as the GEPR were bred specifically for an audiogenic seizure response to this type of maximal stimulus.[4] Other investigators utilize white noise generators producing an audiogenic stimulus in this general intensity and frequency range.[39] Still other investigators utilize pure tones in this intensity range as the audiogenic stimulus.[9,10]

Some investigators have examined audiogenic stimulus parameters in individual species or strains. For example, Faingold et al.[40] reported GEPR-9s were maximally susceptible to audiogenic seizures at a stimulus frequency of 12 kHz. DBA/2 mice were reported to be maximally susceptible to audiogenic seizures at a stimulus frequency of 20 kHz.[41] In O'Grady mice, the optimal audiogenic frequency at minimum acoustic intensity was found to be 13 kHz.[19,20] However, when the stimulus intensity was increased, the frequency range capable of inducing 100% incidence of audiogenic seizures in O'Grady mice broadened.[19] Working at a fixed frequency of 22 kHz, O'Grady mice demonstrated a linear increase (0% to 100%) in audiogenic seizure susceptibility between 66.0 and 84.5 dB.[20] Collectively, these findings are consistent with the concept that a general audiogenic stimulus ranging between 90 and 120 dB, consisting of a wide frequency range (10 to 20 kHz), should be considered a maximal audiogenic stimulus.

A typical audiogenic seizure apparatus utilized with mice consists of a battery jar, 30 cm in diameter by 46 cm in height, with an electric bell mounted in the lid.[41] The battery jar is enclosed in a sound attenuating box with an observation window. A typical audiogenic seizure apparatus utilized with rats consists of a cylindrical metal chamber, 40 cm in diameter by 50 cm in height (Figure 6.1).[4] The cylinder is enclosed in a sound attenuating wooden box with a hinged lid. Two electric fire bells, a light, and an observation window are located in the lid.

G. Seizure Repetition

1. Postictal Refractoriness

One obvious result of repeated audiogenic seizures would be the production of a refractory period in which an animal would not exhibit a seizure when reexposed to the acoustic stimulus. Refractory periods are generally considered to be long in mice susceptible to audiogenic seizures.[42] For example, DBA/2 mice retested for audiogenic seizure susceptibility 60 min after an initial tonic-clonic audiogenic seizure exhibited only a 74% incidence of tonic seizures.[39] In contrast, postictal refractory periods are extremely short in the GEPR.[43] Three minutes following the initial audiogenic seizure, 100% of GEPR-3s exhibit their generalized clonic seizure (ARS of 3) when sound stimulated. GEPR-9s require 12 min for recovery following an audiogenic seizure before they exhibit a 100% incidence of their characteristic tonic extensor convulsion (ARS of 9). Recovery of latencies to the running episode and convulsion are more prolonged in the GEPR. Complete restoration of the latency to the running episode requires 24 min in both GEPR-3s and GEPR-9s. Complete restoration of the latency to convulsion requires 64 min in GEPR-3s and 24 min in GEPR-9s.

2. Kindling of Forebrain Seizures in Audiogenic Susceptible Rats

Audiogenic seizures in rats are considered to be of brainstem origin.[2] Initiation of audiogenic seizures in the GEPR appears to occur in the inferior colliculus within the brainstem.[44] Although epileptiform discharges can be recorded from the inferior colliculus of the GEPR,[44] cortical epileptiform discharges cannot be recorded from GEPRs during their initial audiogenic seizure.[45] However, repetition of audiogenic seizures on a daily basis resulted in the appearance of cortical epileptiform activities and additional seizure responses in both GEPR-3s and GEPR-9s.[45] In GEPR-3s, a pattern of facial and forelimb clonus superimposed upon rearing and falling appeared after a clonic audiogenic seizure. These forebrain seizures were accompanied by spike and wave discharges on the cortical EEG. These behavioral and electrographic observations in GEPR-3s are paralleled by reports in Wistar-derived audiogenic susceptible rats experiencing daily audiogenic seizures.[29,30] GEPR-9s developed post-tonic clonic seizures involving both the forelimbs and hindlimbs following repeated audiogenic seizures.[45] Post-tonic clonic seizures in GEPR-9s did not involve rearing or facial clonus. Repeated brainstem convulsions appear to be capable of inducing forebrain seizures in the GEPR.

3.   Electrical Kindling in GEPRs

Electrical kindling of limbic seizures (see Chapter 3) is accelerated in the GEPR. Savage et al.[46] reported that angular bundle kindling is accelerated in audiogenic seizure-naïve GEPR-3s and GEPR-9s as compared to nonepileptic Sprague-Dawley control rats. Acceleration of limbic kindling was by far greatest in GEPR-9s which required fewer trials to reach each stage of kindling than either GEPR-3s or Sprague-Dawley controls. GEPR-3s required fewer trials than Sprague-Dawley controls only in reaching class 5 kindled seizures. In a study comparing amygdala kindling rates in audiogenic seizure-naïve and seizure-experienced GEPR-9s with nonepileptic Sprague-Dawley control rats, Coffey et al.[47] found that both seizure-experienced and seizure-naïve GEPR-9s required fewer trials to reach stage 5 seizures than the control rats. Audiogenic seizure-experienced GEPR-9s required fewer trials than the naïve GEPR-9s to reach stage 5 seizures. It is interesting to note that brainstem seizure experience (audiogenic seizures) is capable of facilitating limbic kindling. In addition, the acceleration of kindling reported by both Coffey et al.[47] and Savage et al.[46] in seizure-naïve GEPR-9s demonstrates a role for genetic vulnerability in the kindling of limbic seizures. Coffey et al.[47] also reported that a high number of GEPR-9s exhibited severe brainstem seizures following a kindling stimulation, with the majority exhibiting a running episode that terminated in a full tonic extensor convulsion (ARS of 9). These authors suggested that amygdala kindling in GEPRs resulting in generalized brainstem seizures could serve as a model of partial seizures secondarily generalized in humans.

H.   Selective Breeding in the GEPR

As DBA/2 mice are readily available through established vendors, a description of selective breeding is not necessary for the purpose of this chapter. Although the GEPR has recently become commercially available through Harlan Sprague-Dawley, Inc, Indianapolis, IN, it is currently available only in limited supply. A number of laboratories breed GEPRs in-house from breeding stock obtained from the core GEPR colonies maintained by Phillip C. Jobe at the University of Illinois College of Medicine at Peoria. In-house production ensures adequate supply of subjects as well as enables investigators to conduct experiments with seizure-naïve GEPRs or to conduct developmental studies. However, the high degree of consistent audiogenic seizure responsiveness of the two GEPR colonies requires strict adherence to GEPR breeding protocols.

All GEPR breeders maintain a unique breeder identification code which consists of a combination of numbers and letters which indicate generation of inbreeding and inclusion in a specific litter. Experienced technicians know at a glance the code of a given breeder's mate(s). Pups born to a specific breeder are identified by her breeder code and whether they were members of her first, second, third, or fourth litter delivered. Pups are weaned at 21 d of age, sexed and either ear tagged or ear punched to establish a specific identity. Pups of one sex are housed together under their mother's breeder code and their litter number. In this manner, an individual GEPR's identity can be maintained for the life of the individual.

Animal handlers and technicians should be advised that GEPR-9s at the age of weaning are very sensitive to audiogenic seizures. When they are placed in a new cage, the noise of dropping rat chow into a wire cage top can cause the entire litter to experience audiogenic seizures, often leading to status epilepticus and death. This sensitivity dissipates by 30 d of age.

Initial audiogenic seizure screening begins when GEPRs reach 45 d of age. By this age members of both colonies can be expected to exhibit their characteristic adult seizures.[28] Testing for audiogenic seizure susceptibility is performed as described above in Section II.A. GEPRs are screened for audiogenic seizure susceptibility once a week at 7-d intervals for a total of three screens.[4] Latency to running, latency to convulsion, and seizure severity are recorded for each individually identified member of a specific litter. This 3-week screening process produces GEPRs of known identity and known audiogenic seizure responsiveness that can be utilized in research. It also provides information regarding seizure responsiveness of entire litters required for the selection of new breeders.

Every generation of breeding involves brother-sister pairing for specific seizure traits. As such, the evolution of the GEPR colonies is ongoing. To be selected as a GEPR-3 breeder, every member of the litter must exhibit an ARS of 3 (a single running episode terminating in a generalized clonic convulsion) on each of the three weekly sound stimulations.[4] To be selected as a GEPR-9 breeder, every member of the litter must exhibit an ARS of 9 (a single running episode terminating in full tonic extension) on each of the three weekly sound stimulations. Litters with uncharacteristically long latencies to running or convulsion are not utilized. In the GEPR-9 colony, selective breeding also includes breeding for increasingly short latencies to running and convulsion. Some members of the current GEPR-9 colonies immediately begin their tonic extension simultaneously with the onset of the bell (personal communication, P.C. Jobe). Another criterion for selection is litter size. GEPR breeders are selected from litters of eight or more pups. This is believed necessary to maintain the reproductive vigor of such highly inbred lines. GEPRs are typically paired two females to each male and two such breeding triplets are selected from each eligible litter. This ensures that the loss of one male will not result in the loss of the female breeders from his litter. Occasionally a GEPR-9 that is completely nonsusceptible to audiogenic seizures (ARS of 0) is identified in the screening procedure. Under no circumstances should any littermates of this animal be utilized as GEPR-9 breeders. Adherence to this breeding protocol will result in the production of GEPR-3s and GEPR-9s with the high degree of seizure responsiveness described above.

A control colony of Sprague-Dawley descent has been bred for resistance to audiogenic seizures.[4] Members of this colony are raised and identified just as the GEPRs described above. They also undergo the three weekly screenings for audiogenic seizure susceptibility. To be eligible as control breeders, every member of the litter must be completely nonsusceptible to audiogenic seizures (ARS of 0) on all three screens.

III. Interpretation

A. Sites of Seizure Origin

Audiogenic seizures are generally considered to be of brainstem origin.[2] The following discussion is based upon observations in rats, but many parallel observations exist in mice. Multiple sites within the brainstem have been implicated in the generation of audiogenic seizures in the GEPR. One site has been implicated in the initiation of audiogenic seizures and is considered to be the focal site of audiogenic seizure origin. Other sites have been implicated in the elaboration or propagation of the complete pattern of audiogenic convulsion. These propagation sites promote generalization of the audiogenic seizure initiated at the focal site.

1. Audiogenic Seizure Initiation/Focal Seizure Activity

An increasing body of evidence supports the inferior colliculus as the audiogenic seizure focus, that site responsible for seizure initiation in the GEPR. Bilateral lesions of the inferior colliculus completely suppressed audiogenic seizures in rats.[48] Lesions of other auditory nuclei produced lesser attenuation of audiogenic seizures in rats.[49] Following initiation of the audiogenic stimulus, Ludvig and Moshe[44] reported initial electrographic seizure activity occurred in the inferior colliculus simultaneously with the wild running phase of the audiogenic seizure. The wild running phase constitutes the behavioral correlate of focal audiogenic seizure activity in the brainstem.

Seizure initiation appears to at least partially involve excitatory amino acid neurotransmission in the inferior colliculus. Faingold et al.[50] demonstrated that bilateral microinjection of the excitatory amino acid antagonist 2-amino-7-phospho-heptanoate (APH) into the inferior colliculus completely abolished audiogenic seizures in GEPR-9s. Further, the inferior colliculus was much more sensitive to this effect than other auditory nuclei. This parallels a report in which focal microinjection of the excitatory amino acid NMDA into the inferior colliculus of nonsusceptible Sprague-Dawley rats induced audiogenic susceptibility.[51] Finally, increased aspartate levels have been reported in the inferior colliculus of GEPRs following an audiogenic seizure.[52]

Deficits in GABAergic inhibition in the inferior colliculus also appear to play a critical role in seizure initiation in the GEPR. Faingold et al.[53] reported GABA-mediated inhibition in the inferior colliculus of the GEPR was less than in nonepileptic rats. Bilateral microinjection of the GABA-A agonist muscimol into the inferior colliculus abolished audiogenic seizures in GEPR-9s.[54] Bilateral microinjection of the GABA-B agonist baclofen or the GABA transaminase inhibitor gabaculine into the inferior colliculus also completely suppressed audiogenic seizures in GEPR-9s.[55] In nonsusceptible rats, bilateral microinjection of the GABA antagonist bicuculline induced susceptibility to audiogenic seizures.[51]

Clearly excessive excitatory amino acid and deficient GABAergic neurotransmission are implicated in focal seizure generation in the inferior colliculus of the GEPR in response to intense acoustic stimulation. Yang et al.[56] found six separate electrophysiological differences in inferior collicular neurons between GEPR-9s and

Sprague-Dawley controls that could individually or collectively promote seizures in the GEPR. Finally, two reports offer some insight into the origin of the heightened excitability of the GEPR inferior colliculus. Reigel et al.[25] reported a peripheral hearing impairment in immature GEPRs that preceded the developmental onset of audiogenic seizure susceptibility and proposed that the development of the audiogenic seizure focus in the inferior colliculus was a compensatory response for reduced peripheral auditory input. Consistent with this hypothesis, transient peripheral hearing impairment induced in immature Sprague-Dawley rats by kanamycin resulted in the production of an audiogenic seizure focus that was unmasked following monoamine depletion.[37]

2. Audiogenic Seizure Propagation

Browning et al.[57] demonstrated that bilateral lesions of midbrain or pontine reticular formation abolished all of the tonic components of audiogenic seizures in GEPR-9s. In the same study, bilateral lesions of the midbrain reticular formation abolished the clonic component of audiogenic seizures in GEPR-3s, leaving the running component intact. Bilateral lesions of the substantia nigra abolished most of the tonic components of audiogenic seizures in GEPR-9s.[49] Millan et al.[58] found that bilateral microinjections of the excitatory amino acid antagonist APH into the substantia nigra, pontine reticular formation, or the midbrain reticular formation completely suppressed audiogenic seizures in GEPR-9s, including running. It appears that seizure activity initiated in the inferior colliculus requires activation of these brainstem structures for full expression (or propagation) of the audiogenic seizure. The demonstration of Millan et al.[58] that APH microinjection into the substantia nigra, pontine reticular formation, or midbrain reticular formation completely suppressed wild running suggests that the existence of an audiogenic focus in the inferior colliculus alone is not sufficient for even a minimal audiogenic seizure (running). Even the focal seizure event requires some degree of propagation for its expression. This is consistent with two reports in which Reigel and colleagues[37,59] demonstrated that innately hearing impaired Sprague-Dawley rats and kanamycin audiogenic primed Sprague-Dawley rats failed to exhibit audiogenic seizure susceptibility. These Sprague-Dawley rats did become audiogenic seizure susceptible after monoamine depletion induced by RO4-1284. It would appear that these nonsusceptible rats possessed the focal audiogenic mechanism, but lacked the brainstem seizure propagation mechanism inherent in the GEPR and necessary for the expression of audiogenic seizures. Monoamine depletion activated the brainstem propagation mechanism in these previously nonsusceptible animals.

By far, the two neurotransmitters that have been most implicated in the regulation of seizure susceptibility and severity in the GEPR are norepinephrine and serotonin. Widespread deficits in noradrenergic[2,60] and serotonergic[2,61,62] function have been proposed to be responsible for audiogenic seizure susceptibility and severity in the GEPR. These neurochemical abnormalities may at least partially serve as the basis for the brainstem seizure propagation mechanism of the GEPR. Interestingly, Reigel and Lin[63] reported that thresholds for flurothyl-induced tonic seizures were lower in developing GEPR-9 pups than nonsusceptible Sprague-Dawley pups at ages prior to the onset of audiogenic seizure susceptibility. Further, regional deficits in norepinephrine and serotonin content were also detected in GEPR-9s at these ages.[64] Thus, the proposed brainstem seizure propagation mechanism functionally and neurochemically preceded the development of the audiogenic seizure focus and audiogenic seizure susceptibility in the GEPR. Clearly, focal and propagation mechanisms are separate entities in the GEPR, but both are necessary for the expression of audiogenic seizures.

B. Evaluation of Anticonvulsant Drug Activity

In general, all clinically effective antiepileptic drugs are effective against audiogenic seizures in mice and rats. Phenytoin, phenobarbital, valproate, ethosuximide, trimethadione, diazepam, clonazepam, and lorazepam have all been reported to protect against audiogenic seizures in DBA/2 mice.[65] Broad spectrum anticonvulsant effects have also been reported in Frings[21] and Rb mice.[66] Carbamazepine, phenytoin, valproate, ethosuximide, phenobarbital, and clonazepam have been demonstrated to produce anticonvulsant effects in both GEPR-3s and GEPR-9s.[4,12] Such broad spectrum anticonvulsant responsiveness led Chapman and coworkers[65] to conclude that protection against audiogenic seizures (in mice) is a sensitive screen for anticonvulsant effects, but lacks the ability to determine clinical efficacy against specific seizure disorders. Techniques are described below for data analysis in the GEPR that do offer prediction of clinical efficacy that could be applied to audiogenic seizure susceptible mice.

The consistent audiogenic seizure responsiveness of the GEPR makes it ideally suited for the evaluation of anticonvulsant drug effects. GEPRs that have undergone the audiogenic screening procedure described above can be utilized for anticonvulsant experiments.[4,12] To be eligible for anticonvulsant experiments, GEPR-3s must exhibit a seizure rated at an ARS of 3 on each of the three weekly screens.[4] Eligible GEPR-9s must exhibit an ARS of 5, 7, or 9 on the first weekly screen and an ARS of 9 on the last two weekly screens. All GEPR-3s and GEPR-9s exhibiting this seizure history can be expected to exhibit an ARS of 3 and 9, respectively, on subsequent audiogenic screens.[4] Any reduction of seizure severity is considered to be due to the drug being tested in the GEPR.

Anticonvulsant testing occurs 1 week after the third audiogenic screen. In the majority of GEPR anticonvulsant studies, an anticonvulsant effect is defined as any reduction of seizure severity or ARS score.[12] Any seizure rated below an ARS of 3 in a GEPR-3 or below an ARS of 9 in a GEPR-9 would be considered an anticonvulsant effect. The anticonvulsant response utilized under this procedure is in effect a minimally detectable anticonvulsant response. The effective dose fifty (ED_{50}) is calculated for this minimally detectable response according to the method of Litchfield and Wilcoxon.[67] The advantage of such a technique is that it minimizes the number of animals required to generate an ED_{50} for a particular drug. It is a quick, simple, and easy screen for general anticonvulsant efficacy. The disadvantage of such a technique is that no information is generated about the effect of the drug on each component of the audiogenic seizure.

Data gathered in this manner can also be manipulated to distinguish between drugs effective in generalized tonic-clonic, simple and complex partial seizures from those effective in absence seizures.[4] When the ED_{50} for the minimally detectable anticonvulsant effect in GEPR-3s (vertical axis) is plotted against the ED_{50} in GEPR-9s (horizontal axis) for clinically effective antiepileptic drugs, two distinct clusters of drugs appear (Figure 6.5). The first includes drugs effective against generalized tonic-clonic and partial seizures. The second cluster includes drugs effective against absence seizures. The second cluster occurs at doses from 1 to 2 orders of magnitude greater than the first cluster for both GEPR-3s and GEPR-9s. Also included in the first cluster are a number of tricyclic antidepressants. This model predicts that these antidepressants would be effective in generalized tonic-clonic or partial epilepsy. This remains to be determined. It is important to note that valproate, clinically effective in generalized tonic-clonic, partial, and absence seizures, is located between ethosuximide and the first cluster. Thus, this data manipulation may be capable of predicting broad spectrum clinical efficacy as well. Similar data manipulation could be performed between different mouse strains susceptible to audiogenic seizures.

Focal seizure mechanisms (or seizure initiation mechanisms) and seizure propagation/generalization mechanisms appear to involve separate neurochemical and anatomical substrates in the GEPR. If one attempts to completely suppress audiogenic seizures in a GEPR-9 with phenytoin, the components of the audiogenic seizure are eliminated as a function of increasing dose in the following order: hindlimb extension, forelimb extension, generalized clonus, and finally the running episode.[12] Wild running, the focal event in an audiogenic seizure, is more resistant to the anticonvulsant effects of phenytoin than are the seizure components related to seizure generalization in the GEPR. We are in the process of developing a method of anticonvulsant assessment in the GEPR that distinguishes between the relative antifocal and antigeneralization properties of a drug. This model should allow such comparisons across different drugs.

To accomplish this, quantal dose-response curves are determined for suppression of hindlimb extension, forelimb extension, generalized clonus, and wild running (complete seizure suppression) in GEPR-9s. Then mean ARS scores are plotted for each dose utilized in the generation of these four quantal dose-response curves, generating a graded dose-response curve (Figure 6.6). Finally, linear regression analysis is performed on the ARS scores and the doses falling within the linear portion (ED_{16} to ED_{84}) of the hindlimb extension dose-response score (Figure 6.7). A similar regression analysis is performed at the linear portion of the wild running dose-response curve (Figure 6.7). Thus, two dosage intercepts are obtained, one for suppression of the focal event (wild running) and one for suppression of the generalization event (hindlimb extension). One can then determine the ratio of the wild running dosage intercept and hindlimb extension dosage intercept and obtain a focal/generalization ratio that can be compared across different drugs. The smaller the focal/generalization ratio is determined to be, the greater the antifocal properties of a drug would be.

The same drugs used to treat generalized tonic-clonic seizures are used to treat partial seizures with or without generalization. However, partial seizures can be

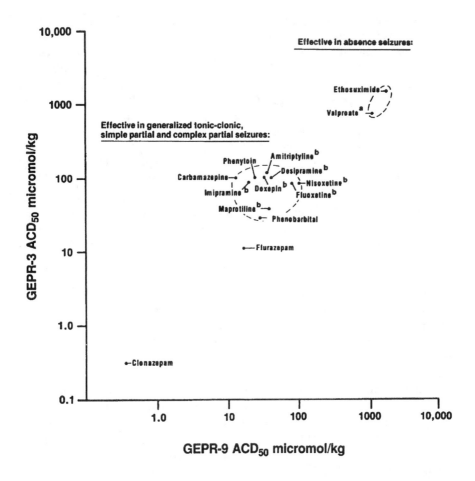

FIGURE 6.5

Relative potency distributions in GEPR-3s and GEPR-9s of clinically effective antiepileptic drugs. Each point reflects the ED_{50} for a minimally detectable anticonvulsant response. ACD_{50} on the figure legends refers to the anticonvulsant dose fifty that appeared on the original published figure that is reproduced here. Note the two distinct clusters of drugs in the figure. The first cluster includes drugs clinically effective against generalized tonic-clonic and partial seizures. The second cluster includes drugs effective against absence seizures. [a]Valproic acid, clinically effective against generalized tonic-clonic, partial, and absence seizures is located between ethosuximide and the first cluster. [b]The first cluster also includes a number of tricyclic antidepressants predicted by the model to be effective against generalized tonic-clonic and partial seizures. (From Reigel, C. E. et al., *Life Sci.*, 39, 763, 1986. With permission.)

resistant to treatment with these agents in some patients. A drug with a lower focal/generalization ratio in the GEPR than existing medications might be effective in these refractory partial seizure patients. This model also offers prediction of a novel type of antiepileptic drug, a drug with a focal/generalization ratio of 1. The wild running regression line becomes a continuation of the hindlimb extension regression line. Such a drug would be a pure antifocal agent in the GEPR. We would expect the slope of the single regression line to be steep and the nature of the

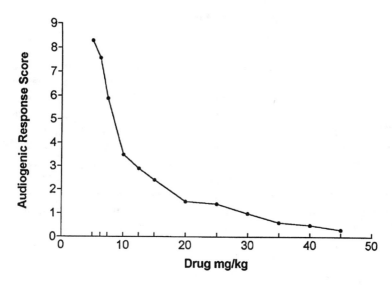

FIGURE 6.6

A graded anticonvulsant dose-response curve for complete suppression of audiogenic seizures in GEPR-9s for a hypothetical drug.

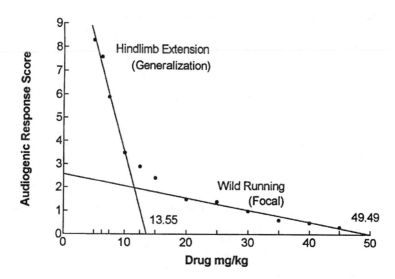

FIGURE 6.7

Regression analysis of hindlimb extension and wild running components of the hypothetical graded dose-response curve depicted in Figure 6.6. In this model, hindlimb extension represents seizure generalization and wild running represents focal seizure activity. Linear regression was performed on ARS scores and the doses falling within the linear portion (ED_{16} to ED_{84}) of the hindlimb extension dose-response curve, generating a generalization dosage intercept of 13.55 mg/kg. Linear regression was also performed on ARS scores and the doses falling within the linear portion (ED_{16} to ED_{84}) of the wild running dose-response curve, generating a focal dosage intercept of 49.49 mg/kg. The focal/generalization ratio for this hypothetical antiepileptic drug is 3.65.

anticonvulsant response to be either complete audiogenic seizure suppression or no effect at all, as a function of dose.

A similar method comparing generalized clonus to wild running dosage intercepts can be performed in the GEPR-3. Studies utilizing drugs clinically effective in generalized tonic-clonic and partial seizures are currently underway in both GEPR-9s and GEPR-3s in our laboratory. Preliminary results are promising, suggesting that phenobarbital possesses a lower focal/generalization ratio than phenytoin in GEPR-9s. There is no reason why this procedure could not be applied to mouse strains susceptible to audiogenic seizures. In fact, the population of partial seizure patients refractory to current antiepileptic drugs might actually consist of multiple populations, each optimally treated by a different, yet undiscovered drug. Such drugs might be discovered in specific mouse or rat strains. In conclusion, this rationale supports the need to screen potential anticonvulsant drugs in animals genetically predisposed to epilepsy in addition to traditional screening methods.

References

1. Collins, R. L., Audiogenic seizures, in *Experimental Models of Epilepsy: A Manual for the Laboratory Worker*, Purpura, D. P., Penry, J. K., Tower, D. M., Woodbury, D. M., and Walker, R., Eds., Raven Press, New York, 1972, 347.
2. Jobe, P. C., Mishra, P. K., and Dailey, J. W., Genetically epilepsy-prone rats: actions of antiepileptic drugs and monoaminergic neurotransmitters, in *Drugs for Control of Epilepsy: Actions on Neuronal Networks Involved in Seizure Disorders*, Faingold, C. L. and Fromm, G. H., Eds., CRC Press, Boca Raton, 1992, 253.
3. Seyfried, T. N., Glaser, G. H., Yu, R. K., and Palayoor, S. J., Inherited convulsive disorders in mice, in *Advances in Neurology*, Vol. 44, *Basic Mechanisms of the Epilepsies: Molecular and Cellular Approaches*, Delgado-Escueta, A. V., Ward, A. A., Jr., Woodbury, D. M., and Porter, R. J., Eds., Raven Press, New York, 1986, 115.
4. Reigel, C. E., Dailey, J. W., and Jobe, P. C., The genetically epilepsy-prone rat: an overview of seizure-prone characteristics and responsiveness to anticonvulsant drugs, *Life Sci.*, 39, 763, 1986.
5. Chapman, A. G. and Meldrum, B. S., Epilepsy-prone mice: genetically determined sound-induced seizures, in *Neurotransmitters and Epilepsy*, Jobe, P. C. and Laird, H. E., III, Eds., Humana, Clifton, NJ, 1987, 9.
6. Seyfried, T. N., Audiogenic seizures in mice, *Fed. Proc.*, 38, 2399, 1979.
7. Dailey, J. W., Reigel, C. E., Mishra, P. K., and Jobe, P. C., Neurobiology of seizure predisposition in the genetically epilepsy-prone rat, *Epilepsy Res.*, 3, 3, 1989.
8. Deckard, B. S., Lieff, B., Schlesinger, K., and DeFries, J. C., Developmental patterns of seizure susceptibility in inbred strains of mice, *Dev. Psychobiol.*, 9, 17, 1976.
9. Seyfried, T. N., Yu, R. K., and Glaser, G. H., Genetic analysis of audiogenic seizure susceptibility in C57BL/6J × DBA/2J recombinant inbred strains of mice, *Genetics*, 94, 701, 1980.
10. Seyfried, T. N. and Glaser, G. H., Genetic linkage between the *AH* locus and a major gene that inhibits susceptibility to audiogenic seizures in mice, *Genetics*, 99, 117, 1981.

11. Jobe, P. C., Picchioni, A. L., and Chin, L., Role of brain norepinephrine in audiogenic seizure in the rat, *J. Pharmacol. Exp. Ther.*, 184, 1, 1973.

12. Dailey, J. W. and Jobe, P. C., Anticonvulsant drugs and the genetically epilepsy-prone rat, *Fed. Proc.*, 44, 2640, 1985.

13. Browning, R. A., Effect of lesions on seizures in experimental animals, in *Epilepsy and the Reticular Formation: The Role of the Reticular Core in Convulsive Seizures*, Fromm, G. H., Faingold, C. L., Browning, R. A., and Burnham, W. M., Eds., Alan R. Liss, New York, 1987, 137.

14. Collins, R. L., A new genetic locus mapped from behavioral variation in mice: audiogenic seizure prone (ASP), *Behav. Genet.*, 1, 99, 1970.

15. Schreiber, R. A., Lehmann, A., Ginsburg, B. E., and Fuller, J. L., Development of susceptibility to audiogenic seizures in DBA/2 and Rb mice: toward a systematic nomenclature of audiogenic seizure levels, *Behav. Genet.*, 10, 537, 1980.

16. Seyfried, T. N., Genetic heterogeneity for the development of audiogenic seizures in mice, *Brain Res.*, 271, 325, 1983.

17. Schlesinger, K., Boggan, W., and Freedman, D. X., Genetics of audiogenic seizures. I. Relation to brain serotonin and norepinephrine in mice, *Life Sci.*, 4, 2345, 1965.

18. Seyfried, T. N., Glaser, G. H., and Yu, R. K., Developmental analysis of regional brain growth and audiogenic seizures in mice, *Genetics*, 88, S90, 1978.

19. Alexander, G. J. and Gray, R., Induction of convulsive seizures in sound sensitive albino mice: response to various signal frequencies, *Proc. Exp. Biol. Med.*, 140, 1284, 1972.

20. Alexander, G. J. and Alexander R. B., Linear relationship between stimulus intensity and audiogenic seizures in inbred mice, *Life Sci.*, 19, 987, 1976.

21. Swinyard, E. A., Castellion, A. W., Fink G. B., and Goodman, L. S., Some neurophysiological characteristics of audio-seizure-susceptible mice, *J. Pharmacol. Exp. Ther.*, 140, 375, 1963.

22. Castellion, A. W., Swinyard, E. A., and Goodman, L. S., Effect of maturation on the development and reproducibility of audiogenic seizures in mice, *Exp. Neurol.*, 13, 206, 1965.

23. Simler, S., Ciesielski, L., Maitre, M., Randrianarisoa, H., and Mandel, P., Effect of sodium n-dipropylactetate on audiogenic seizures and brain γ-aminobutyric acid level, *Biochem. Pharmacol.*, 22, 1701, 1973.

24. Faingold, C. L., Walsh, E. J., Maxwell, J. K., and Randall, M. E., Audiogenic seizure severity and hearing deficits in the genetically epilepsy-prone rat, *Exp. Neurol.*, 108, 55, 1990.

25. Reigel, C. E., Randall, M. E., and Faingold, C. L., Developmental hearing impairment and audiogenic seizure (AGS) susceptibility in the genetically epilepsy-prone rat, *Soc. Neurosci. Abstr.*, 15, 46, 1989.

26. Reigel, C. E., Bell, K. D., Randall, M. E., and Faingold, C. L., Monoaminergic and auditory indices in non-audiogenic seizure (AGS) susceptible genetically epilepsy-prone rats (GEPRs), *Soc. Neurosci. Abstr.*, 17, 172, 1991.

27. Thompson, J. L., Carl, F. G., and Holmes, G. L., Effects of age on seizure susceptibility in genetically epilepsy-prone rats (GEPR-9s), *Epilepsia*, 32, 161, 1991.

28. Reigel, C. E., Jobe, P. C., Dailey, J. W., and Savage, D. D., Ontogeny of sound-induced seizures in the genetically epilepsy-prone rat, *Epilepsy Res.*, 4, 63, 1989.

29. Vergnes, M., Kiesmann, M., Marescaux, C., Depaulis, A., Micheletti, G., and Warter, J. M., Kindling of audiogenic seizures in the rat, *Int. J. Neurosci.*, 36, 167, 1987.
30. Marescaux, C., Vergnes, M., Kiesmann, M., Depaulis, A., Micheletti, G., and Warter, J. M., Kindling of audiogenic seizures in Wistar rats: an EEG study, *Exp. Neurol.*, 97, 160, 1987.
31. Henry, K. R., Audiogenic seizure susceptibility induced in C57Bl/6J mice by prior auditory exposure, *Science*, 158, 938, 1967.
32. Iturrian, W. B. and Fink, G. B., Effect of age and condition-test interval (days) on an audio-conditioned convulsive response in CF #1 mice, *Dev. Pyschobiol.*, 1, 230, 1968.
33. Fuller, J. L. and Collins, R. L., Temporal parameters of sensitization for audiogenic seizures in SJL/J mice, *Dev. Psychobiol.*, 1, 185, 1968.
34. Schreiber, R. A., Effects of stimulus intensity and stimulus duration during acoustic priming on audiogenic seizures in C57BL/6J mice, *Dev. Psychobiol.*, 10, 77, 1977.
35. Norris, C. H., Cawthon, T. H., and Carroll, R. C., Kanamycin priming for audiogenic seizures in mice, *Neuropharmacol*ogy, 16, 375, 1977.
36. Pierson, M. G. and Swann, J. W., The sensitive period and optimal dosage for induction of audiogenic seizure susceptibility in the Wistar rat, *Hearing Res.*, 32, 1, 1988.
37. Reigel, C. E. and Aldrich, W. M., Kanamycin-induced audiogenic seizure susceptibility requires monoamine depletion in Sprague-Dawley (SD) rats, *Soc. Neurosci. Abstr.*, 16, 781, 1990.
38. Pierson, M. G. and Swann, J., Ontogenetic features of audiogenic seizure susceptibility induced in immature rats by noise, *Epilepsia*, 32, 1, 1991.
39. Willott, J. F. and Henry, K. R., Roles of anoxia and noise-induced hearing loss in the postictal refractory period for audiogenic seizures in mice, *J. Comp. Physiol. Psychol.*, 90, 373, 1976.
40. Faingold, C. L., Travis, M. A., Gehlbach, G., Hoffman, W. E., Jobe, P. C., Laird, H. E., and Caspary, D.M., Neuronal response abnormalities in the inferior colliculus of the genetically epilepsy-prone rat, *Electroencephalogr. Clin. Neurophysiol.*, 63, 296, 1986.
41. Schreiber, R. A., Stimulus frequency and audiogenic seizures in DBA/2J mice, *Behav. Genet.*, 8, 341, 1978.
42. Reid, H. M. and Collins, R. L., Recovery of susceptibility to audiogenic seizure in mice, *Epilepsia*, 34, 18, 1993.
43. Reigel, C. E., Dailey, J. W., Ferrendelli, J. A., and Jobe, P. C., Duration of postictal refractoriness in the genetically epilepsy-prone rat (GEPR), *Soc. Neurosci. Abstr.*, 11, 1314, 1985.
44. Ludvig, N. and Moshe, S. L., Different behavioral and electrographic effects of acoustic stimulation and dibutyryl cyclic AMP injection into the inferior colliculus in normal and in genetically epilepsy-prone rats, *Epilepsy Res.*, 3, 185, 1989.
45. Naritoku, D. K., Mecozzi, L. B., Aiello, M. T., and Faingold, C. L., Repetition of audiogenic seizures in genetically epilepsy-prone rats induces cortical epileptiform activity and additional seizure behaviors, *Exp. Neurol.*, 115, 317, 1992.
46. Savage, D. D., Reigel, C. E., and Jobe, P. C., Angular bundle kindling is accelerated in rats with a genetic predisposition to acoustic stimulus-induced seizures, *Brain Res.*, 376, 412, 1986.

47. Coffey, L. L., Reith, M. E. A., Chen, N. H., Mishra, P. K., and Jobe, P. C., Amygdala kindling of forebrain seizures and the occurrence of brainstem seizures in genetically epilepsy-prone rats, *Epilepsia*, 37, 188, 1996.
48. Kesner, R. P., Subcortical mechanisms of audiogenic seizures, *Exp. Neurol.*, 15, 192, 1966.
49. Browning, R. A., Neuroanatomical localization of structures responsible for seizures in the GEPR: lesion studies, *Life Sci.*, 39, 857, 1986.
50. Faingold, C. L., Millan, M. H., Boersma, C. A., and Meldrum, B. S., Excitant amino acids and audiogenic seizures in the genetically epilepsy-prone rat. I. Afferent seizure initiation pathway, *Exp. Neurol.*, 99, 678, 1988.
51. Millam, M. H., Meldrum, B. S., and Faingold, C. L., Induction of audiogenic seizure susceptibility by focal infusion of excitant amino acid or bicuculline into the inferior colliculus of normal rats, *Exp. Neurol.*, 91, 634, 1986.
52. Chapman, A. G., Faingold, C. L., Hart, G. P., Bowker, H. M., and Meldrum, B. S., Brain regional amino acid levels in seizure susceptible rats: changes related to sound-induced seizures, *Neurochem. Int.*, 8, 273, 1986.
53. Faingold, C. L., Gelbach, G., Travis, M. A., and Caspary, D. N., Inferior colliculus response abnormalities in genetically epilepsy-prone rats and evidence for a deficit of inhibition, *Life Sci.*, 39, 869, 1986.
54. Browning, R. L., Lanker, M. L., and Faingold, C. L., Injections of noradrenergic and GABAergic agonists into the inferior colliculus: effects on audiogenic seizures in genetically epilepsy-prone rats, *Epilepsy Res.*, 4, 119, 1989.
55. Faingold C. L., Marcinczyk, M. J., Casebeer, D. J., Randall, M. E., Arneric, S. P., and Browning, R. L., GABA in the inferior colliculus plays a critical role in control of audiogenic seizures, *Brain Res.*, 640, 40, 1994.
56. Yang, L., Evan, M. S., and Faingold, C. L., Inferior colliculus neuronal membrane and synaptic properties in genetically epilepsy-prone rats, *Brain Res.*, 660, 232, 1994.
57. Browning, R. A., Nelson, D. K., Mogharreban, N., Jobe, P. C., and Laird, H. E., Effect of midbrain and pontine tegmental lesions on audiogenic seizures in the genetically epilepsy-prone rats, *Epilepsia*, 26, 175, 1985.
58. Millan, M. H., Meldrum, B. S., Boersma, C. A., and Faingold, C. L., Excitant amino acids and audiogenic seizures in the genetically epilepsy-prone rat. II. Efferent seizure propagating pathway, *Exp. Neurol.*, 99, 687, 1988.
59. Reigel, C. E. and Faingold, C. L., Innately hearing impaired Sprague-Dawley rats exhibit audiogenic seizure susceptibility following monoamine depletion, *Proc. West. Pharmacol. Soc.*, 36, 267, 1993.
60. Jobe, P. C., Mishra, P. K., Browning, R. A., Wang, C., Adams-Curtis, L. E., Ko, K. H., and Dailey, J. W., Noradrenergic abnormalities in the genetically epilepsy-prone rat, *Brain Res. Bull.*, 35, 493, 1994.
61. Dailey, J. W., Mishra, P. K., Ko, K. H., Penny, J. E., and Jobe, P. C., Serotonergic abnormalities in the central nervous system of seizure-naive genetically epilepsy-prone rats, *Life Sci.*, 50, 319, 1992.
62. Stanick, M. A., Dailey, J. W., Jobe, P. C., and Browning, R. A., Abnormalities in brain serotonin concentration, high-affinity uptake, and tryptophan hydroxylase activity in severe-seizure genetically epilepsy-prone rats, *Epilepsia*, 37, 311, 1996.

63. Reigel, C. E. and Lin, B. K., Differential neonatal ontogeny of heightened seizure sensitivity in two strains of genetically epilepsy-prone rats, *Soc. Neurosci. Abstr.*, 19, 1470, 1993.

64. Reigel, C. E., Whitehead, H., Lovering, A. T., and Lin, B. K., Neonatal norepinephrine content in two strains of genetically epilepsy-prone rats, *Soc. Neurosci. Abstr.*, 20, 405, 1994.

65. Chapman, A. G., Croucher, M. J., and Meldrum, B. S., Evaluation of anticonvulsant drugs in DBA/2 mice with sound-induced seizures, *Arzneim.-Forsh.*, 34, 1261, 1984.

66. Lehmann, A. G., Psychopharmacology of the response to noise, with special reference to audiogenic seizure in mice, in *Physiological Effects of Noise*, Welch, B. L. and Welch, A. S., Eds., Plenum Press, New York, 1970, 227.

67. Litchfield, J. T. and Wilcoxon, F., A simplified method of evaluating dose-effect experiments, *J. Pharmacol. Exp. Ther.*, 96, 99, 1949.

Chapter

7

Models of Focal Epilepsy in Rodents

Charles R. Craig

Contents

I. Introduction

Experimental seizure models in laboratory animals have played a prominent role in epilepsy research over the years. Among these models are those that utilize the placement of metals to the brains of animals to produce seizures. The species most often employed have been monkeys, cats, or rats, although other species have also been studied. An advantage of metals over many other substances, such as organic chemicals, is that metals tend not to be absorbed by the body, and not to be changed by metabolism and/or other chemical reactions. Therefore, they tend not to diffuse from the area of injection and to persist at the site of implantation or administration for at least several days or weeks.

A large number of metals, as well as other substances, have been shown to produce seizures when applied directly to the brains of experimental animals. The first such report was by Kopeloff et al.[1] They observed that application of alumina cream to the cerebral cortex of monkeys resulted in the development of seizures. Aluminum, in the form of the metal or as alumina cream, has been widely employed to study mechanisms of seizures: it does not appear to be epileptogenic to rodents,[2] however, and this has curtailed its usage somewhat. Kopeloff[3] studied the effects of cobalt, nickel, and antimony powder, in addition to alumina, applied intracerebrally to mice. She substantiated the ability of cobalt and nickel to also produce seizures in this species. Kopeloff also found that antimony produced death soon after its administration and substantiated the lack of any epileptic effects of alumina cream in mice. Subsequently, this group studied the epileptogenic effects of implantation of several pure metal pellets[4] or metallic powders[5] to the cerebral cortices of monkeys. They concluded that aluminum, cobalt, nickel, bismuth, antimony, vanadium, iron, molybdenum, zirconium, mercury, tungsten, tantalum, lead, and beryllium were all capable of causing spontaneous seizures or of causing epileptogenic activity in the electroencephalogram (EEG) of the monkey; however, in most cases, only one monkey was employed for each metal. About the same time, Blum and Liban[6] demonstrated the epileptogenic effects of tungstic acid in cats and Donaldson et al.[7] showed that zinc was capable of causing seizures in rats.

The use of aluminum or its salts became the metal of choice to study focal experimental epilepsy in primates. For those investigators interested in using rodents, cobalt or nickel appeared to be the metals of choice. Subsequently, cobalt evolved as the metal of choice for studying epileptogenesis in rats[8] and mice; it has also been shown to produce seizures in the cat,[9] monkey,[5] and the gerbil.[10]

There are several reasons for the choice of cobalt. First, seizures are produced in virtually all animals in which cobalt has been applied to the brain. While seizures are prominent, the animals otherwise show very little in the way of toxicity. During the period of the most intense seizure activity, the animals are hyperactive, but as the seizure activity subsides, they become indistinguishable from controls. The time course for the development of seizure activity after cobalt implantation is particularly desirable for many types of investigations. Very few seizures occur during the first 5 d after implantation of cobalt in the rat, but the frequency increases rapidly thereafter, reaching a peak by day 7 or 8. There is a period of 7 to 10 d during which seizure activity is maximal; this is followed by a period of decreased seizure activity and after about 2 weeks after cobalt implantation, seizure activity is again virtually absent.[11,12] This time course makes the cobalt model particularly desirable when studying biochemical parameters prior to the onset of seizures, during peak seizure activity and, again, after cessation of convulsions. With the use of cobalt rods, the metal remains localized to the site of implantation and only minimal diffusion, metabolism, or transport from the site of implantation occurs. There is, however, recent evidence that some cobalt ions may be transported via axons to the thalamus.[13] The area of the lesion is also quite circumscribed and available for histochemical, microscopic, or other types of analysis. The cobalt rods can even be removed later to see if the productions of the epileptic state can be reversed or terminated by removal of the stimulus.

Although cobalt-experimental epilepsy, particularly in rodents, possesses several characteristics, making it a desirable model to employ, it has not been universally accepted as a valid model of seizure disorders. In fact, a review of the literature indicates that its use has declined in recent years.

Another model is that of iron as an epileptogenic agent. The fact that iron produces an experimental epilepsy in laboratory animals may be of relevance to human posttraumatic epilepsy. It is known that there frequently is extravasation of blood during head trauma; the development of posttraumatic epilepsy is a frequent sequela of severe head trauma. It has been postulated that the deposition of iron from hemoglobin and its subsequent sequestration as hemosiderin into brain cells as a consequence of the blood loss may be a cause of seizures seen in this condition.[14-16] If this is the case, then iron-induced epilepsy could be an ideal model of posttraumatic epilepsy; certainly, iron can produce seizures when applied to brain tissue of experimental animals.

Penicillin G has also been widely studied as an epileptogenic agent. The seizures produced by the direct application of penicillin G to the cortex occur very rapidly, persist for 3 to 5 h, and disappear. Penicillin G does not appear to cause any morphological changes and its epileptic activity is presumably related to the capacity it has to reduce the neuronal inhibition produced by the amino acid transmitter, gamma-aminobutyric acid (GABA).

II. Methodology

A. Cobalt as a Focal Epileptogenic Agent

1. Form of Cobalt to Use

Cobalt can be employed as the metal, sometimes as powdered cobalt and sometimes in the form of cobalt wire, or as a solution of either cobaltous or cobaltic chloride. In this discussion, we will concentrate on the use of cobalt wire as the epileptogenic agent (the paper by Hattori et al.[17] describes a method utilizing cobalt chloride if one is interested). The most recent purchase of cobalt wire that this author made was from Aldrich Chemical Company, Inc., St. Louis, MO. For rats, the best size for producing seizures is wire 1.0 mm in diameter. For some purposes, one may choose cobalt that is thinner; wire 10 mil in diameter is available and will also produce seizures.

2. Preparation of Experimental Animals

Several anesthetic agents have been used; however, this investigator has generally employed Innovar®. Innovar® is a commercial mixture containing 0.05 mg fentanyl and 2.5 mg droperidol per milliter. It is best administered subcutaneously in the nape of the neck in a dose of about 0.5 ml/adult rat. There appears to be some strain differences and in some studies, the author has been successful administering 0.25 ml per animal while at other times, 0.7 ml or so was required. This procedure produces surgical anesthesia in about 15 min that ordinarily persists for about an hour. There is a very low incidence of death with this agent. If necessary, the administration of an opioid antagonist, such as naloxone, may be useful to reverse the effects of the fentanyl component of Innovar.

Either male or female rats may be used, since there appear to be no differences in the response of either gender to the application of cobalt. Female rats are generally employed if long-term electroencephalography is contemplated, since they do not grow as rapidly as males and, therefore, there is less likelihood of pulling electrodes out of the skull as the skull is developing.

The rat is placed in a stereotaxic frame as soon as it is anesthetized. For most studies, it is not necessary to be particularly concerned about using specific stereotaxic coordinates. The bregma is a particularly useful landmark for application of cobalt to the cerebral cortex of rats. The cerebral cortex directly beneath the bregma and at least 2 to 4 mm lateral, rostral, and caudal is essentially motor cortex. The stereotaxic frame is very useful for holding the rat's head in a fixed position so that the surgery can be easily accomplished. The skull is exposed by using a scalpel and blunt dissection. The entire skull should be exposed and, if the animal is to be prepared for electromyographic tracings, the temporalis muscle must be exposed as well. Earlier studies involved application of the cobalt to either left or right cerebral hemisphere; however, for the maximum number of seizures to occur, bilateral placement of cobalt into both left and right cerebral cortices is indicated.[18] With a dental drill, holes are made in the cortex 2 or 3 mm to the left and right of the saggital suture, 2 mm rostral or caudal to the bregma (depending on the choice for the placement of cobalt). One should be careful not to allow the dental drill to penetrate into the brain tissue, since it can cause significant damage. When the hole is drilled into the skull, the dura mater and pia mater should be carefully punctured with a needle. A cobalt rod, 1 mm in diameter and 1 to 2 mm in length, should carefully be inserted into the exposed cortex (Figure 7.1 illustrates the placement of the cobalt). The hole should be covered with a piece of gelfoam to prevent excessive bleeding. If neither electroencephalography nor electromyography is planned, the skin should be sutured carefully and the animals placed in individual cages to recover for subsequent studies (for greater detail on intracranial implant surgery, see Chapter 3 on kindling). There have been no rigorous studies to determine if there is an optimal cortical location for the production of seizures. Depending upon the nature of the study, one area may be favored over others. However, if the production of seizure activity is the primary aim, any location ±3 or 4 mm in any direction from the bregma should be satisfactory. In a given study, however, the same location should be used in all rats.

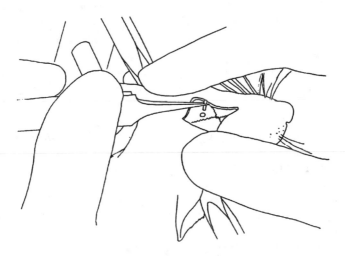

FIGURE 7.1
This figure depicts the placement of a cobalt rod to a rat's cerebral cortex. The rat has been anesthetized and placed in a stereotaxic frame. The skull has been exposed and an opening made in the skull in the area of the bregma with a dental drill. The dura mater and pia mater have been punctured with a needle. The technician will carefully insert the cobalt rod into the opening, cover the opening with a small piece of gelfoam, and suture the skin.

The administration of cobalt to the brain results in the formation of an area of necrosis that is roughly proportional to the amount of cobalt inserted;[19] this is a reason to choose a smaller diameter cobalt wire. It is likely that the area of epileptogenic neurons lies immediately outside the area of necrosis, with these neurons exhibiting altered electrophysiological and neurochemical activity. The area of necrosis is quite clearly demarcated and recording from neurons in the "normally appearing cortex" should be relatively easy, but to date, has not been done in any systematic fashion. Similarly, studies can be carried out to measure ions, enzymes, and putative neurotransmitters in the same tissues. A limited number of such studies have been carried out.[20-23] Recently Van Ostrand and Cooper[13] have combined cobalt epilepsy in the rat with a study of 2-deoxyglucose uptake and have made some interesting observations. They found that the area around the site of cerebral cobalt application in rats demonstrated hypermetabolism and that there was also a region of hypermetabolism in the thalamus.

3.   Preparation for Electrocorticographic Measurements

There are certain advantages to measuring electrocorticographic (ECoG) activity following cobalt placement to the brain. Although seizures are frequently observed following administration of cobalt, it is difficult to quantify their occurrence, unless there are facilities for continuous video monitoring of the animals. The electrocorticogram (ECoG) offers a reliable means of determining seizure activity in a conscious animal throughout the time period chosen: this interval has been up to several weeks. The author for many years used a system in which the rats were tethered to a mercury swivel by means of a cable connected to a headpiece. The original source of the swivels no longer exists, but Grass Instrument Co., West Warwick, RI should be contacted for a current supplier. Figure 7.2 depicts a rat prepared for electrocorticographic and electromyographic (EMG) recording.

Figure 7.3 shows a typical ECoG tracing of a seizure induced by administration of cobalt. It can be noted that it is possible to differentiate sleep from wakefulness, even at the compressed chart speed of 25 mm/min. A complete discussion of methods currently available for electroencephalographic recording of rats is beyond the scope of this chapter. If one is planning to do rat electroencephalography, he or she should contact Grass Instruments to find out what telemetry methods are currently available.

If one wishes to record ECoG and/or EMG activity, the surgical preparation should be accomplished at the same time that cobalt is applied. In addition to holes for the insertion of cobalt, holes are also drilled bilaterally over the frontal and parietal cortices for placement of stainless steel screws that will be used as electrodes for the electrocorticogram. The screw in the right parietal region is used only as an anchor to help hold the headpiece on. Stainless steel wires are also sewn into the temporalis muscle to record the electromyogram. All electrodes (three ECoG and two EMG) are fastened to a 7-pin electrical connector (obtained from Vantage Electronics, Waltham, MA) which is attached to the skull with acrylic dental cement. The dental cement dries very quickly and within an hour ECoG and EMG activity can be collected; the cable is plugged into an electrical connector and into a Grass polygraph via a mercury swivel.

As mentioned, currently there is equipment available that allows for intermittent or even continuous monitoring of a freely moving unrestrained rat, without the necessity of the rat being connected through a mercury swivel.

4.   Choice of Appropriate Controls

There is some question as to what would constitute an ideal control to the administration of a cobalt rod, since many other metals would be expected to also produce an epileptogenic state when administered to the cortex. The insertion of the cobalt rod produces tissue damage as well as some bleeding.

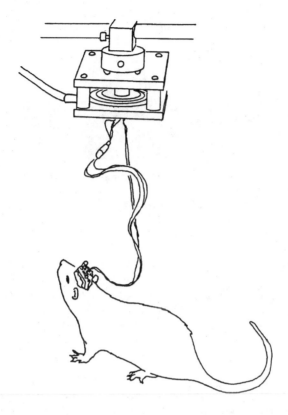

FIGURE 7.2

Artist's conception of a rat that has been implanted with a cobalt rod in the cerebral cortex. For the recording of electrocorticogram (ECoG) activity, screw electrodes have been placed in the cortex in the left frontal, right frontal, left parietal, and right parietal areas. Muscle electrodes have been inserted into the left and right temporalis muscles for recording of the electromyogram (EMG). Following implantation of cobalt into the cerebral cortex and connection of the electrodes, all of the electrodes are inserted into a headpiece which is held in position on the skull with acrylic dental cement. Two hours after the surgery is completed, the rat is placed in an individual recording cage and is connected to a cable, and via a mercury swivel to a Grass polygraph (Grass Instrument Co., West Warwick, RI) for the recording of ECoG and EMG activity. The ECoG and EMG activity of these unrestrained and freely moving rats can be recorded continuously for 7 to 21 d. To conserve chart paper, a slow speed of 25 mm/min is generally employed.

We have used two different agents as controls. First, a glass rod, equivalent in length and diameter, will control for the tissue damage produced by the cobalt rod: glass produces neither significant necrosis nor seizures. Another control that has been employed is the use of a material that produces a degree of necrosis roughly equivalent to that seen with cobalt, but that appears to be nonepileptogenic. We have employed copper rods of the same size as cobalt. Copper produces a significant amount of necrosis, but does not appear to produce seizure activity. A sham operation, in which a hole is drilled in the skull and the dura and pia mater punctured, is a control that should also be carried out. According to Van Ostrand and Cooper,[13]

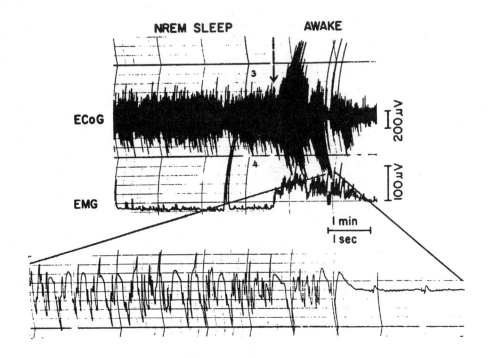

FIGURE 7.3

ECoG tracings collected during the appearance of a completely generalized seizure occurring at the end of sleep in a cobalt-epileptic rat. ECoG tracings at the slow chart speed of 25 mm/min is depicted; a section of ECoG at the speed of 25 mm/s is shown to illustrate the nature of the epileptic spikes. (From Craig, C. R. and Colasanti, B. K., in *Drugs for Control of Epilepsy: Actions on Neuronal Networks Involved in Seizure Disorders,* Faingold, C. L. and Fromm, G. H., Eds., CRC Press, Boca Raton, 1992. With permission.)

who also used copper as a control, there was no indication of hypermetabolism at the site of copper implantation or in any other areas of the brain.

5.   Methods to Determine Epileptogenicity

For many studies, it may not be necessary to determine precisely the level of seizure activity present, particularly if one uses a standard amount of cobalt (e.g., cobalt rods, 1 mm in diameter and 1 to 2 mm in length). If one has the facilities, it is relatively easy to monitor the EEG for evidence of epileptic activity (e.g., the presence of epileptic spiking) and even the incidence of overt seizures as evidenced by the ECoG (see Figure 7.3). In the absence of electroencepalographic monitoring, careful observation of the rats, particularly if cobalt has been administered bilaterally, should allow one to observe head jerks, clonic movements of the forelimbs, and even the occurrence of generalized convulsions. Such activity will be most prominent about 6 to 7 d after application of the cobalt and should persist at a relatively constant level for another 6 or 7 d.

    Another method that can be employed to establish that the rats are indeed epileptic and that can give a measure of the degree of epileptogenicity, is to determine the threshold for seizures induced by a convulsant agent, such as pentylenetetrazol. The implantation of cobalt produces a state in which the rats exhibit an increased sensitivity to most central nervous system convulsants (pentylenetetrazol, picrotoxin, or bicuculline). The method of Levine et al.[24] involves administering an intraperitoneal dose of pentylenetetrazol, 15 mg/kg, every 15 min until a generalized convulsion occurs. The threshold is then expressed as the time period after pentylenetetrazol in which an observed seizure occurs: the time required for seizures to occur should be decreased in rats rendered epileptic by cobalt. This method was utilized in the study of Hartman et al.,[25] in which control rats usually exhibited a generalized convulsion only after three injections of pentylenetetrazol while animals treated with cobalt 5 to 7 d earlier convulsed after the first injection. One must not attempt to determine the pentylenetetrazol seizure threshold in the same rat more than one time, since a type of "chemical kindling" can occur.[26] With repeated administrations of pentylenetetrazol no more frequently than one time per week, the seizure threshold significantly declines. This is not unlike the phenomenon of kindling that is more commonly produced by electrical stimulation to the brain and that is one of the most widely employed models for seizure studies. However, the administration of pentylenetetrazol in the manner discussed 6 or 7 d after implantation of cobalt should clearly indicate that the cobalt treatment was effective in producing an epileptic state. The methods to determine epileptic activity after cobalt implantation should be equally effective in dealing with any other form of focal epilepsy and will not be discussed again in relation to either focal iron or penicillin epilepsy in the rat.

B.   Iron as a Focal Epileptogenic Agent

1.   Form of Iron to Use

Iron was initially tested in the form of metallic iron. This form is, at best, only minimally epileptogenic.[25] Ionic iron, usually in the form of ferric chloride ($FeCl_3$) appears to be the preferred form.[14,27]

2.   Preparation of Experimental Animals

The usual amount of ferric chloride to produce epileptic discharges appears to be in the order of 5 µl of a 100 mM solution of $FeCl_3$.[27] The solution is applied to a small area of the cerebral cortex exposed following surgery, as described above for cobalt. The dura and pia mater are punctured to allow the direct instillation to the cerebral cortex. The solution is administered slowly (0.5 µl/min) via a syringe. Rats prepared similarly, but that, instead of receiving the iron salt, have received a solution of physiological saline, also slowly administered, serve as controls.

3.   Description of Seizure Activity

The first manifestation of an effect of the iron takes the form of spike and wave discharges from the electrocorticogram. These can be expected to begin within the first 24 to 48 h. Behavioral seizures also occur.[14] According to the authors, the behavioral convulsions took the form of an interruption of exploratory activity, rhythmic twitching of the vibrissae and neck musculature, and piloerection. Occasionally turning movements contralateral to the injected hemisphere were also observed. The electrocorticographic abnormalities persist for at least 90 d.[27]

C. Penicillin as a Focal Epileptogenic Agent

Penicillin has been widely used to produce seizures in rats and cats. Currently, penicillin is usually employed parenterally in high doses to produce seizures.[28,29] When used parenterally, the dose is very high, generally in the order of 1.2 million units per kilogram in the rat.[29] On the other hand, when penicillin is applied directly to the cortex, much lower quantities are required.

1. Form of Penicillin to Use

Like ferric chloride, when penicillin is used to produce focal epilepsy, it is commonly administered as a liquid. Sodium benzylpenicillin is dissolved in a mock cerebrospinal fluid (CSF) solution and the pH is adjusted to 7.3–7.4 immediately before filling the syringe. A small amount of fast green dye may be added to the penicillin solution to assure identification of the penicillin focus, if this is desirable.[30] This study by Collins[30] determined the amount of penicillin required to produce epileptic-like activity and characterized the time course for the appearance and disappearance of epileptic spiking. He found that 10 units of penicillin was the threshold dose for producing repetitive spike discharges while 300 units was required to reliably produce evidence of afterdischarge on the EEG.

2. Preparation of Experimental Animals

Because of the short duration of penicillin, in order to observe ECoG and behavioral changes, it is necessary to paralyze the animals with a substance such as d-tubocurarine instead of relying on anesthesia. This also requires that the laboratory be set up to maintain respiration during the period of observation.

3. Description of Seizure Activity

The primary indication of epileptic activity in the paralyzed animals is the appearance of repetitive ECoG spiking activity. Collins reported that the half-life of penicillin, using radiolabeled penicillin, was about 15 min and that after 45 to 60 min only 5% to 10% of the original dose remained. The disappearance of penicillin from the brain correlated nicely with the behavioral events and the EEG evidence of epileptic-like activity.

III. Interpretation

A search of the literature reveals that many fewer studies of focal epilepsy in experimental animals are being conducted at the present time than was the case a decade or so ago. If the use of focal models of epilepsy is declining, it might be instructive to consider some reasons why this is the case.

It is certainly much more expensive to conduct experimentation with laboratory animals now than it was a decade ago. The cost of animals, animal care, and the cost in time of complying with all of the regulations that are now required before laboratory studies can be carried out makes it much more expensive to conduct such studies. The explosion of knowledge in the area of molecular biology and the perception that, in order to be funded, grant applications must include a significant amount of molecular biology experiments may have pushed many investigators away from studies concerned with whole animals and into areas of *in vitro* research.

A more important consideration, however, is whether the focal models of experimental epilepsy have led to advances in knowledge about convulsive disorders that their proponents had hoped; and whether it is likely that they will prove to be important in our search for new understanding of the mechanisms of the epilepsies.

A. Advantages of Focal Models of Epilepsy in Experimental Animals, Particularly in the Rat

1. Seizures Produced Resemble those of Human Epilepsy

Primary features of focal models of epilepsy in the rat are that an epileptogenic focus can be induced at a specific anatomic site and that behavioral seizures, as well as EEG spiking, can be reliably produced that resemble the human counterpart, in that they tend to be intermittent and generalized (particularly apparent after cobalt). In this respect, focal seizure models more closely resemble the human counterpart than many other models that are used. Although the stimulus (metal or penicillin) is obviously not a causative factor in human epilepsy (an exception may be iron), the clinical syndrome is very similar to certain types of human epilepsy.

2. Focal Models of Epilepsy Have a Reliable Time Course

There is ordinarily a clearly defined onset, a prolonged time of consistent seizure activity, and a period in which seizure activity is declining, and ultimately ceases. This makes certain types of studies easy to plan and to carry out. In the case of penicillin, the duration of seizure activity is terminated by the half-life of the penicillin in the brain; it leaves the brain by diffusion and probably active transport. Metals persist in the site, but calcification may occur[19] and this calcification can seal off the area of the lesion and terminate the seizures.

3. Focal Models Are Adaptable to a Variety of Studies

Measurements of enzyme activity, transmitter levels, or a large number of other parameters can be measured and related to the behavioral state of the animal, prior to seizure onset, during periods of active seizure activity or at a time when seizures no longer occur. A large number of such studies have been done that demonstrate the possibilities; a few examples will be indicated here. Goldberg et al.[20] demonstrated a decrease in activity of the cholinergic enzymes, choline acetyltransferase and cholinesterase, in cerebral cortex adjacent to the site of cobalt implantation at 7 d with no changes in the contralateral cortices. Levels of activity returned to control values by 21 d. Hoover et al.[18] demonstrated that acetylcholine levels were maximally depressed in the ipsilateral cortex 7 d following implantation of cobalt, with recovery to control levels by 21 d. Ross and Craig[23] showed that GABA and its synthesizing enzyme, GAD, were both depressed at 7 d with recovery by 21 d when measured in the ipsilateral cortex. Esclapez and Trottier[31] found decreased densities of GABA-positive cells and terminals that appeared in early states of cobalt-induced epilepsy became more pronounced at a time of maximal seizure activity and returned to control levels at a time coinciding with extinction of seizures in this model. Ribak[32] has also done extensive studies with the GABAergic system using alumina cream and cobalt-induced focal epilepsy; he has reviewed his studies as well as those of others[32] finding a decrease in GABAergic terminals in focal models of epilepsy. Witte[33] conducted a study with ion channels to see what factors are responsible for the afterpotentials associated with penicillin-induced focal epilepsy. He showed that both $GABA_A$ and $GABA_B$ receptors were likely involved. Hattori et al.[17] studied the accumulation of cyclic AMP elicited by either adenosine or 2-chloroadenosine in brain slices of rats rendered epileptic by cobalt chloride. They observed a significant increase in cyclic AMP accumulation only in the primary epileptic area of the cortex adjacent to the injection of the cobalt chloride. The increased accumulation was observed as early as 8 d and had returned to control at 40 to 50 d; the time course for accumulation of cyclic AMP paralleled the development of epileptic activity. It is interesting that most studies that have demonstrated changes in neurotransmitters, enzymes, or second messengers have seen maximum activity (either decreases or increases) at the time of peak seizure activity and that most changes have returned to control levels by 20 to 30 d.

B. Some Indications that Focal Epilepsy may not Be the Best Choice of a Model to Study

A major validation of the use of focal models as a legitimate seizure model would be if anticonvulsant drugs show significant protection against the induced convulsions at dose levels that are effective in human epilepsy. Further validation would be achieved if anticonvulsant drugs, known to work in a particular manner, would predictably affect the seizures in the focal model.

Although anticonvulsant drugs are able to block seizures in cobalt-induced epilepsy in the rat, relatively high doses are usually required. The doses are also generally higher than those that show anticonvulsant activity in traditional screening procedures such as maximal electroshock or subcutaneous pentylenetetrazol seizure tests. The doses required are also higher than those used in human epilepsy. Most of the major anticonvulsant drugs have been tested, including phenytoin,[34] valproic acid,[35] ethosuximide,[36] clonazepam,[37] and carbamazepine.[38]

There may be valid reasons why anticonvulsant drugs do not appear any more effective than they do in the cobalt focal epilepsy model. First, this is a very severe seizure state, particularly if cobalt is applied bilaterally, as it was in some of the studies. Second, there is a great deal of variability in the number of seizures that occur in the cobalt model, thus making quantification and demonstration of statistical significance difficult to show without employing a large number of rats. Last, it is difficult to maintain steady anticonvulsant blood levels throughout a several day study. A new approach has been suggested by the use of a substance such as 2-deoxyglucose in cobalt-epileptic rats.[13,39] The area of hypermetabolism, as evidenced by "dark patches" (indicative of increased uptake of 2-deoxyglucose) may be less variable than other measures of epileptic-like activity. An ideal anticonvulsant compound may be expected to decrease the hypermetabolism associated with the seizure activity without altering normal brain metabolism. This hypothesis is interesting and is testable.

On the other hand, studies with anticonvulsant drugs may have revealed that there are significant differences in the mechanisms whereby the seizures are generated in idiopathic human epilepsy and those induced by cobalt, iron, or other materials in experimental animals. If this is the case, perhaps the decreased use of focal experimental epilepsy as a tool for studying human epilepsies may be justified.

C. Summary

In summary, the application of certain metals and nonmetals to the cerebral cortices of laboratory animals, particularly of rodents, leads to the production of a reproducible seizure state that lends itself well to a variety of studies. Studies utilizing focal epilepsy may significantly increase our knowledge of seizure mechanisms and, ultimately, our knowledge of human seizure disorders.

References

1. Kopeloff, L. M., Barrera, S. E., and Kopeloff, N., Recurrent convulsive seizures in animals produced by immunologic and chemical means, *Am. J. Psychiat.*, 98, 881, 1942.
2. Servit, Z. and Sterc, J., Audiogenic epileptic seizures evoked in rats by artificial epileptogenic foci, *Nature*, 181, 1475, 1958.
3. Kopeloff, L. M., Experimental epilepsy in the mouse, *Proc. Soc. Exp. Biol. Med.*, 104, 500, 1960.
4. Chusid, J. G. and Kopeloff, L. M., Epileptogenic effects of pure metals implanted in motor cortex of monkeys, *J. Appl. Physiol.*, 17, 696, 1962.
5. Chusid, J. G. and Kopeloff, L. M., Epileptogenic effects of metal powder implants in motor cortex of monkeys, *Int. J. Neuropsychiat.*, 3, 24, 1967.
6. Blum, B. and Liban, E., Experimental baso-temporal epilepsy in the cat. Discrete epileptogenic lesions produced in the hippocampus or amygdaloid by tungstic acid, *Neurology*, 10, 546, 1960.
7. Donaldson, J., St.-Pierre, T., Minnich, J., and Barbeau, A., Seizures in rats associated with divalent cation inhibition of Na^+-K^+-ATPase, *Can. J. Biochem.*, 49, 1217, 1971.

8. Dow, R., Fernandez-Guardiola, A., and Manni, E., The production of cobalt experimental epilepsy in the rat, *Electroencephalogr. Clin. Neurophysiol.*, 14, 399, 1962.
9. Henjyoji, E. Y. and Dow, R. S., Cobalt-induced seizures in the cat, *Electroencephalogr. Clin. Neurophysiol.*, 19, 152, 1965.
10. Payan, H. M., Cobalt experimental epilepsy in gerbils, *Exp. Med. Surg.*, 28, 163, 1970.
11. Colasanti, B. K., Hartman, E. R., and Craig, C. R., Electrocorticogram and behavioral correlates during the development of chronic cobalt experimental epilepsy in the rat, *Epilepsia*, 15, 361, 1974.
12. Trottier, S., Lindwall, O., Chauvel, P., and Bjorklund, A., Facilitation of focal cobalt-induced epilepsy after lesions of the noradrenergic locus coeruleus system, *Brain Res.*, 454, 308, 1988.
13. Van Ostrand, G. and Cooper, R. M., (^{14}C)2-deoxyglucose autoradiographic technique provides a metabolic signature of cobalt-induced focal epileptogenesis, *Epilepsia*, 35, 939, 1994.
14. Willmore, L. J., Hurt, R. W., and Sypert, G. W., Epileptiform activity initiated by pial iontophoresis of ferrous and ferric chloride on rat cerebral cortex, *Brain Res.*, 152, 406, 1978.
15. Willmore, L. J. and Rubin, J. J., Antiperoxidant pretreatment and iron-induced epileptiform discharges in the rat: EEG and histopathologic studies, *Neurology*, 31, 63, 1981.
16. Sypert, G. W., Metallic salts and epileptogenesis, in *Physiology and Pharmacology of Epileptogenic Phenomena*, Klee, M. R., Lux, H. D., and Speckmann, E.-J., Eds., Raven Press, New York, 1982, 81.
17. Hattori, Y., Moriwaki, A., Hayashi, Y., and Hori, Y., Involvement of adenosine-sensitive cyclic AMP-generating systems in cobalt-induced epileptic activity in the rat, *J. Neurochem.*, 61, 2169, 1993.
18. Hoover, D. B., Craig, C.R., and Colasanti, B. K., Cholinergic involvement in cobalt-induced epilepsy in the rat, *Exp. Brain Res.*, 29, 501, 1977.
19. Hunt, W. A. and Craig, C. R., Alterations in cation concentration and Na-K ATPase activity during the development of cobalt-induced epilepsy in the rat, *J. Neurochem.*, 20, 559, 1973.
20. Goldberg, A. M., Pollock, J. J., Hartman, E. R., and Craig, C. R., Alterations in cholinergic enzymes during the development of cobalt-induced epilepsy in the rat, *Neuropharmacology*, 11, 253, 1972.
21. Cenedella, R. J. and Craig, C. R., Changes in cerebral cortical lipids in cobalt induced epilepsy, *J. Neurochem.*, 20, 743, 1973.
22. Craig, C. R. and Hartman, E. R., Concentration of amino acids in the brain of cobalt-epileptic rat, *Epilepsia*, 14, 409, 1973.
23. Ross, S. M. and Craig, C. R., Gamma-aminobutyric acid concentration, L-glutamate-1-decarboxylase activity, and properties of the gamma-aminobutyric acid postsynaptic receptor in cobalt epilepsy in the rat, *J. Neurosci.*, 1, 1388, 1981.
24. Levine, S., Payan, H., and Strebel, R., Metrazol thresholds in experimental allergic encephalomyelitis, *Proc. Soc. Exp. Biol.*, 113, 901, 1963.
25. Hartman, E. R., Colasanti, B. K., and Craig, C. R., Epileptogenic properties of cobalt and related metals after direct application to the cerebral cortex of the rat, *Epilepsia*, 15, 121, 1974.

26. Craig, C. R. and Colasanti, B. K., A study of pentylenetetrazol kindling in rats and mice, *Pharmacol. Biochem. Behav.,* 31, 867, 1989.
27. Moriwaki, A, Hattori, Y., Nishida, N., and Hori, Y., Electrocorticographic characterization of chronic iron-induced epilepsy in rat, *Neurosci. Lett.,* 110, 72, 1990.
28. Fariello, R. G., Parenteral penicillin in rats: an experimental model of multifocal epilepsy, *Epilepsia,* 17, 217, 1976.
29. Sullivan, H. C. and Osorio, I., Aggravation of penicillin-induced epilepsy in rats with locus cereuleus lesions, *Epilepsia,* 32, 591, 1991.
30. Collins, R. C., Metabolic response to focal penicillin seizures in rat: spike discharge vs. afterdischarge, *J. Neurochem.,* 27, 1473, 1976.
31. Esclapez, M. and Trottier, S., Changes in GABA-immunoreactive cell density during focal epilepsy induced by cobalt in the rat, *Exp. Brain Res.,* 76, 369, 1989.
32. Ribak, C. E., Epilepsy and the cortex anatomy, in *Cerebral Cortex,* Peters, A., Ed., Plenum Press, 1991, chap. 10.
33. Witte, O. W., Afterpotentials of penicillin-induced epileptiform neuronal discharges in the motor cortex of the rat in vivo, *Epilepsy Res.,* 18, 43, 1994.
34. Craig, C. R., Chiu, P., and Colasanti, B. K., Effects of diphenylhydantoin and trimethadione on seizure activity during cobalt experimental epilepsy in the rat, *Neuropharmacology,* 15, 485, 1976.
35. Emson, P. C., Effects of chronic treatment with amino-oxyacetic acid or sodium n-dipropylacetate on brain GABA levels and the development and regression of cobalt epileptic foci in rats, *J. Neurochem.,* 27, 1489, 1976.
36. Scuvee-Moreau, J., Lepot, M., Brotchi, J., Gerebtzoff, M. A., and Dresse, A., Action of phenytoin, ethosuximide and of the carbidopa-L-dopa association in semi-chronic cobalt-induced epilepsy in the rat, *Arch. Int. Pharmacodyn.,* 230, 92, 1977.
37. Colasanti, B. K. and Craig, C. R., Reduction of seizure frequency by clonazepam during cobalt experimental epilepsy, *Brain Res. Bull.,* 28, 329, 1992.
38. Craig, C. R. and Colasanti, B. K., Reduction of frequency of seizures by carbamazepine during cobalt experimental epilepsy in the rat, *Pharmacol. Biochem. Behav.,* 41, 813, 1992.
39. Krenz, N. R. and Cooper, R. M., A combined cobalt and ^{14}C 2-deoxyglucose approach to antiepileptic drug asssessment, *Int. J. Neurosci.,* 86, 55, 1996.

Chapter 8

Evaluation of Associated Behavioral and Cognitive Deficits in Anticonvulsant Drug Testing

Piotr Wláz and Wolfgang Löscher

Contents

0-8493-3362-8/98/$0.00+$.50

I. Introduction

Concerns about the adverse effects of antiepileptic drugs have always existed. By definition, the ideal antiepileptic drug would interrupt seizure activity without causing any unwanted effects. Unfortunately, neither requirements have been fulfilled. Seizure control can be achieved only in about 80% of epileptic patients,[1] whereas adverse effects are frequently reported in a wide spectrum ranging from mild central manifestations such as sedation to fatal hepatoxicity and aplastic anemia.[2,3] Therefore, there is a critical need for systematic evaluation at the preclinical level of not only the anticonvulsant effects but also of the side effects induced by novel drugs.[4] This chapter focuses on how to evaluate behavioral and cognitive side effects of antiepileptic drugs using relatively simple methods that can be utilized routinely in the search for new anticonvulsant agents. This chapter will also consider the evaluation of new drug combinations, since it is known that polytherapy of epilepsy is saddled with greater risk of undesirable side effects.[5]

Behavioral measures are very sensitive to even slight modification of the equipment used and are subjective in appearance. Thus, it is not surprising that data obtained with the same drug in different laboratories may differ dramatically. Apart from the technical side of drug testing, the human factor also introduces potential variability in the data. And therefore, before going into detailed descriptions of the techniques employed in our laboratory we will concisely review factors that should be kept in mind during anticonvulsant drug testing.

II. Methodology

A. Laboratory Conditions

The reliability and reproducibility of behavioral experiments in animals largely depends on many environmental conditions that frequently do not receive adequate attention. In the following we will briefly enumerate the most important factors that can significantly bias the outcome of behavioral studies.

1. Ambient temperature. Temperature inside the laboratory should essentially match that of the vivarium and be independent of external temperature fluxes. The most appropriate environmental temperature range for rats is 18 to 23°C and for mice 20 to 25°C. Relative humidity should be kept constant within the range of 50% to 70%. During transportation of animals from the home colony to the laboratory, draughts and sudden temperature and humidity changes should be avoided.

2. Sound and light. When possible, the laboratory should be acoustically isolated and illuminated by artificial light of the same intensity as in the vivarium during the light period. This is often not the case, so at least all windows and doors should be closed during experiments. This will also help to keep temperature, humidity, and sound conditions constant. Other sources of sound, such as radios and loud conversation, should be eliminated. To reduce seasonal variation in light intensity in a laboratory with windows, blinds and artificial light should be used. The vivarium should be

illuminated artificially so that the light-dark cycle is independent of the external light conditions. Usually a 12 h/12 h light/dark cycle is employed, although this might not be optimal, as discussed in Chapter 12.

3. Time of experiments. Because of marked circadian and diurnal influences on various behavioral measures in animals, which is obviously connected with the nocturnal activity of rodents, all experiments should be performed during the same time of day, usually between 8 AM and 12 PM. Experiments can be performed in the afternoon but the entire study must be done at that time of day. Otherwise, due to the introduction of an additional variable the results may not be comparable. When animals from commercial breeders are used, animals should be acclimated after arrival for at least 1 week before experimentation to compensate transportation stress-induced alterations in behavior.

4. Laboratory personnel. Persons who are involved in a given project and work with a given group of animals generally should not be replaced, as the animals become familiar with them. Before experiments have started animals should be handled on a few occasions. The number of handling sessions depend on the study purpose. For ordinary behavioral tests, like those used in antiepileptic drugs testing, four to five handling sessions are recommended. For more specialized studies where, for instance, subsequent neurotransmitter levels in the brain are measured, more occasions may be necessary. It has been shown that handled and nonhandled animals differ as regards brain biochemistry[6] and the variability of data from previously handled animals is expected to be lower. All persons working in the laboratory should always wear a unicolored laboratory coat.

5. Odors. Most animals have a much more sensitive sense of smell than humans. Accordingly, excessive use of cosmetics by persons conducting experiments is inappropriate. Blood also has a specific smell that produces clear excitation or anxiety behavior in animals. Therefore, manipulations such as blood sampling or decapitation of animals should be done in other rooms by other persons.

6. Bedding material. Replacement of bedding (washed sawdust) in animal cages should be scheduled on the basis of currently running investigations. Ideally, the animal cages should be covered with fresh bedding at least 1 to 2 d before the experimental session. Freshly changed bedding (devoid of specific and familiar animal smells) induces unpredictable behaviors and causes high variability in seizure threshold in mice and rats.[6a]

B. Motor Impairment

Acute toxicity from antiepileptic drugs in laboratory animals almost invariably is manifested by signs of neurological deficit, such as sedation, hypo- or (less often) hyperlocomotion, ataxia, abnormal gait, reduced or inhibited righting reflexes, muscle relaxation, and cognitive deficits.[4] These effects are commonly referred to as neurotoxicity which, however, is somewhat misleading because some of these pharmacological actions are utilized clinically. These include the sedative and hypnotic effects caused by benzodiazepines and barbiturates.[7] Neurological deficits associated with antiepileptic drug administration can be detected and quantified by using relatively simple methods. Motor impairment can be measured quantitatively by rotarod, chimney test, and inverted screen test. The data obtained in these tests may differ since, for instance, in the chimney test muscle strength and coordination are certainly of primary importance, whereas sense of balance and coordination appear to be crucial to perform in the rotarod test. Although, both tests give almost the same values for a particular drug, the striking exception to this rule is diazepam (see Tables 8.1 and 8.2).[8] In mice diazepam is more potent in the chimney test than in the rotarod test.[8] In contrast, rats are more susceptible to diazepam when tested in the rotarod test,[8] which possibly reflects differences in the muscle relaxant action of this antiepileptic drug in the two species. Very valuable information can be provided by palpation and by a direct observation of the animals in the open field. Cognitive deficits, as measured by the ability of animals to acquire and retrieve information, can be measured by passive avoidance procedures, which will be discussed later in this chapter.

Quantal data obtained in tests used to quantify motor impairment (rotarod, chimney, and inverted screen test) are then used to calculate median minimal "neurotoxic" dose (TD_{50}, i.e., the dose of an anticonvulsant drug which induces minimal neurological deficit in 50% of the animals). This measure in conjunction with the ED_{50} value obtained during anticonvulsant testing permits calculation of protective indices (see Table 8.3 and below). Both values are commonly obtained using the method of Litchfield and Wilcoxon.[9]

1.   Rotarod

This commonly used test for evaluation of "neurotoxicity" or neurological deficit in laboratory animals was originally described by Dunham and Miya[10] and is based on the assumption that an animal with normal motor efficiency (and not reduced muscular tone) is able to maintain its equilibrium on a rotating rod. This test is especially useful for rats and mice, and several more or less substantial modifications exist since many behavioral laboratories run this test using equipment built in-house. A clear disadvantage of this situation is lack of any standardization of this test. Basically, the rotarod consists of a metal rod coated with rubber or polypropylene foam to provide friction and to prevent animals from slipping off the rod. The rod is driven by a motor and the rotational speed can be regulated. The diameters of the rotating rods used in our laboratory are 5 and 2.5 cm, and the number of revolutions per minute is set at 8 and 6 rpm for rats and mice, respectively.[8] However, these parameters are by no means obligatory and each laboratory must adjust these values depending on the animal strain and weight. The distance between the drum and the floor of the test apparatus is approximately 30 cm for rats and 15 cm for mice. When it is too low, the incidence of intentional jumping off the rod will increase; when it is too high, a possibility of injury to the animal falling from the rod will occur. Usually, the rod is divided in several identical sections to allow the testing of more animals at a time. These sections should be separated from each other by opaque discs of sufficient diameter to prevent animals from climbing up the divisions during the test. The initial position of the animal on the rod seems to be important. When the degree of motor impairment is low to moderate, the animal will assume the most comfortable position, but when the impairment is marked, the animal can fall off the rod immediately after placement. More consistent results are obtained when the

TABLE 8.1

Anticonvulsant and "Neurotoxic" Potencies of Standard Anticonvulsants after Intraperitoneal Administration in Male Mice

Drug	Vehicle used	Time of tests (min)	Doses effective to increase seizure thresholds (mg/kg i.p.)				Anticonvulsant ED_{50}s (mg/kg i.p.)		Neurotoxic TD_{50}s (mg/kg i.p.)	
			MES threshold model		IV PTZ model		MES test (50 mA)	s.c. PTZ test (80 mg/kg)	Rotarod test	Chimney test
			TID_{20}	TID_{50}	TID_{20}	TID_{50}				
Phenobarbital (as sodium salt)	Saline	30	2.9	4.0	7.0	11.4	24 (21–28)	15.0 (12.3–18.4)	80 (75–86)	82 (76–88)
Carbamazepine	10% GF or 30% PEG 400	15	1.2	1.5	n.e.	n.e.	8 (6.7–9.6)	n.e.	33 (26–40)	34 (25–47)
Phenytoin	Saline (with NaOH)	120	4.9	5.4	n.e.	n.e.	11.5 (10.5–12.7)	n.e.	50 (44–57)	50 (41–62)
Valproate (as sodium salt)	Saline	5	50	69	85	113	320 (286–385)	160 (125–205)	430 (391–473)	385 (363–408)
Ethosuximide	Saline	30	n.e.	n.e.	89	148	n.e.	120 (99–149)	505 (459–556)	440 (405–478)
Diazepam	Saline (with HCl)	15	1.6	2.7	0.1	0.31	23 (21–26)	0.29 (0.21–0.41)	4.7 (3.8–5.7)	2.25 (1.56–3.24)
Clonazepam	PEG 400 (10–30%)	15	0.37	0.65	0.015	0.032	n.e.	0.031 (0.018–0.054)	1.2 (0.87–1.5)	0.48 (0.37–0.62)

Note: All tests were carried out at the time of maximum anticonvulsant activity. The following four seizure tests were used: (1) the threshold for maximal (tonic hindlimb extension) electroshock seizures (MES), (2) the threshold for the initial myoclonic twitch induced by IV infusion of pentylenetetrazol (PTZ), (3) the MES test with fixed, supramaximal current stimulation (50 mA), and (4) the s.c. PTZ seizure test with a fixed dose of PTZ (80 mg/kg) and generalized clonic seizures of at least 5 s in the 30 min following PTZ as an end point. In the electroshock models, electrical stimuli were applied via transauricular electrodes. In the threshold tests, the doses of anticonvulsants increasing the threshold by 20% or 50% ($TID_{20/50}$) were calculated from dose-response curves. In the other tests, ED_{50}s or TD_{50}s were calculated from dose-response curves and are given with confidence limits for 95% probability. Inactivity of a drug in a model or too weak efficacy for determination of ED_{50} is indicated by "n.e." (not effective). All drugs were administered as solutions to yield comparative drug adsorption. The vehicles used for preparation of drug solutions had no effects on seizure thresholds or other tests used. PEG 400, polyethylene glycol 400; GF, glycofurol. All doses of drugs refer to the free acid or base.

From Löscher, W. and Nolting, B., *Epilepsy Res.*, 9, 1, 1991. With permission.

TABLE 8.2
Anticonvulsant and "Neurotoxic" Potencies of Standard Anticonvulsants after Intraperitoneal Administration in Female Rats

Drug	Vehicle used	Time of tests (min)	Anticonvulsant ED$_{50}$s (mg/kg i.p.)		Neurotoxic TD$_{50}$s (mg/kg i.p.)	
			MES test (150 mA)	s.c. PTZ test (90 mg/kg)	Rotarod test	Chimney test
Phenobarbital	Saline	60	18	41	58	47
(as sodium salt)			(13–25)	(33–50)	(51–66)	(39–56)
Carbamazepine	PEG 400	30	6	n.e.	37	30
	30%		(4.9–7.4)		(31–34)	(24–37)
Phenytoin	Saline	30	13	n.e.	140	145
	(with NaOH)		(11–14)		(113–172)	(116–181)
Valproate	Saline	15	140	195	275	285
(as sodium salt)			(110–178)	(157–242)	(239–322)	(254–319)
Ethosuximide	Saline	30	n.e.	140	390	440
				(128–153)	(351–433)	(386–502)
Diazepam	Saline	15	15	1.8	2.8	4.8
	(with HCl)		(12–19)	(1.2–2.8)	(1.4–5.6)	(3.5–6.6)
Clonazepam	PEG 400	15	n.e.	0.082	1.6	1.1
	(10–30%)			(0.061–0.11)	(1.1–2.3)	(0.73–1.7)

Note: All tests were carried out at the time of maximum anticonvulsant activity. End points in the seizure tests were tonic hindlimb extension in the MES test with supramaximal stimulation (150 mA; transauricular application) and a generalized seizure of at least 5 s in the s.c. PTZ test with administration of 90 mg/kg (observation time 30 min). ED$_{50}$s and TD$_{50}$s were calculated from dose-response curves and are given with confidence limits for 95% probability. Inactivity of a drug in a model or too weak activity for determination of ED$_{50}$ are signified by "n.e." (not effective). All drugs were administered as solutions to yield comparative drug absorption. For clonazepam this limited the dosages that could be tested (higher doses could have been injected only as suspensions) so that indication of "n.e." for this drug does not exclude that the drug might have been active in the respective test at higher doses. The vehicles used for drug solutions had no effect on seizure thresholds or other tests used. PEG 400, polyethylene glycol 400. All doses of drugs refer to the free acid or base.

From Löscher, W. and Nolting, B., *Epilepsy Res.*, 9, 1, 1991. With permission.

rod rotates against or toward the animal rather than away from the animal. In every case the animals have to be placed gently on the rod in the same manner and in the same direction.

In contrast to naïve (untrained) mice, which are usually able to remain on the rotating rod for several minutes, naïve rats require at least two or three training sessions before the regular experiments are started.

At appropriate time points during an experiment, vehicle- or drug-treated animals are placed on the rod. Animals that are not able to maintain their equilibrium on the rod for 1 min are again put on the rod a further two times. Only animals that are unable to stay on the rod during three sequential 1-min trials are considered to exhibit neurological deficit. This procedure, however, can be performed in several ways. For instance, rotational speed and time of test can be changed. Also, the end

TABLE 8.3
Protective Indices Calculated for the Different Models Used for Anticonvulsant Drug Evaluation in Mice and Rats

Protective indices (PI)

	Mice								Rats			
	MEST model		IV PTZ threshold model		MES test		s.c. PTZ test		MES test		s.c. PTZ test	
Drug	PI (rotarod)	PI (chimney)	PI (rotarod)	PI (chimney)	PI (rotarod)	PI (chimney)	PI (rotarod)	PI (chimney)	PI (rotarod)	PI (chimney)	PI (rotarod)	PI (chimney)
Phenobarbital	20	21	11	12	3.3	3.4	5.3	5.4	3.2	2.6	1.4	1.1
Carbamazepine	22	23	n.e.	n.e.	4.1	4.3	n.e.	n.e.	6.1	5.0	n.e.	n.e.
Phenytoin	9.3	9.3	n.e.	n.e.	4.3	4.3	n.e.	n.e.	11	11	n.e.	n.e.
Valproate	6.2	5.6	5.1	4.5	1.3	1.2	2.7	2.4	1.9	2.0	1.4	1.5
Ethosuximide	n.e.	n.e.	5.7	4.9	n.e.	n.e.	4.2	3.7	n.e.	n.e.	2.8	3.1
Diazepam	1.7	0.8	47	23	0.2	0.1	16	7.8	0.2	0.3	1.6	0.4
Clonazepam	1.8	0.7	80	32	n.e.	n.e.	39	15.5	n.e.	n.e.	20	13

Note: Protective indices were calculated by dividing the TD_{50} determined in the rotarod or chimney test by effective doses of the respective drugs determined in the different seizure models (see Tables 8.1 and 8.2). For the MES threshold test (MEST model), the dose increasing the threshold for tonic hindlimb extension by 50% (TID_{50}) was used for calculations, whereas the IV PTZ threshold test, the dose (TID_{20}) increasing the threshold for the initial myoclonic twitch by 20% was taken. For the supramaximal MES tests and the s.c. PTZ test, anticonvulsant $ED_{50}s$ were used for calculation of PI. Insufficient anticonvulsant activity for PI calculation is indicated by "n.e." (not effective).

From Löscher, W. and Nolting, B., *Epilepsy Res.*, 9, 1, 1991. With permission.

point can be modified in that the time to fall can be measured or it can be measured when the rotational speed of the rod is constantly increasing, which may increase the sensitivity of the method of neurological deficit and produce less variable data.[11]

Modern rotarods (e.g., Basile Rota-Rod Treadmills; Stoelting, Wood Dale, Illinois) possess built-in timers that are all stopped by the animal itself when it falls down and presses the pad located underneath the rod. The speed of rotation can be programmed so that it may vary (usually accelerate) over time.

In addition to using the rotarod test for determination of "neurotoxic" or neurological deficit potency, it can also be used to determine the time course of adverse drug effects, simply by using the rotarod test in the same group of animals after various time points following drug administration.

2.   Chimney Test

In this test, first described by Boissier and colleagues,[12] the inability of an animal to climb backward up through a plastic or glass tube within a given period of time is an indication of neurologic impairment. The tube is usually made of transparent plastic and the dimensions depend on the tested animal species and their body weight. The dimensions for rats (200 to 300 g) and mice (25 to 30 g) are recommended to be 5.5 cm/50 cm and 3 cm/25 cm (inner diameter/length), respectively.[8] Time of test is arbitrary and should be chosen *a priori*. A range of 30 to 60 s is long enough to detect a neurologic deficit. A nonimpaired mouse or rat usually fulfills this criterion in 10 to 20 s. Animals do not require prior training for this test. It is not advisable to measure the time it takes the animal to reach the upper end of the tube, since results thus obtained are markedly variable. Quantal data (the animal either reached the end of the tube or it did not) are more appropriate.

The tube is placed horizontally on the bench top and a rat or mouse is directed so that it enters the tube (it might have to be lightly pushed in). When the animal reaches the opposite end of the tube, the observer immediately stands the tube vertically, so that the animal is in an upside-down position. This obviously unnatural position triggers an escape response and the animal starts to climb backwards up the tube. Because defecation and urination occur very frequently during this test and they both adversely affect the animal's traction (and thus may influence the results), the tube should be cleaned after each individual animal. It should be stressed that this ethological test involves complex and coordinated movements of the whole body rather than just the paws and thus the diameter of the tube seems to be important.

3.   Inverted Screen

In this simple test described by Coughenor and coworkers,[13] also known as the wire mesh test or horizontal screen test, a single untrained drug-treated animal (usually a mouse) is put upon the grid or screen (consisting of parallel steel rods 2 mm in diameter localized 1 cm apart; several groups just use commercial metal grid covers from rodent cages). The screen is then slowly rotated by 180°. The number of mice that either fall off the screen or are not able to climb to the top of the screen within a preset time period is a measure of motor impairment in this test. The duration of this test is arbitrary and varies between 5 s (when falling off the screen is used as an end point[14]) and 60 s (in which case the inability to climb to the top of the screen serves as an end point[15]). Apart from a quantal ("all-or-none") measure in this test, a scoring system can be used as well. Maxwell and coworkers[16] described the following grading system for the inverted screen test: 0, the mouse climbs to the top; 1, the mouse fails to reach the top but holds onto the screen; and 2, the mouse falls from the screen during 60 s of the test duration. This test has an obvious advantage over the two modifications described above since it combines both end points for which the data can be calculated independently. Based on the percentage of impaired mice, TD_{50} values and their confidence limits for 95% probability are calculated.[9]

4.   Open Field

In this test the animals (rats and mice) are evaluated for ataxia and changes in general behavior. The open field consists of a circular arena of 1 to 1.5 m diameter with walls high enough (30 cm) to prevent the animals from escaping. The open field floor and walls are usually painted black with matte paint and lit by dim and diffuse light. An animal is placed gently in the center of the open field, facing away from the observer, and observed for a period of 1 to 2 min.

Ataxia can be rated as follows: 1, slight ataxia in hind legs (tottering of hind quarters); 2, more pronounced ataxia with dragging of hind legs; 3, further increase of ataxia and more pronounced dragging of hind legs; 4, marked ataxia with only occasional loss of balance during forward locomotion; 5, very marked ataxia with frequent loss of balance during forward locomotion; and 6, permanent loss of righting reflexes but animal still attempts to move forward.[17] It should be noted that open field observation of locomotion may be more sensitive to detect motor impairment than more commonly used tests, such as the rotarod test. Thus, a drug-treated animal might pass the rotarod test as normal, but show moderate ataxia in the open field. An added advantage of open field observation is that behavioral abnormalities other than those resulting from motor impairment can be detected, too.

Sedation can be assessed separately from ataxia according to a four-point system: 1, slightly reduced forward locomotion; 2, reduced locomotion with rest periods in between (partly with closed eyes); 3, reduced locomotion with more frequent rest periods; and 4, no forward locomotion, animal sits quietly with closed eyes.[17]

Ataxia is often considered a consequence of central drug action. However, it should be realized that a disturbance of muscle coordination may result from a drug action upon muscle tone. The reduced muscle tone can be easily measured by the resistance to fingers pressed gently into the abdomen. Even though this examination is very subjective, when performed by an experienced observer it is a reliable and sensitive tool to differentiate between drugs that cause ataxia by muscle relaxation from those in which ataxiogenic action resides centrally. Therefore, when removing animals from the open field, abdominal muscle tone can be evaluated by palpation and rated according to the scoring system described by Löscher and Hönack: 0, normal muscle tone; 1, equivocal muscle relaxation; 2, unequivocally reduced muscle tone; and 3, markedly reduced muscle tone.[18]

C. Alterations in General Behavior

In addition to motor impairment and sedation, observation of drug-treated animals can disclose complex alterations in general behavior, which may be overlooked if only simple tests such as rotarod, chimney, or inverted screen are used for detection of "neurotoxicity." Detection of altered behavior by visual observation requires a trained observer. Drug-induced behavioral alterations should be separated from behavioral alterations induced by mere handling (including injection) and new environment (e.g., laboratory, open field). Thus, there are two important prerequisites for such observations: (1) animals should be adapted to both the testing environment and handling by training sessions prior to the drug trial (see above), and (2) age-matched vehicle-controls should be used together with the drug-treated groups. The observer should not be aware of which groups are drug-treated and which are vehicle-treated, i.e., the observation should be performed in a blinded fashion.

In our laboratory, we use the following protocol. For examination of behavioral drug effects, the animals are removed from their home cages and placed alone in plastic cages (590 × 380 × 190 mm high) without sawdust flooring or grid covers, the cages being placed on a table. The animals are continuously observed for alterations in behavior for up to 3 h after intraperitoneal injection of vehicle or drug. For comparative evaluation of experiments with different drugs, behavioral alterations determined at suitable time points (depending on time of peak effect and duration of action) after drug administration can be scored (see below). For all observations, rigorous observational protocols are used, using a ranked intensity scale for each altered behavior: 0, absent; 1, equivocal; 2, present; and 3, intense. Examples for drug-induced behavioral alterations scored in this respect include hyperlocomotion, stereotyped behaviors, such as head weaving (swaying movements of the head and upper torso from side to side for at least one complete cycle; i.e., left-right-left), stereotyped sniffing, biting, licking or grooming, reciprocal forepaw treading ("piano playing"), stereotyped rearing, hyperexcitability (as indicated by increased reactions to noise or handling), tremor, "wet dog" shakes, abduction of hind limbs, reduction of righting reflexes, flat body posture, circling, Straub tail (i.e., raised tail as a sign of central excitation), and piloerection. Ataxia and sedation are scored by a modified intensity scale as described above, placing the animals briefly in an open field (in which it is easier and less variable to score locomotion than in the cage). Behavioral alterations other than those described above are separately recorded.

We use group sizes of five animals per drug and vehicle. A maximum of 10 animals can be continuously observed by an experienced investigator. In the case of behavioral alterations, experiments are repeated once and, if both experiments give similar results, the data are pooled for statistical evaluation.

It is mandatory to observe the animals not only at one fixed time point (e.g., 30 or 60 min) after drug injection, but continuously, starting immediately after the injection. Otherwise, adverse behavioral effects of short duration may be overlooked. Furthermore, since animals with chronic brain dysfunction, such as that resulting from epileptogenesis, may exhibit altered responsiveness to adverse behavioral effects of drug, these tests should include groups of epileptic (e.g., kindled rats) as well as normal (healthy) laboratory rodents (see below).

In addition to observation of animals in new (clean) plastic cages and open field, the animals should also be observed in their home cage. These studies should be performed in separate experiments when a particularly interesting or novel drug is found. This may be particularly valuable in the case of drug-induced sedation, which may be observable in the normal environment of the animal, i.e. in the home cage in the vivarium, but not in the new environment (e.g., open field, empty new cage), because the stress of the observation procedure counteracts the sedative drug effect. Automatic motility and activity meters also may be helpful in this respect, because they allow measurements in the absence of the experimenter e.g., during the dark phase. However, even if sophisticated systems for measurement of behavioral alterations are used, such automated systems cannot replace observation by an experienced investigator.

D. Body Temperature

Body temperature should be carefully monitored after both vehicle and test drug application, because changes in body temperature influence seizure susceptibility.[19] Following transportation of animals from the vivarium to the laboratory, the animals should be allowed some time (e.g., 0.5 h) for acclimatization. This is very important since body temperature without this acclimatization period usually is higher by at least 0.5°C due to the stress of the move. Thus, measurement of body temperature before and after drug injection without acclimatization results in a false positive (hypothermic) drug effect, because the predrug value is erroneously high.

Body temperature is commonly measured in mice and rats by using electronic (thermocouple) thermometers (e.g., as provided by Technical & Scientific Equipment Ltd., Bad Homburg, Germany, or by Columbus Instruments, Columbus, Ohio). The probe (sensor) of the thermometer should be lubricated by immersion in paraffin oil or water and inserted in the rectum to a depth of 1 to 2 cm in mice and 2 to 3 cm in rats. Between the measurements the probe is immersed in oil or water which is maintained at the approximate body temperature in order to avoid inertia of the thermocouple and thus to facilitate speed of measurement. The thermoelectric thermometers work reasonably fast so that 20 to 30 s usually is a sufficient period of time to obtain a correct temperature reading.

To allow conclusions concerning whether the change in body temperature has any influence on the anticonvulsant action of a drug, body temperature should be measured at least three times during both vehicle and drug application: 1, before vehicle (and drug) administration (after acclimatization, see above); 2, in the middle of the pretreatment time; and 3, at the point where anticonvulsant activity is actually tested. For practical reasons, this last measurement is taken just before the appropriate behavioral and/or convulsive test. Because temperature will show some time-dependent fluctuations due to handling-induced stress, the data obtained from vehicle and drug sessions should be compared using two-way repeated measures analysis of variance (ANOVA) with appropriate *post hoc* test to determine if the drug apart from other factors had any significant influence upon body temperature.

The physiological body temperature of laboratory animals depends on numerous internal as well as external factors, such as age, sex, strain, stressful events, ambient temperature, and number of animals per cage, so it is mandatory to use vehicle controls in each drug trial.

Particularly in the case of a hypothermic drug effect, it may be advisable to repeat the experiment under conditions that prohibit the fall in body temperature (e.g., at elevated ambient temperature) to ensure that an anticonvulsant drug effect was not secondary to hypothermia.

E. Tests for Memory Function

It is not universally clear how disruptive anticonvulsant therapy is to the finer intellectual processes of the epileptic patient.[20] This issue is very complicated to study in human subjects because of difficulties in forming homogeneous groups of seizure type, antiepileptic drug used, duration of treatment, social issues, and age of patients.[21] In addition, the results are always difficult to interpret since epilepsy, as a chronic brain dysfunction, has obviously negative impact upon the cognitive capacity of a patient.[22] Furthermore, by ameliorating the memory-disrupting action of epilepsy, antiepileptic drugs may actually be perceived as beneficial to the patient's intellectual efficiency, thus masking their own impairing effects. Nevertheless, many studies have provided some evidence that antiepileptic drugs do impair cognitive functions.[23]

It is thus surprising that the cognitive deficit induced by antiepileptic drugs is only rarely studied in detail. Such laboratory tests like the one-trial passive avoidance procedures and maze tests (as exemplified by a Y-maze test) have shown that doses of anticonvulsant drugs that produce effects and/or plasma concentrations comparable with those seen in clinical conditions in patients may produce impairments of short-term and long-term memory in rodents.[24-27] For this reason, simple tests that are able to detect adverse cognitive effects of antiepileptic drugs need to be included in the battery of routinely performed preclinical tests in rodents. The passive avoidance task is believed to offer information pertaining to long-term memory[28] and spontaneous alternation in the Y-shaped maze can be regarded as a measure involving spatial working memory.[29]

1. Passive Avoidance

Similar to the rotarod, most passive avoidance apparatuses are built in-house because of their generally uncomplicated construction, but there are also commercial products (e.g., the passive avoidance test system provided by Technical & Scientific Equipment Ltd., Bad Homburg, Germany, or by Columbus Instruments, Columbus, Ohio). An apparatus consists of two compartments ($10 \times 15 \times 15$ cm each for mice and $30 \times 30 \times 30$ cm each for rats) separated by a guillotine door of appropriate size. The whole apparatus is made of nontransparent plastic or wood. One compartment is painted black and covered with a lid (to prevent animals from jumping out upon stimulation) while the other one is painted white and brightly illuminated. The light should be emitted by a fluorescent bulb, since a regular bulb produces excessive heat. The floor of the white compartment is made of the same material as the walls and is also painted white, while the floor of the dark compartment is made of stainless steel rods. The 3 mm diameter rods are separated by about 0.5 cm for mice and 1.5 cm for rats and are connected to a shock generator.

During the acquisition trial an animal is placed into the white, illuminated box facing the wall opposite to the door and then, after a certain time interval (e.g., 10 s) the door is opened. Once the animal enters the dark compartment with all four paws, the door is closed and a footshock is delivered. Usually animals enter the dark compartment within 10 to 30 s. Animals that enter the dark box with a considerably longer latency should be excluded from further testing. Because a footshock is used as a reinforcer for this task, the shock parameters are critical. The footshock duration is usually 1 to 5 s and the current intensities vary between 0.1 and 0.8 mA for mice and 0.4 and 1.0 mA for rats. The stimulus duration is relatively easy to determine, while the current intensity is not. The footshock strength depends on the sex and strain of the animals, as well as on the cleanliness of the grid floor (feces block and urine facilitates current passage). Therefore, several preliminary trials have to be performed using some reference drugs that are known to produce memory impairment, like the anticholinergic drug scopolamine, in order to find the current intensity best suited for the purpose. Current intensities that are too high will likely result in no influence of test substances on memory ("ceiling" effect). With weak current intensities even low doses of antiepileptic drugs will produce memory impairment ("floor" effect). In either case it will not be possible to construct a dose-response curve. The time that each animal spends in the illuminated box before entering the dark box is measured using a stopwatch. Twenty-four hours later (retention or retrieval trial), each animal is placed again in the illuminated box in the same way as during the acquisition trial and the latency to enter the dark box is noted. The time the animal is allowed to stay in the illuminated box without entering the dark box or the cut-off time is arbitrary and usually is set at 120 to 300 s. The measure of long-term memory in this test is the time spent by an animal in the illuminated box during the acquisition trial subtracted from that spent during the retention trial (the greater the difference, the better the task was remembered). Usually, the data obtained do not conform to the requirements imposed by the theory for parametric data because of high variability (e.g., lack of normal distribution) and an *a priori* setting of cut-off time. Accordingly, nonparametric statistics should be used. Also, the aforementioned high dispersion of experimental data forces the researcher to use larger groups of animals (10 to 15 per group).

One of the very early questions an investigator is faced with during the planning of a passive avoidance experiment is whether to administer the drug before or after footshock exposure. It appears most reasonable that the animals should be dosed after the shock trial since the drug itself can change the experimental outcome of the acquisition trial due to indirect effects, through modulation of perception, emotion, and motivation. For instance, when a drug produces analgesic or even just clear sedative effects, the pain threshold will increase, thus rendering a high probability of drawing false conclusions. To eliminate this draw-back, drugs can be administered after the shock application. However, one should keep in mind that mnemonic consolidation of the shock event takes place over a relatively short period of time (single hours) after the shock trial. Consequently, drugs that affect memory formation processes should be active only if given within this short period of time. Conversely, substances that adversely affect information retrieval will only be active when administered shortly (e.g., 0.5 to 1 h) before the retention (retrieval) trial. Ghelardini and coworkers[30] recently described a modification of the step-through passive avoidance procedure that could be used without concerns about changes in drug-induced pain perception. In their version of the test, the electrifiable grid floor is substituted by a pitfall floor. Thus, an animal entering the dark compartment receives a nonpainful punishment consisting of fall into a cold-water bath (10°C).[30]

A second passive avoidance procedure is the so called step-down passive avoidance. In this variation, a small platform is located in the center of the electrifiable grid floor. The platform is made of an insulating and easy-to-clean material (e.g., plastic or wood). The dimensions of this platform are about 12 × 12 × 5 cm for rats and 4 × 4 × 2 cm for mice. Similarly to the step-through passive avoidance, the latency to step down from the pedestal is measured on the first day (acquisition trial) and on the second day (retention or retrieval trial) when the animals are punished upon stepping down on the grid. Parameters of the stimulus are basically similar to those utilized in the step-through passive avoidance. How-ever, the grid floor may need to be enlarged. Interestingly, although these two types of passive avoidance would seem to give similar results, this may not be the case. For instance, while competitive and noncompetitive N-methyl-D-aspartate (NMDA) antagonists impaired acquisition of the step-through passive avoidance paradigm, they actually improved retention performance in the step-down passive avoidance situation when the same doses were used.[31] Because step-through pas-sive avoidance is used much more frequently and thus could be considered a standard method, results obtained by using step-down passive avoidance should be interpreted cautiously.

2. Spontaneous Alternation in a Y-Maze Test

A modified Y-maze test suitable for memory testing in mice consists of three identical compartments 10 × 10 × 10 cm each.[24] The maze has no ceiling or floor and is simply put on a sheet of paper which is replaced after an animal has been tested to avoid gustatory bias in the responsiveness of the next animal. Mice are placed individually in the maze for 8 min. In this test mice explore the maze systematically, entering each arm sequentially.[32] The total number of arm entries (a measure of locomotor activity) and alternation behavior (consecutive entries into all three arms with no repetitions) are noted. Test compounds may alter locomotor activity and reduce the total number of arm entries. Alternation behavior in this test is expressed as a percentage of the total arm entries.

F. Influences of Epileptogenesis on Drug Adverse Effects

It is not known with certainty whether epileptogenesis alters a patient's responsiveness to antiepileptic agents. The discouraging experience with the competitive NMDA antagonist SDZ EAA-494 (d-CPP-ene)[33] has brought our attention to the fact that drugs that are well tolerated by normal (healthy) volunteers can be unacceptably neurotoxic to epileptic patients. Some parallels can be drawn between laboratory and clinical evaluation of side effects of antiepileptic drugs. At both levels of drug development, toxicity and tolerance trials are performed in normal laboratory animals and healthy volunteers, respectively. At the laboratory level, however, it would be more appropriate (and possible!) to use epileptic animals, ideally with the type of seizures against which the drug is expected to be effective. Indeed, it has been shown that amygdala-kindled rats are more prone to show distinct neuropsychological abnormalities in response to administration of NMDA antagonists than age-matched nonkindled rats.[18] Thus, the selective population toxicity could have been correctly predicted in the laboratory,[18,34] and then confirmed during the subsequent clinical trial,[33] that epileptogenesis, as mimicked by the kindling process, can lower the threshold for precipitation of some drug-induced neurological side effects.[34]

Further, it has been shown that some of the new prototype anticonvulsant drugs may also induce more pronounced adverse effects in kindled than in normal rats.[17] This suggests that such studies may be more predictive of potential neurotoxicity. Another typical example of differences induced by epileptogenesis is that HA-966, a low-efficacy partial agonist at the glycine-insensitive site at the NMDA receptor (glycine$_B$ receptor), was devoid of significant electroencephalographic activity in normal rats,[35] whereas kindled rats injected with this substance showed pronounced paroxysmal EEG activation,[36] pointing to functional differences between normal and epileptic brains.

In related recent experiments we have demonstrated that neurotoxicity from NMDA antagonists in stroke patients could also be predicted using simple observational procedures in animals with middle cerebral artery (MCA) occlusion.[45] MCA occluded rats showed more adverse effects (hyperlocomotion, stereotyped behavior) than age-matched, sham-lesioned animals, again demonstrating that chronic brain dysfunction may increase drug adverse effect potential and that side effects and tolerance studies on novel drugs should also be conducted in relevant animal models.

G. Evaluation of Drug Combinations

The clinical efficacy of newly developed antiepileptic drugs is usually evaluated in add-on trials, i.e. by adding the new drug to the existing medication with standard antiepileptic drugs in patients with chronic epilepsy. The combination of two or more drugs often produces therapeutic or toxic effects which may be quite different from what would be expected from the known pharmacological actions of the single compounds involved.[37] Thus, during preclinical evaluation, a novel antiepileptic drug should not only be tested alone but also in combination with standard antiepileptic drugs in order to study drug interactions in terms of anticonvulsant activity and adverse effects.[37] For this purpose, the same test procedures described above for single drug experiments can be used.[37] However, in view of the fact that for preclinical evaluation of drug combinations each drug of a two-drug combination has to be tested at different dose levels, the necessary number of animal experiments is so high that polypharmacy testing should start with simple models for neurotoxicity testing, such as the rotarod test.[37] We have recently described a suitable test strategy for evaluation of antiepileptic drug combinations.[37]

III. Interpretation

A. Role of Behavioral and Cognitive Deficit Testing

Although behavioral pharmacology often works with relatively simple methods, including visual observation of animals' behaviors, it is irreplaceable in antiepileptic drug development. As shown in this chapter, behavioral pharmacology is much more complex than generally thought, and necessitates carefully planned experimental protocols and experience which cannot be replaced by automatic measurement systems. For instance, the failure of NMDA antagonists in the clinic could have been foreseen by simple observation of behavioral alterations in suitable animal models, such as kindling, which is a model of difficult-to-treat partial epilepsy for which these drugs were developed and clinically evaluated. Of course, it is not possible to use all of the tests described in this chapter during early phases of drug development, or as part of a screening battery. However, depending on the characteristics of a novel drug, they could be subsequently used during the more advanced phases of preclinical drug development, allowing more sophisticated methods of behavioral pharmacology to be used, some of which were out of the scope of this review.[38] Table 8.4 shows an example of a battery of seizure models and models for detection of adverse effects as used in our laboratory. Examples of data from acute drug administration obtained with the test hierarchy summarized in Table 8.4 are illustrated in Tables 8.1 to 8.3. As shown in Table 8.4, we propose to use both models with fixed (suprathreshold) seizure stimuli, e.g., the traditional maximal electroshock (MES) and subcutaneous pentylenetetrazol (s.c. PTZ) tests, and models in which the individual seizure threshold is determined after drug administration.[4] It has been shown that drugs that are inactive in the traditional models may be effective in the threshold models, thus reducing the risk that a potentially interesting new anticonvulsant is missed.[4] Examples in this regard are clinically effective drugs such as vigabatrin and levetiracetam. A further advantage of the threshold models is that calculation of protective indices on the basis of anticonvulsant activity in such models may more reliably predict

TABLE 8.4
A Test Hierarchy Proposed for Evaluation of Antiepileptic Drugs

(1) Models of primary generalized seizures:

   (a) Maximal electroshock seizure (MES) test with tonic hindlimb seizures induced by stimulation via corneal and/or transauricular stimulation in mice (50 mA) and rats (150 mA)

   (b) s.c. pentylenetetrazol (PTZ) seizure test with clonic convulsions in mice (80 mg/kg) and rats (90 mg/kg); doses are CD_{97} for seizure induction, which may vary among strains

   (c) Threshold for tonic seizures induced by electrical stimulation in mice and rats via corneal and/or transauricular electrodes

   (d) Thresholds for myoclonic, clonic, and tonic seizures induced by IV infusion of PTZ in mice

(2) Models of partial seizures with secondary generalization:

   (a) Threshold for induction of afterdischarges (ADT) induced by electrical stimulation of the amygdala in fully amygdala-kindled rats; recording of seizure severity (focal and secondarily generalized seizures), seizure duration, and afterdischarge duration at threshold current

   (b) Suprathreshold stimulation of amygdala-kindled rats; recording of seizure severity (focal and secondarily generalized seizures), seizure duration, and afterdischarge duration at suprathreshold current (e.g., 500 μA)

(3) Models for detection of motor impairment and other adverse effects:

   (a) Rotarod and chimney test in mice and rats, including kindled rats

   (b) Open field behavior in mice and rats, including kindled rats

(4) Models for chronic efficacy testing:

   Chronic drug experiments in mice (MES, PTZ) and fully amygdala-kindled rats, including studies on tolerance and dependence liability

(5) Models for detection of antiepileptogenic effects:

   Chronic drug administration during kindling development

(6) Further (more specialized) models if test drug looks promising:
   e.g., models for drug-resistant seizure types, genetic animal models of epilepsy, seizure models in higher mammals (e.g., PTZ-induced seizures in dogs), models for detection of antipsychotic drug effects, tests for memory function

Based on data by Löscher and Schmidt.[4]

the dose ratios between adverse and beneficial drug effects in patients than indices based on models such as the MES test (see below).

As shown in Table 8.4, not only single drug administration but also repeated drug administration is used in this test hierarchy. This is because behavioral and cognitive deficits may change during chronic drug administration, e.g., by development of tolerance. Chronic drug studies are also needed to examine whether drug dependence develops. Experimental methods for the study of tolerance and dependence are beyond the topic of this chapter, but the interested reader is referred to some recent reports illustrating that a battery of chronic experiments is needed in this respect.[39-41]

B. Protective Indices

The protective index (PI) is a numerically expressed measure of the relative safety of a drug, i.e., separation between the anticonvulsant (desired) and neurotoxic (undesired) effects of a given antiepileptic drug.[8] This value is calculated according to the following equation:

$$PI = TD_{50}/ED_{50}$$

where TD_{50} is a dose that produces neurotoxicity or neurological deficit in a given 50% of animals, and ED_{50} is a dose that protects 50% of animals from occurrence of a given convulsive end point. The higher the PI value, the less relative toxicity of the drug. Both values used for PI calculation are commonly obtained from the quantal dose-response relationship of the respective effects by the method of Litchfield and Wilcoxon.[9]

   TD_{50} is usually determined by using the tests already described for neurological impairment, such as the chimney, rotarod, or inverted screen test. Other undesirable effects (for example, memory impairments) can also serve to calculate protective indices. Anticonvulsive tests that are usually used to determine ED_{50} include MES- and PTZ-induced seizures.[8] However, again by no means should the repertoire of convulsive tests be restricted to these two. Any other test can be used. When tests with fixed (maximal or supramaximal) seizure stimuli (MES and PTZ) are used, the ED_{50} of drugs is used to calculate PI.[8] For threshold tests such as MES threshold or intravenous (IV) PTZ threshold tests a dose that increases the threshold (threshold increasing dose; TID) by 50% or 20% (TID_{50} and TID_{20}, respectively) can be used as determined by log-linear regression analysis from dose-effect experiments.[8] For the MES threshold model the TID_{50} is used for calculation of PIs, whereas the TID_{20} is used in the case of the IV PTZ seizure threshold model, because TID_{20}s determined in this model are more predictive of therapeutic plasma levels in humans than TID_{50}s.[8]

   Examples of PIs determined for mice and rats using rotarod and chimney tests to quantify motor impairment and electrically and PTZ-induced seizures are shown in Table 8.3. Apparently, despite high standardization of relatively uncomplicated experimental procedures, the PI for a given drug and species may vary quite markedly. The reasons for this are innumerable. Of greatest importance are the route of administration of a drug (oral vs. parenteral), its formulation (solution vs. suspension),[42] pretreatment time (peak neurotoxic effect does not necessarily parallel peak anticonvulsant effect),[43] and modifications of the convulsive and behavioral tests. Often marked differences in the results obtained in various laboratories illustrate the importance of even small variation in the procedures (e.g., Löscher and Nolting[8] vs. Stagnitto et al.[15]).

   In the initial description of the Antiepileptic Drug Development (ADD) Program of the Epilepsy Branch of the National Institutes of Health (NIH; Bethesda, Maryland) it was proposed that only drugs displaying PIs of at least 5 can proceed from initial screening to further stages of anticonvulsant evaluation.[44] Drugs that have a PI less than or equal to 1 (i.e., $ED_{50} \geq TD_{50}$) are considered as nonselective and thus should not be subjected to a more specialized evaluation. Because some interesting drugs may be excluded from further tests for not having a favorable toxicity profile (for discussion see Löscher and Nolting[8]) a PI of 2 has been proposed as less prohibitive and thus less likely to discard a potentially interesting drug.[8] Furthermore, as shown in Table 8.3, PIs calculated anticonvulsant activity on seizure thresholds may more reliably characterize the true therapeutic ratio of a drug than PIs calculated on the basis of anticonvulsant activity in a test with fixed seizure stimuli, such as the traditional MES and PTZ tests.

Research using behavioral techniques has continued to evolve over the about four decades since the emergence of neuropsychopharmacology as a scientific discipline.[38] As the field of neuropsychopharmacology continues its seemingly inevitable progression towards more molecular analyses, it will be of continued importance to maintain the experimental and conceptual rigor that has characterized the study of behavior as it has developed within the broader context of this field.[38] Although the actions of anticonvulsant drugs can be studied at many different levels, it is inevitable that a thorough analysis of risk-benefit ratio will eventually address issues of a behavioral nature. As shown in this chapter, use of a simple test such as the rotarod test alone is certainly not sufficient in this regard. Furthermore, in order to avoid underestimation of a drug's potency to induce behavioral and cognitive deficits, animals with chronic brain dysfunction as induced by epileptogenesis should be involved in the evaluation of associated behavioral and cognitive deficits in anticonvulsant drug testing.

References

1. Schmidt, D. and Morselli, P., Eds., *Intractable Epilepsy: Experimental and Clinical Aspects,* Raven Press, New York, 1986.
2. Dreifuss, F. E. and Langer, D. H., Hepatic considerations in the use of antiepileptic drugs, *Epilepsia,* 28 (Suppl. 2), S23, 1987.
3. Pennell, P. B., Ogaily, M. S., and Macdonald, R. L., Aplastic anemia in a patient receiving felbamate for complex partial seizures, *Neurology,* 45, 456, 1995.
4. Löscher, W. and Schmidt, D., Which animal models should be used in the search for new antiepileptic drugs? A proposal based on experimental and clinical considerations, *Epilepsy Res.,* 2, 145, 1988.
5. Reynolds, E. H., Polytherapy, monotherapy, and carbamazepine, *Epilepsia,* 28 (Suppl. 3), S77, 1987.
6. Warenycia, M. W., Kombian, S. B., and Reiffenstein, R. J., Stress-induced increases in brainstem amino acid levels are prevented by chronic sodium hydrosulfide treatment, *Neurotoxicology,* 11, 93, 1990.
6a. Wláz, P., unpublished observations.
7. Schallek, W. and Schlosser, W., Neuropharmacology of sedatives and anxiolytics, *Mod. Probl. Pharmacopsychiatry,* 14, 157, 1979.

8. Löscher, W. and Nolting, B., The role of technical, biological and pharmacological factors in the laboratory evaluation of anticonvulsant drugs. IV. Protective indices, *Epilepsy Res.,* 9, 1, 1991.

9. Litchfield, J. T., Jr. and Wilcoxon, F., A simplified method of evaluating dose-response experiments, *J. Pharmacol. Exp. Ther.,* 86, 99, 1949.

10. Dunham, N. W. and Miya, T. S., A note on a simple apparatus for detecting neurological deficit in mice and rats, *J. Am. Pharm. Assoc.,* 46, 208, 1957.

11. Sanger, D. J., Morel, E., and Perrault, G., Comparison of the pharmacological profiles of the hypnotic drugs, zaleplon and zolpidem, *Eur. J. Pharmacol.,* 313, 35, 1996.

12. Boissier, J.-R., Tardy, J., and Diverres, J.-C., Une nouvelle methode simple pour explorer l'action 'tranquillisante': le test de la cheminee, *Med. Exp.,* 3, 81, 1960.

13. Coughenor, L. L., McLean, J. R., and Parker, R. B., A new device for the rapid measurement of impaired motor function in mice, *Pharmacol. Biochem. Behav.,* 6, 351, 1977.

14. Yamaguchi, S. and Rogawski, M. A., Effects of anticonvulsant drugs on 4-aminopyridine-induced seizures in mice, *Epilepsy Res.,* 11, 9, 1992.

15. Stagnitto, M. L., Palmer, G. C., Ordy, J. M., Griffith, R. C., Napier, J. J., Becker, C. N., Gentile, R. J., Garske, G. E., Frankenheim, J. M., Woodhead, J. H., White, H. W., and Swinyard, E. A., Preclinical profile of remacemide: a novel anticonvulsant effective against maximal electroshock seizures in mice, *Epilepsy Res.,* 7, 11, 1990.

16. Maxwell, D. M., Brecht, K. M., Doctor, B. P., and Wolfe, A. D., Comparison of antidote protection against soman by pyridostigmine, HI-6 and acetylcholinesterase, *J. Pharmacol. Exp. Ther.,* 264, 1085, 1993.

17. Hönack, D. and Löscher, W., Kindling increases the sensitivity of rats to adverse effects of certain antiepileptic drugs, *Epilepsia,* 36, 763, 1995.

18. Löscher, W. and Hönack, D., Anticonvulsant and behavioral effects of two novel competitive N-methyl-D-aspartic acid receptor antagonists, CGP 37849 and CGP 39551, in the kindling model of epilepsy. Comparison with MK-801 and carbamazepine, *J. Pharmacol. Exp. Ther.,* 256, 432, 1991.

19. Bowker, H. M. and Chapman, A. G., Adenosine analogues. The temperature-dependence of the anticonvulsant effect and inhibition of 3H-D-aspartate release, *Biochem. Pharmacol.,* 35, 2949, 1986.

20. Dodrill, C. B., Problems in the assessment of cognitive effects of antiepileptic drugs, *Epilepsia,* 33 (Suppl. 6), S29, 1992.

21. Vining, E. P., Cognitive dysfunction associated with antiepileptic drug therapy, *Epilepsia,* 28 (Suppl. 2), S18, 1987.

22. Aldenkamp, A. P., Alpherts, W. C., Dekker, M. J., and Overweg, J., Neuropsychological aspects of learning disabilities in epilepsy, *Epilepsia,* 31 (Suppl. 4), S9, 1990.

23. Devinsky, O., Cognitive and behavioral effects of antiepileptic drugs, *Epilepsia,* 36 (Suppl. 2), S46, 1995.

24. Parada Turska, J. and Turski, W. A., Excitatory amino acid antagonists and memory: effect of drugs acting at N-methyl-D-aspartate receptors in learning and memory tasks, *Neuropharmacology,* 29, 1111, 1990.

25. Pietrasiewicz, T., Czechowska, G., Dziki, M., Turski, W. A., Kleinrok, Z., and Czuczwar, S. J., Competitive NMDA receptor antagonists enhance the antielectroshock activity of various antiepileptics, *Eur. J. Pharmacol.,* 250, 1, 1993.

26. Zarnowski, T., Kleinrok, Z., Turski, W. A., and Czuczwar, S. J., The NMDA antagonist procyclidine, but not ifenprodil, enhances the protective efficacy of common antiepileptics against maximal electroshock-induced seizures in mice, *J. Neural Trans. Gen. Sect.,* 97, 1, 1994.

27. Wláz, P., Rolinski, Z., and Czuczwar, S. J., Influence of D-cycloserine on the anticonvulsant activity of phenytoin and carbamazepine against electroconvulsions in mice, *Epilepsia,* 37, 610, 1996.

28. Venault, P., Chapouthier, G., de Carvalho, L. P., Simiand, J., Morre, M., Dodd, R. H., and Rossier, J., Benzodiazepine impairs and beta-carboline enhances performance in learning and memory tasks, *Nature,* 321, 864, 1986.

29. Sarter, M., Bodewitz, G., and Stephens, D. N., Attenuation of scopolamine-induced impairment of spontaneous alteration behaviour by antagonist but not inverse agonist and agonist beta-carbolines, *Psychopharmacology,* 94, 491, 1988.

30. Ghelardini, C., Gualtieri, F., Romanelli, M. N., Angeli, P., Pepeu, G., Giovannini, M. G., Casamenti, F., Malmbergaiello, P., Giotti, A., and Bartolini, A., Stereoselective increase in cholinergic transmission by r-(+)-hyoscyamine, *Neuropharmacology,* 36, 281, 1997.

31. Mondadori, C., Weiskrantz, L., Douglas, R. S., and Isaacson, R. L., NMDA receptor blockers facilitate and impair learning via different mechanisms. Homogeneity of single trial response tendencies and spontaneous alternation in the T-maze, *Behav. Neural Biol.,* 16, 87, 1965.

32. Douglas, R. S. and Isaacson, R. L., Homogeneity of single trial response tendencies and spontaneous alternation in the T-maze, *Psych. Rep.,* 16, 87, 1965.

33. Sveinbjornsdottir, S., Sander, J. W., Upton, D., Thompson, P. J., Patsalos, P. N., Hirt, D., Emre, M., Lowe, D., and Duncan, J. S., The excitatory amino acid antagonist D-CPP-ene (SDZ EAA-494) in patients with epilepsy, *Epilepsy Res.,* 16, 165, 1993.

34. Löscher, W. and Hönack, D., Responses to NMDA receptor antagonists altered by epileptogenesis, *Trends Pharmacol. Sci.,* 12, 52, 1991.

35. Tortella, R. C. and Hill, R. G., EEG seizure activity and behavioral neurotoxicity produced by (+)-MK801, but not the glycine site antagonist L-687,414, in the rat, *Neuropharmacology,* 35, 441, 1996.

36. Wláz, P., Ebert, U., and Löscher, W., Low doses of the glycine/NMDA receptor antagonist R-(+)-HA-966 but not D-cycloserine induce paroxysmal activity in limbic brain regions of kindled rats, *Eur. J. Neurosci.,* 6, 1710, 1994.

37. Löscher, W. and Wauqier, A., Use of animal models in developing guiding principles for polypharmacy in epilepsy, in *Rational Polypharmacy,* Leppik, I. E., Ed., *Epilepsy Res. Suppl.,* 11, Elsevier, Amsterdam, 1996, 61.

38. Barrett, J. E. and Miczek, K. A., Behavioral techniques in preclinical neuropsychopharmacology research, in *Psychopharmacology — The Fourth Generation of Progress,* Bloom, F. E. and Kupfer, D. J., Eds., Raven Press, New York, 1995, 65.

39. Rundfeldt, C., Wláz, P., Hönack, D., and Löscher, W., Anticonvulsant tolerance and withdrawal characteristics of benzodiazepine receptor ligands in different seizure models in mice. Comparison of diazepam, bretazenil and abecarnil, *J. Pharmacol. Exp. Ther.*, 275, 693, 1995.

40. Löscher, W., Rundfeldt, C., Hönack, D., and Ebert, U., Long-term studies on anticonvulsant tolerance and withdrawal characteristics of benzodiazepine receptor ligands in different seizure models in mice. I. Comparison of diazepam, clonazepam, clobazam, and abecarnil, *J. Pharmacol. Exp. Ther.*, 279, 561, 1996.

41. Löscher, W., Rundfeldt, C., Hönack, D., and Ebert, U., Long-term studies on anticonvulsant tolerance and withdrawal characteristics of benzodiazepine receptor ligands in different seizure models in mice. II. The novel imidazoquinazolines NNC 14-0185 and NNC 14-0189, *J. Pharmacol. Exp. Ther.*, 279, 573, 1996.

42. Löscher, W., Nolting, B., and Fassbender, C. P., The role of technical, biological and pharmacological factors in the laboratory evaluation of anticonvulsant drugs. I. The influence of administration vehicles, *Epilepsy Res.*, 7, 173, 1990.

43. Wláz, P., Baran, H., and Löscher, W., Effect of the glycine/NMDA receptor partial agonist, D-cycloserine, on seizure threshold and some pharmacodynamic effects of MK-801 in mice, *Eur. J. Pharmacol.*, 257, 217, 1994.

44. Krall, R. L., Penry, J. K., White, B. G., Kupferberg, H. J., and Swinyard, E. A., Hepatic considerations in the use of antiepileptic drugs, *Epilepsia*, 19, 409, 1978.

45. Löscher, W., Wláz, P., and Szabo, L., Focal ischemia enhances the adverse effect potential of *N*-methyl-D-aspartate receptor antagonists in rats, *Neurosci. Lett.*, in press.

Chapter 9

Gene Targeting Models of Epilepsy: Technical and Analytical Considerations

Laurence H. Tecott

Contents

I. Introduction

A. Gene Targeting Technology

A mutational approach has proven to be invaluable to investigators examining the roles of gene products in complex biological processes within prokaryotic and cultured eukaryotic cells. Recently, it has become possible to apply this approach to a mammalian system. Gene targeting procedures enable the precise (site-specific) introduction of a mutation to one of the estimated 100,000 murine genes. Typically, mutations have been designed to eliminate gene function, resulting in the generation of "knockout" or "null mutant" mice. The introduction of mutations that produce more subtle alterations in gene function has also been achieved. Two major developments have made gene targeting experiments feasible: (1) the generation of embryonic stem (ES) cells, and (2) the elucidation of techniques to achieve gene targeting in mammalian cells.

Initially, the generation of chimeric mice (animals whose tissues contain mixtures of cells derived from two genetic backgrounds) was achieved through the introduction of cells derived from teratocarcinomas (embryonal carcinoma cells) into early-gestation mouse embryos. However, these chimeras often displayed low rates of embryo colonization, restricted patterns of differentiation, and the development of tumors. A major advance was provided in the early 1980s by the establishment of embryonic stem cell lines derived from early-gestation mouse embryos.[1,2] In 1984, it was demonstrated that the injection of ES cells into blastocysts resulted in the formation of chimeras with high rates of colonization. Moreover, these chimeras demonstrated germ line transmission of ES cell genetic material to their offspring.[3]

The precise introduction of planned mutations into the genome was pioneered in the 1970s and 1980s in yeast. Gene targeting in yeast was achieved by a process of homologous recombination, involving recombination between homologous regions of a targeting vector and a native genomic locus. Homologous recombination in mammalian cells was first achieved in 1985, with the targeting of a neomycin resistance gene into the β-globin locus of erythroleukemic and carcinoma cell lines.[4] Relative to yeast, gene targeting in mammals was a more challenging proposition; whereas the majority of recombination events in yeast were homologous, such events were rare in mammalian cells. Thus, in mammalian cells, the vast majority of recombination events were nonhomologous, resulting in the apparently random integration of targeting vectors throughout the genome. This necessitated the development of selection strategies to enrich for homologous recombinant clones (see Section II.B). Homologous recombination was soon achieved in ES cells,[5,6] demonstrating the feasibility of the gene targeting approach.

The major impact of gene targeting approaches in the biomedical sciences is highlighted by the exponential growth in the number of mutant mouse strains generated in the last 8 years.[7] In the majority of cases, mutations are designed to eliminate the function of the targeted gene (null mutation). Many of these knockout (null mutant) strains have provided useful mouse models of human disease. In the last 3 years, several knockout strains have been found to exhibit epilepsy syndromes, providing new insights into the pathophysiology of epileptic disorders.

B. Gene Knockout Models of Epilepsy

Much evidence has accumulated to indicate that genetic factors contribute significantly to the susceptibility to idiopathic human epilepsies.[8,9] However, the identification of epilepsy genes has been hindered by several factors that commonly complicate human genetic studies, including issues of diagnostic classification, heterogeneous phenotypes, small family sizes, variable expressivity, and polygenic inheritance.[8,10] Therefore, epilepsy-prone rodent strains have been frequently used to examine the impact of genetic factors on seizure susceptibility.

Most intensively studied have been a number of genetically epilepsy-prone mouse strains arising from spontaneous mutations. Several of these syndromes are attributable to single gene mutations, including those of the *tottering, stargazer,* and *lethargic* strains. By contrast, complex patterns of inheritance underlie the audiogenic seizures of DBA/2 mice and the handling-induced seizures of EL mice. These models have been useful for testing anticonvulsant compounds and these strains have provided tools for studies examining mechanisms through which genes impact seizure susceptibility. However, this task has been complicated by the lack of information regarding the identities of the responsible genes.

The advent of gene targeting technology has produced a dramatic increase in the number of mutant epilepsy models. The first of these, serotonin 5-HT$_{2C}$ receptor null mutant mice, was reported in 1995.[11] Subsequently, at least nine additional transgenic mouse epilepsy models have been reported. This cohort of models illustrates the great diversity of genes that are involved in the regulation of neuronal network excitability. Some of the targeted gene products play direct roles in synaptic transmission, such as the 5-HT$_{2C}$ and glutamate GluR-B receptors, the GLT-1 glutamate transporter, and synaptic vesicle proteins, synapsins I and II.[12-14] Others play indirect roles in synaptic transmission, such as Ca^{2+}/calmodulin-dependent protein kinase II (CaMKII), a protein kinase that is critical to neuronal signaling,[15] the GAD65 isoform of glutamic acid decarboxylase, a GABA synthetic enzyme,[16] and nonspecific alkaline phosphatase, which regulates a cofactor required for GABA synthesis.[17] In contrast with the above models, other mutations of genes implicated in neuronal excitability are associated with a surprising absence of seizures. Examples include mice lacking the potassium channel Kv 3.1[18] and the GABA β3 receptor subunit.[19]

In some cases, epilepsy syndromes have resulted from disruptions of genes that have not been previously implicated in neurotransmission. One particularly serendipitous example of this is the recent report of the *jerky* mouse.[20] These animals resulted from an effort to generate transgenic mice bearing the SV40 large T antigen. The offspring of one particular transgenic founder mouse displayed handling-induced behaviors manifested by whole body jerks and generalized clonic seizures. Chronic recordings revealed large-amplitude interictal spikes in the dentate gyrus and spike-and-wave patterns in the neocortex. Molecular and genetic analysis revealed that the phenotype resulted not from the introduction of the SV40 antigen *per se*, but from the disruption produced by its insertion. The transgene had integrated into a novel gene termed *jerky*, producing an unintended *jerky* knockout. The increased seizure susceptibility of mice heterozygous for this genetic perturbation revealed it to be a dominant mutation. *Jerky* was found to be ubiquitously expressed in all tissues examined. The gene encodes a putative 60-kDa protein of 509 amino acids that is homologous to the mouse centromere binding protein and to the yeast autonomously replicating sequence-binding protein, indicating a possible function as a DNA binding protein. The mechanism through which the disruption of this protein alters seizure susceptibility remains to be determined.

Given estimates that 30,000 genes are expressed in the mammalian brain,[21] it is likely that the proliferation of new mutant mouse strains will include many that are relevant to the epilepsies. The first members of this new wave of epilepsy models illustrate the wide variety of genes that participate in the regulation of neuronal network excitability. It is likely that, in the future, an abundance of mechanisms involved in the regulation of excitability will be discovered. An advantage of pursuing this work in gene knockout models is that the genetic lesions are known, providing molecular points of reference for these studies. Furthermore, these models will provide candidate genes for studies aimed at uncovering the genetic bases of seizure susceptibility in humans.

II. Methodology

A number of excellent sources exist for detailed descriptions of the techniques required for the generation of gene targeted mouse strains.[22-24] In this section, an overview of the basic concepts and considerations in the generation of mutant strains will be presented.

A. Embryonic Stem Cells

ES cells are derived from 3.5-d-old mouse embryos, at the blastocyst stage of development. Blastocysts are cultured individually under conditions that permit the proliferation of the inner cell mass cells; those cells that would normally become the fetus. These cells are then disaggregated, and individual ES cell clones are grown on feeder layers. Under optimal conditions, ES cells retain the ability to contribute to all of the tissues of the developing fetus. The derivation of ES cells was pioneered using embryos derived from the 129 strain of mice, a strain that is commonly used in studies of early embryonic development. Although neurological function and behavior have not been extensively characterized for this strain, most ES cell lines in current use are 129-derived.

Rather than generating their own ES cell lines, most investigators acquire them from other investigators. A number of considerations in the selection of ES cell lines are worth noting. The first is the ES cell genetic background. There are multiple substrains of 129 which have resulted from outcrossings that have led to marked genetic variability among the substrains.[25] This has implications for homologous recombination frequencies (see the next section) and the interpretation of phenotypic differences (see Section III.B). Some of these issues will be resolved as ES cell lines generated from other strains (e.g., C57BL/6) become available (ES cells may be stored and shipped). Another consideration involves the predisposition of ES cell lines to lose their totipotency (ability to contribute to all of the tissues of an adult animal) in proportion to the number of prior cell passages.[26] Presumably, this reflects an accumulation of mutations and genetic rearrangements. Obviously, cells from low passage numbers are favored, and caution should be exercised in using cells for which such information is unavailable. In many cases, cells are found that are not euploid (with other than their expected complement of 40 chromosomes). It is therefore advisable to karyotype a newly acquired ES cell line prior to use. ES cell lines with less than two thirds of the cells euploid are unlikely to produce germ line-transmitting chimeric mice. Finally, it is desirable that cells of a similar passage have a prior history of contributing to the generation of such animals.

ES cells are most commonly cultured on feeder cell layers which serves the dual functions of providing a substrate for ES cell attachment and providing factors that prevent ES cell differentiation. The two most commonly used types of feeder cells are the STO fibroblast cell line (a mouse fibroblast cell line demonstrated to support the growth of ES cells[2]) and primary mouse embryonic fibroblasts (MEFs). MEFs are typically obtained from the mechanical dissociation of cells from 14-d-old mouse embryos. Embyros from strains bearing the neomycin resistance gene are optimal. Both STO and MEF cells must be mitotically inactivated prior to use. This is accomplished using mitomycin C or by irradiation. The optimal feeders are generally considered to be those on which the ES cell lines were initially derived. It is believed that leukemia inhibitory factor (LIF) released from feeders contributes to their ability to inhibit ES cell differentiation. Many investigators therefore supplement their ES cell media with LIF from commercial sources or from the conditioned medium of Chinese hamster ovary (CHO) cells. In most instances, the benefit of such supplementation is unclear.

B. Targeting Vectors

A targeting construct is generated which typically consists of a long target gene sequence into which a loss-of-function mutation has been engineered (Figure 9.1). Most targeting constructs are designed so that homologous recombination events occur in which recombination at the target locus results in replacement of native target sequences with construct sequences. As previously mentioned, in mammalian cells, fragments of DNA preferentially integrate into the genome at random locations, at rates that greatly exceed those of homologous recombination. Therefore, targeting constructs are designed for use in selection strategies that enrich for ES clones in which homologous recombination has occurred. The most commonly used strategy is the positive-negative selection strategy.[27] In this scheme, a portion of a protein-coding exon is replaced by sequences that confer resistance to the drug neomycin. This mutation serves two functions: (1) it inactivates the gene product and (2) in the presence of neomycin or geneticin it provides a marker that permits the selection of cells that have integrated the construct. This exogenous DNA fragment is flanked by regions of DNA that are homologous to the native gene. Adjacent to one of these homologous regions is a gene fragment encoding Herpes simplex virus thymidine kinase. Thymidine kinase expression is not toxic under normal conditions; however, in the presence of the nucleoside analogs gancyclovir (9-(1,3-dihydroxy-2-propoxymethyl) guanine) or FIAU (1-(2-deoxy, 2-fluoro-b-d-arabino-furanosyl) 5-iodouracil), the enzyme produces toxic metabolites. Thus, treatment

HOMOLOGOUS RECOMBINATION

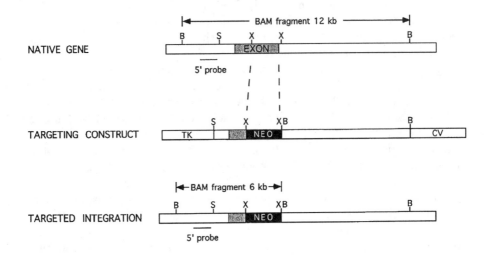

FIGURE 9.1

Homologous recombination using a targeting construct designed for gene replacement in a positive-negative selection strategy. In this example, a deletion is made by excising an XbaI (X) gene fragment corresponding to a protein-coding portion of an exon. It is replaced in the targeting construct by a neomycin resistance cassette (Neo), and an exogenous BamHI (B) restriction site is introduced in the process. In a targeted integration event, construct sequences replace the corresponding region of the native gene. The Neo cassette, a nonhomologous sequence, is included in the integration because it is flanked on both sides by regions of homology. In contrast, the thymidine kinase cassette (TK) and cloning vector (CV) are excluded because they are not flanked on both ends by homologous sequences. Southern blot screening for homologous recombination may be achieved using a probe corresponding to a 5′ flanking region of the integration site. Following BamHI digestion of genomic DNA, the wild-type allele will be indicated by a 12-kb fragment and the mutant allele by a 6-kb fragment.

with FIAU or gancyclovir will kill cells that express this gene. The importance of this effect is described below (Section II.C).

For reasons that are poorly understood, targeting efficiencies vary widely. Although much of this variability results from unknown factors, two contributing factors have been identified: length of homology and genetic background. A positive correlation is believed to exist between targeting efficiency and the length of homology between targeting vector sequences and those of the native gene.[28,29] Conversely, targeting efficiency does not appear to be markedly sensitive to the length of gene fragments that are deleted in targeting constructs. Another factor to consider in optimizing targeting efficiency is the use of a genomic fragment that is isogenic (derived from the same strain) with the ES cell line to be used. It has been suggested that strain polymorphisms, which are most prevalent in intronic regions (gene sequences not found in mRNA), may negatively impact targeting efficiency. In several cases the use of isogenic constructs has been shown to enhance efficiency relative to nonisogenic constructs for the same gene.[30] Most ES cell lines in common use are derived from genomic fragments obtained from strain 129 libraries. Due to the marked genetic variability among 129 substrains, the optimal situation would be for both the targeting construct and ES cell line to be derived from the same substrain. However, it should be noted that many gene targeting attempts have been successfully achieved with nonisogenic DNA constructs.

C. Transfection, Selection, and Screening

The targeting construct is typically linearized and introduced into ES cells by electroporation. In this step, cells are subjected to an electrical current that facilitates the internalization and integration of the DNA construct. The cells are then plated onto feeder layers. One day following electroporation, double drug selection is applied with geneticin and FIAU or gancyclovir. Those cells that had failed to incorporate the targeting construct are killed by the addition of geneticin to the culture medium (positive selection). The majority of the surviving cells have incorporated the entire DNA construct (including the thymidine kinase gene) at random sites throughout the genome. By contrast, during homologous recombination, non-homologous regions of the construct that are not flanked by homologous sequences are excluded from the integration event. Therefore, homologous recombinant clones will not contain the thymidine kinase gene. Thus, the addition of a second drug, gancyclovir, will selectively kill cells that have randomly incorporated the construct (negative selection), thereby enriching for targeted clones. In our experience, this strategy typically provides a 2- to 30-fold enrichment for targeted clones. These clones are expected to be heterozygous for the targeted allele (except when targeting X-linked genes in male ES cell lines).

After approximately 10 d of drug selection, ES cell colonies are visible to the unaided eye. Commonly, individual clones are picked and placed into U-bottomed 96-well plates, where they are treated with trypsin, dispersed, and then plated onto feeders in flat-bottomed multiwell plates. The individual clones are subsequently grown for DNA isolation and portions are frozen pending screening. Screening for homologous recombination events is performed by either polymerase chain reaction (PCR) or Southern blot analysis. PCR strategies typically involve the use of one primer that is homologous to the neomycin resistance cassette and another that is homologous to an external genomic region that flanks the insertion site. Advantages of PCR-based strategies include rapidity, applicability to low-quality DNA, and suitability for screening pooled samples. On the other hand, PCR screening strategies that require the generation of long fragments may be difficult to implement. Furthermore, some investigators experience difficulties with the consistency and reliability of these strategies. For some investigators, screening by Southern blot analysis provides greater reliability; however, these strategies are often more cumbersome. The probe and restriction enzyme to be used must be carefully selected so that mutant and wild-type alleles are readily distinguishable. Strategies involving the use of a probe corresponding to a genomic region external to the targeting construct are typically used. Once putative targeted clones are identified, they should be subjected to a thorough restriction analysis to confirm the organization of the recombinant allele.

D. Generation of Germ Line Chimeras

Following the isolation of targeted ES cell clones, chimeras are made using either blastocyst injection or, more recently, aggregation techniques (Figure 9.2). In the blastocyst injection technique, ES cells are microinjected into the fluid-filled blastocoel cavity of 3.5-d-old embryos at the blastocyst stage. The blastocyst injection apparatus commonly includes an inverted microscope and micromanipulation equipment. The blastocyst is positioned using a holding pipette, and 10 to 15 ES cells are introduced into the embryo using an injection pipette. Both the holding and injection pipettes are fashioned from drawn glass capillary tubing using a microforge. The injected embryos are then surgically transferred into the uterus of pseudopregnant female mice. Pseudopregnant females are generated by matings with vasectomized males. The act of copulation initiates the endocrine changes of pregnancy, providing a suitable uterine environment for the survival and implantation of the transferred embyros. These animals will then give birth to chimeric mice, which are partly derived from the injected ES cells, and partly derived from the host embryo. For example, ES cells derived from a 129 substrain with an agouti coat color are often injected into embryos derived from black C57BL/6 mice, resulting in chimeras with coats containing black and agouti patches. The extent to which the ES cells have colonized the animal may be roughly approximated by the extent of the agouti contribution to the coat. It is also notable that marked strain effects exist in the extent to which ES cell progeny colonize host embryos. For this reason, most investigators use the proven combination of 129-derived ES cells and C57BL/6 embryos. For investigators who are experienced with these techniques, over 30% of injected embryos can give rise to germ line competent chimeras.

Blastocyst injection methods require a significant degree of expertise and access to microinjection setups that some investigators find prohibitively expensive. Recently, alternative aggregation techniques have been developed for the production of chimeric mice.[31] These methods involve the harvest of morula-stage embryos, the removal of the zona pellucida, and coculture with ES cells. When 10- to 15-cell clumps of ES cells are cultured in apposition to such morulae overnight, they become integrated into the embryo inner cell mass. Embryos that have incorporated ES cells are removed from culture and introduced into pseudopregnant females by uterine transfer. Using these techniques, over 10% of transferred embryos may develop into germ line competent chimeras. Aggregation methods are gaining in popularity; aside from a dissecting microscope, no specialized equipment is needed and only basic embryo manipulation skills are required.

As mentioned previously, inspection of the coat of chimeric mice is often used to estimate the extent of the ES cell contribution to the animal. Although there does appear to be a rough correlation between ES cell coat contribution and likelihood of germ line transmission, an animal whose coat is greater than 95% ES cell derived is not guaranteed to be germ line competent. Conversely, there are anecdotal reports of germ line transmission from chimeras that have an ES cell coat contribution of less than 10%. Because most ES cell lines in common use are male (XY), a sex

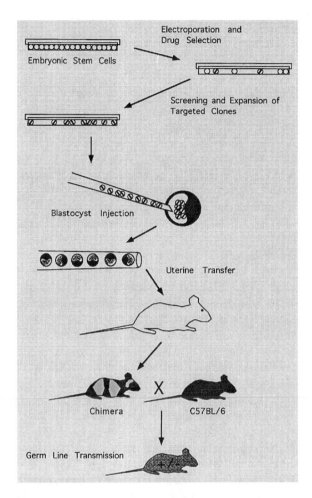

FIGURE 9.2
Procedures required for the generation of mice with targeted mutations. In this example, the coat of the chimera contains black and agouti patches. Germ line transmission of ES cell genetic material is indicated by the generation of completely agouti offspring.

conversion phenomenon may occur, so that the number of male chimeras exceeds that of females. Commonly, chimeras derived from agouti strain 129 ES cells are bred with C57BL/6 mice. Because the agouti allele is dominant, germ line transmission of 129 genetic material is indicated by pups with agouti coats. Such animals have a 50% likelihood of bearing one copy of the targeted allele. The percentage of agouti pups varies widely among chimeras. Typically, genomic DNA is prepared from tail biopsies of these animals, and genotyping is performed by PCR or Southern blot analysis. Heterozygous mutant mice are then crossed to produce homozygous mutant animals that completely lack the normal gene product.

III. Interpretation

A. Epilepsy in 5-HT$_{2C}$ Receptor Mutant Mice

A number of important considerations in the interpretation of mutant phenotypes are illustrated by the analysis of an epilepsy syndrome in serotonin 5-HT$_{2C}$ receptor null mutant mice. The 5-HT$_{2C}$ receptor is widely expressed throughout the central nervous system, and has been implicated in numerous behavioral and physiological actions of serotonin.[32-35] Studies of the functional significance of this receptor have been hindered by a lack of available selective agonist and antagonist drugs. Therefore, to gain further insight into the role of 5-HT$_{2C}$ receptors in central nervous system function, a null mutant mouse strain was generated.[11]

A 5-HT$_{2C}$ receptor genomic fragment was obtained from a strain 129 genomic phage library, and a targeting construct was generated for use in a standard positive-negative selection strategy.[27] The construct contained 8.5 kb of genomic DNA and a 16 base pair insertion encoding stop codons in all three reading frames so as to produce a truncation of the 5-HT$_{2C}$ receptor protein within the fifth transmembrane domain. This would eliminate nearly one half the protein mass, and by analogy with adrenergic receptors[36] would disrupt G protein interactions and ligand binding. Following electroporation of 129-derived ES cells and drug selection, Southern blotting revealed a targeting frequency of 1/45 drug resistant clones. Germ line transmitting chimeras generated by blastocyst injection were bred to produce null mutant mice. Because the 5-HT$_{2C}$ receptor gene is X-linked,[37] males possessing a single mutant allele lack the normal receptor and are termed "hemizygous" for the mutation.

Although the absence of functional 5-HT$_{2C}$ receptor mRNA and protein was confirmed in the hemizygous mutants, no overt abnormalities in appearance or behavior were initially noted. However, it became clear that a minority of knockout mice were subject to spontaneous death beginning at approximately 3 to 4 weeks of age. Regular health checks revealed no animals that appeared ailing or dehydrated, but in several cases animals that appeared healthy were found dead a few hours later. Careful postmortem analyses with particular attention to the brain and cardiovascular structures revealed no clues regarding the cause of death. To gain insight into the cause of death, small groups of mutant mice were subject to continuous videotape monitoring. Review of the tapes revealed the presence of a seizure syndrome, manifested by spontaneous tonic-clonic seizures associated with the enhancement of grooming behaviors. Following most seizure episodes, the mice recovered; however, animals occasionally died within seconds of seizure onset. These findings are consistent with the observation that mice are quite prone to death from seizures. They illustrate that seizure disorders should be included in the differential diagnosis of sudden death in transgenic mice.

Spontaneous seizures were found to be infrequent and sporadic in 5-HT$_{2C}$ receptor mutant mice. To gain a more quantitative measure of the extent to which the mutation had altered seizure susceptibility, the sensitivity of animals to administration of the convulsant drug pentylenetetrazol (Metrazol) was determined.

Loosely restrained animals received a continuous intravenous infusion of pentyl-enetetrazol and exhibited a stereotyped progression of convulsive behaviors. The latencies of animals to display the following responses were determined: (1) first twitch, (2) repetitive tonic-clinic seizure activity, and (3) tonic extension. Knockout mice exhibited a 24% reduction in seizure threshold (latency or cumulative dose required for the first twitch) and a striking 83% reduction in the duration of the tonic-clonic phase of the response relative to wild-type littermates (time between the onset of the tonic-clonic and tonic extension phases). These results suggested that the loss of $5\text{-}HT_{2C}$ receptors leads to both a lowered seizure threshold and to a more rapid progression of seizure activity. A role for $5\text{-}HT_{2C}$ receptors in the regulation of neuronal network excitability was further indicated by the finding that the nonspecific $5\text{-}HT_{2C}$ receptor antagonist mesulergine enhanced pentylenetetrazol sensitivity in wild-type C57BL/6 mice.

Studies of the seizure syndrome of $5\text{-}HT_{2C}$ receptor mutant mice have been complicated by the rarity of their spontaneous seizures. To determine whether seizures could be reliably induced by a noninvasive stimulus, the sensitivity of these animals to sound-induced seizures was determined.[38] Mutant and wild-type animals were exposed to a complex acoustic stimulus consisting of a mixture of high-frequency tones presented at 108 dB. The stimulus was terminated at the first signs of seizure activity, or after 1 min. Whereas none of the wild-type animals displayed behavioral evidence of seizures, the mutants displayed severe audiogenic seizures (AGSs) within seconds of sound presentation. At 2 to 3 s following tone onset, the mutants exhibited the sudden onset of a bout of wild running and erratic leaping. This response persisted for 1 to 2 s, and was immediately followed by a period of extensor rigidity leading to apparent respiratory arrest and death. It was subsequently determined that animals could be resuscitated by artificial ventilation. The resuscitation "device" consisted of a 21-in. length of 0.5-in. inner diameter tygon tubing. Artificial ventilation was begun during the extensor rigidity phase by placing one end of the tubing over the snout of the animal and gently puffing into the other end. Typically, spontaneous ventilations resumed within 1 min, and following a period of postictal lethargy, animals appeared to recover completely.

Expression of the immediate early gene transcription factor c-fos is rapidly induced in response to seizure activity. To identify neural substrates of AGSs in $5\text{-}HT_{2C}$ receptor mutant mice, AGS-induced patterns of c-fos-like immunoreactivity were examined. Staining was observed in subcortical structures associated with auditory processing in a pattern similar to those found in other AGS-susceptible strains. Because no X-linked genes have been implicated in AGS-susceptibility in these other strains, it is likely that AGSs may be produced not only by the absence of $5\text{-}HT_{2C}$ receptors, but also by additional independent genetic mechanisms. This epilepsy syndrome is also the first audiogenic seizure model for which the causative genetic perturbation has been identified. Greater detail concerning sound-induced seizures may be found in Chapter 6.

The observation of an epilepsy syndrome in $5\text{-}HT_{2C}$ receptor mutant mice is consistent with several lines of evidence indicating that serotonin system activity produces anticonvulsant effects in a variety of epilepsy models, including AGSs.[39-42] Despite these findings, the identity of the receptors that contribute to these actions of serotonin has been unclear. Interestingly, several commonly used psychiatric drugs with convulsant side effects have potent 5-HT_{2C} receptor antagonist properties.[43] The above findings indicate that the 5-HT_{2C} receptor may play a significant role in the serotoninergic inhibition of neuronal network excitability, and indicate its potential as a target for anticonvulsant drug development. 5-HT_{2C} receptor null mutant mice provide a useful tool for exploring mechanisms through which serotonin systems regulate brain excitability.

B. Knockout Models of Epilepsy: Caveats and Future Directions

An important consideration in the interpretation of mutant phenotypes is genetic background. It is well known that large variations in behavioral and neural function exist among inbred mouse strains. Moreover, marked behavioral differences have been observed among various 129 substrains.[44] Studies are frequently performed using mice with mixed genetic backgrounds, leading to enhanced phenotypic variability. Although the problem of genetic heterogeneity may be solved by breeding chimeras with strain 129 mice, this strain may not be optimal for studies of central nervous system function. Structural abnormalities of the central nervous system have been observed in a subpopulation of 129 mice. For example, incomplete formation of the corpus callosum is a partially penetrant phenotype observed in some 129 substrains.[45]

Alternatively, many investigators seek to minimize genetic heterogeneity by backcrossing their hybrid mice to another inbred strain. Although such a breeding program can greatly reduce genetic variability, it has been recently pointed out that backcrossing will not readily eliminate 129-derived genes that are tightly linked to the target locus.[46] For example, even after 12 generations of backcrossing to a C57BL/6 background, animals may retain a length of flanking 129-derived genetic material capable of encoding 300 genes. This could complicate data interpretation if significant differences in the relevant phenotype occur between C57BL/6 and 129 animals. Strategies to control for these effects include the use of transgenic approaches to rescue the mutant phenotype by restoring the targeted gene and the use of controls generated by crossing the wild-type progeny of chimeras.[47] An optimal solution to these problems may lie in the ability to derive ES cell lines from other inbred strains, such as the well-characterized C57BL/6 strain.[48] Chimeras generated from C57BL/6 ES cells could be crossed with C57BL/6 mice to yield null mutant and control animals devoid of genetic polymorphisms.

Another important caveat in the analysis of null mutant strains is the potential for abnormal development. Because standard knockout strains lack the targeted gene product throughout development, it is unclear whether a mutant phenotype reflects the normal adult role of a gene product or an indirect consequence of perturbed development. In some cases, aberrant brain structure indicates the occurrence of abnormal development; however, an apparent lack of morphological anomalies does not rule out developmental defects. Such defects can lead to overestimation of the functional significance of a gene in the adult brain. Conversely, developmental compensation for the loss of a functional gene may produce a phenotype that underestimates its significance to adult brain function. When the targeted gene is a receptor molecule, developmental issues may be addressed by determining the extent to which the mutant phenotype may be mimicked by antagonist treatment in wild-type animals. This strategy was used in the evaluation of the 5-HT$_{2C}$ receptor seizure model.

An optimal solution for eliminating the confound of potential developmental perturbations will be the generation of mouse strains in which targeted mutations are induced following a period of normal development. The feasibility of "inducible knockout" strategies that enable temporal control of targeted mutations has been demonstrated through the use of a promoter that is regulated by tetracycline in transgenic mice.[49,50] It is likely that advances over the next several years will enable these approaches to become widely available.

Additional "second generation" knockout strategies address another limitation of the standard technology. The loss of gene expression can be ubiquitous, involving the entire organism. This leads to situations in which the function of a gene product in a particular region is complicated by defects arising from its absence at other sites. It is now possible to overcome this problem by generating animals in which mutations are restricted to particular regions or cell types. "Tissue-specific" knockout strategies have been successfully applied using the loxP-cre site-specific recombination system of bacteriophage P1.[51,52] Such strategies may be ultimately used to restrict mutations to neurons or to subpopulations of neurons. Finally, in many instances it will be advantageous to introduce small mutations into a target locus, rather than generating a null mutation. Strategies utilizing the loxP-cre system may also be applied to produce more subtle alterations in gene function. This technology would have a number of uses, including the ability to simulate many human genetic disorders in mice.

It is clear that gene targeting technology is making a major impact on the understanding of human diseases, including the epilepsies. These procedures enable the study of gene function in the context of the intact organism. The new murine models of epilepsy are revealing that a wide variety of genes may determine the heritability of seizure predisposition. Given the large number of genes that are expressed in the central nervous system, many additional models are likely to be described in the future. A major challenge will be to unravel the mechanisms through which genetic alterations undermine those mechanisms that dampen neuronal network excitability. These efforts will be greatly aided by the advent of more sophisticated technologies that enable the regional and temporal control of gene inactivation.

References

1. Evans, M. J. and Kaufman, M. H., Establishment in culture of pluripotent cells from mouse embryos, *Nature,* 292, 154, 1981.

2. Martin, G. R., Isolation of a pluripotent cell line from early mouse embryos cultured in medium conditioned by teratocarcinoma stem cells, *Proc. Natl. Acad. Sci. U.S.A.,* 78, 7634, 1981.

3. Bradley, A., Evans, M., Kaufman, M. H., and Robertson, E., Formation of germ-line chimaeras from the embryo-derived teratocarcinoma cell lines, *Nature,* 309, 255, 1984.

4. Smithies, O., Gergg, R. G., Boggs, S. S., Koralewski, M. A., and Kuckerlapati, R. S., Insertion of DNA sequences into the human chromosomal b-globin locus by homologous recombination, *Nature,* 317, 230, 1985.

5. Thomas, K. R. and Capecchi, M. R., Site-directed mutagenesis by gene targeting in mouse-embryo-derived stem cells, *Cell,* 51, 503, 1987.

6. Doetschman, T., Gregg, R. G., Maeda, N., Hooper, M. L., Melton, D. W., Thompson, S., and Smithies, O., Targeted correction of a mutant HPRT gene in mouse embryonic stem cells, *Nature,* 330, 576, 1987.

7. Brandon, E. P., Idzerda, R. L., and McKnight, G. S., Targeting the mouse genome: a compendium of knockouts. II, *Current Biol.,* 5, 758, 1995.

8. Andermann, E., Multifactorial inheritance of generalized and focal epilepsy, in *Genetic Basis of the Epilepsies,* Anderson, V. E., Hauser, W. A., Penry, J. K., and Sing, C. F., Eds., Raven Press, New York, 1982, 355.

9. Delgado-Escueta, A. V., Ward, A. A., Woodbury, D. M., and Porter, R. J., New wave of research in the epilepsies, in *Advances in Neurology,* Vol. 44, Delgado-Escueta, A. V., Ward, A. A., Woodbury, D. M., and Porter, R. J, Eds., Raven Press, New York, 1986, 3.

10. Frankel, W. N., Taylor, B. A., Noebels, J. L., and Lutz, C. M., Genetic epilepsy model derived from common inbred mouse strains, *Genetics,* 138, 481, 1994.

11. Tecott, L. H., Sun, L. M., Akana, S. F., Strack, A. M., Lowenstein, D. H., Dallman, M. F., and Julius, D., Eating disorder and epilepsy in mice lacking 5HT2C serotonin receptors, *Nature,* 374, 542, 1995.

12. Brusa, R., Zimmermann, F., Koh, D. S., Feldmeyer, D., Gass, P., Seeburg P. H., and Sprengel, R., Early-onset epilepsy and postnatal lethality associated with an editing-deficient GluR-B allele in mice, *Science,* 270, 1677, 1995.

13. Rosahl, T. W., Spillane, D., Missler, M., Herz, J., Selig, D. K., Wolff, J. R., Hammer, R. E., Malenka, R. C., and Sudhof, T. C., Essential functions of synapsins I and II in synaptic vesicle regulation, *Nature,* 375, 488, 1995.

14. Tanaka, K., Watase, K., Manabe, T., Yamada, K., Watanabe, M., Takahashi, K., Iwama, H., Nishikawa, T., Ichihara, N., Kikuchi, T., Okuyama, S., Kawashima, N., Hori, S., Takimoto, M., and Wada, K., Epilepsy and exacerbation of brain injury in mice lacking the glutamate transporter GLT-1, *Science,* 276, 1699, 1997.

15. Butler, L. S., Silva, A. J., Abeliovich, A., Watanabe, Y., Tonegawa, S., and McNamara, J. O., Limbic epilepsy in transgenic mice carrying a Ca^{2+}/calmodulin-dependent kinase II α–subunit mutation, *Proc. Natl. Acad. Sci. U.S.A.,* 92, 6852, 1995.

16. Kash, S., Johnson, R., Tecott, L. H., Lowenstein, D., Hanahan, D., and Baekkeskov, S., Targeted disruption of the murine glutamic acid decarboxylase (65 kDa isoform - GAD65) gene, *Soc. Neurosci. Abstr.,* 22, 1295, 1996.

17. Waymire, K. G., Hahuren, J. D., Jaje, J. M., Guilarte, T. R., Coburn, S. P., and MacGregor, G. R., Mice lacking tissue non-specific alkaline phosphatase die from seizures due to defective metabolism of vitamin B-6, *Nature Genet.,* 11, 45, 1995.

18. Ho, C. S., Grange, R. W., and Joho, R. H., Pleiotropic effects of a disrupted K$^+$ channel gene: reduced body weight, impaired motor skill and muscle contraction, but no seizures, *Proc. Natl. Acad. Sci. U.S.A.,* 94, 1533, 1997.

19. Culiat, C. T., Stubbs, L. J., Woychik, R. P., Russell, L. B., Johnson, D. K., and Rinchik, E. M., Deficiency of the beta 3 subunit of the type A gamma-aminobutyric acid receptor causes cleft palate in mice, *Nature Genet.,* 11, 344, 1995.

20. Toth, M., Grimsby, J., Buzsaki, G., and Donovan, G. P., Epileptic seizures caused by inactivation of the novel gene, jerky, related to centromere binding protein-B in transgenic mice, *Nature Genet.,* 11, 71, 1995.

21. Milner, R. J. and Sutcliffe, J. G., Gene expression in rat brain, *Nucleic Acids Res.,* 11, 5497, 1983.

22. Hogan, B., Costantini, F., and Lacy, E., *Manipulating the Mouse Embryo: A Laboratory Manual,* Cold Spring Harbor Laboratory, Cold Spring Harbor, NY, 1986.

23. Joyner, A. L., Gene Targeting: a Practical Approach, IRL Press, New York, 1993.

24. Ramirez-Solis, R., Davis A. C., and Bradley, A., Gene targeting in embryonic stem cells, *Methods Enzymol.,* 225, 855, 1993.

25. Simpson, E. M., Linder, C. C., Sargent, E. E., Davisson, M. T., Mobraaten, L. E., and Sharp, J. J., Genetic variation among 129 substrains and its importance for targeted mutagenesis in mice, *Nature Genet.,* 16, 19, 1997.

26. Nagy, A., Rossant, J., Nagy, R., Abramow-Newerly, W., and Roder, J. C., Derivation of completely cell culture-derived mice from early-passage embryonic stem cells, *Proc. Natl. Acad. Sci. U.S.A.,* 90, 8424, 1993.

27. Mansour, S. L., Thomas, K. R., and Capecchi, M. R., Disruption of the proto-oncogene int-2 in mouse embryo-derived stem cells: a general strategy for targeting mutations to non-selectable genes, *Nature,* 336, 348, 1988.

28. Hasty, P., Rivera-Perez, J., Chang, C., and Bradley, A., Target frequency and integration pattern for insertion and replacement vectors in embryonic stem cells, *Mol. Cell. Biol.,* 11, 4509, 1991.

29. Deng, C. and Capecchi, M. R., Re-examination of gene targeting frequency as a function of the extent of homology between the targeting vector and the target locus, *Mol. Cell. Biol.,* 12, 3365, 1992.

30. Te Riele, H., Maandag, E. R., and Berns, A., Highly efficient gene targeting in embryonic stem cells through homologous recombination with isogenic DNA constructs, *Proc. Natl. Acad. Sci. U.S.A.,* 89, 5128, 1992.

31. Wood, S. A., Allen, N. D., Rossant, J., Auerbach, A., and Nagy, A., Non-injection methods for the production of embryonic stem cell-embryo chimaeras, *Nature,* 365, 87, 1993.

32. Molineaux, S., Jessell, T., Axel, R., and Julius, D., 5-HT1c receptor is a prominent serotonin receptor subtype in the central nervous system, *Proc. Natl. Acad. Sci. U.S.A.,* 86, 6793, 1989.

33. Curzon, G. and Kennett, G. A., m-CPP: a tool for studying behavioural responses associated with 5-HT1c receptors, *Trends Pharmacol. Sci.,* 11, 181, 1990.

34. Wright, D. E., Seroogy, K. B., Lundgren, K. H., Davis, B. M., and Jennes, L., Comparative localization of serotonin 1A, 1C, and 2 receptor subtype mRNAs in rat brain, *J. Comp. Neurol.,* 351, 357, 1995.

35. Tecott, L. H., Serotonin receptor diversity: implications for psychopharmacology, in *Review of Psychiatry,* Vol. 15, Dickstein, L. J., Riba, M. B., and Oldham, J. M., Eds., American Psychiatric Press, Washington, D.C., 1996, 331.
36. Kobilka, B., Adrenergic receptors as models for G protein-coupled receptors, *Annu. Rev. Neurosci.,* 15, 87, 1992.
37. Milatovich, A., Hsieh, C. L., Bonaminio, G., Tecott, L. H., Julius, D. J., and Francke, U., Serotonin receptor 1c gene assigned to X chromosome in human (band q24) and mouse (bands D-FG4), *Hum. Mol. Genet.,* 1, 681, 1992.
38. Brennan, T. J., Seeley, W. W., Kilgard, M., Schreiner, C. E., and Tecott, L. H., Sound-induced seizures in serotonin 5-HT2C receptor mutant mice, *Nature Genet.,* 16, 387, 1997.
39. Jobe, P. C., Picchioni, A. L., and Chin, L., Role of brain 5-hydroxytryptamine in audiogenic seizure in the rat, *Life Sci.,* 13, 1, 1973.
40. Browning, R. A., Hoffman, W. E., and Simonton, R. L., Changes in seizure susceptibility after intracerebral treatment with 5,7-dihydroxytrytpamine: role of serotonergic neurons, *Ann. NY Acad. Sci.,* 305, 437, 1978.
41. Dailey, J. W., Slater, K., Crable, D. J., and Jobe, P. C., Anticonvulsant effects of the antidepressant fluoxetine in the gentically epilepsy-prone rat (GEPR), *Fed. Proc.,* 46, 2282, 1987.
42. Hiramatsu, M. K., Ogawa, K., Kabuto, H., and Mori, A., Reduced uptake and release of 5-hydroxytryptamine and taurine in the cerebral cortex of epileptic El mice, *Epilepsy Res.,* 1, 40, 1987.
43. Jenck, F., Moreau, J. L., Mutel, V., and Martin, J. R., Brain 5-HT1C receptors and antidepressants, *Progr. Neuro-Psychopharmacol. Biol. Psychiatr.,* 18, 563, 1994.
44. Owen, E. H., Logue, S. F., Rasmussen, D. L., and Wehner, J. M., Assessment of learning by the Morris water task and fear conditioning in inbred mouse strains and F1 hybrids: implications of genetic background for single gene mutations and quantitative trait loci analyses, *Neuroscience,* 80(4), 1087, 1997.
45. Ozaki, H. S. and Wahlsten, D., Cortical axon trajectories and growth cone morphologies in fetuses of acallosal mouse strains, *J. Comp. Neurol.,* 336, 595, 1993.
46. Gerlai, R., Gene-targeting studies of mammalian behavior: is it the mutation or the background genotype?, *Trends Neurosci.,* 19, 177, 1996.
47. Zimmer, A., Gene targeting and behaviour: a genetic problem requires a genetic solution, *Trends Neurosci.,* 19, 470, 1996.
48. Kawase, E., Suemori, H., Takahashi, N., Okazaki, K., Hashimoto, K., and Nakatsuji, N., Strain differences in establishment of mouse embryonic stem (ES) cell lines, *Int. J. Dev. Biol.,* 38, 385, 1994.
49. Furth, P. A., Onge, L., Boger, H., Gruss, P., Gossen, M., Kistner, A., Bujard, H., and Henninghausen, L., Temporal control of gene expression in transgenic mice by a tetracycline-responsive promoter, *Proc. Natl. Acad. Sci. U.S.A.,* 91, 9302, 1994.
50. Gossen, M., Freundlieb, S., Bender, G., Muller, G., Hillen, W., and Bujard, H., Transcriptional activation by tetracyclines in mammalian cells, *Science,* 268, 1766, 1995.

51. Sternberg, N., Sauer, B., Hoess, R., and Abremski, K., Bacteriophage P1 cre gene and its regulatory region. Evidence for multiple promoters and for regulation by DNA methylation, *J. Mol. Biol.,* 187, 197, 1986.
52. Gu, H., Marth, J. D., Orban, P. C., Mossmann, H., and Rajewsky, K., Deletion of DNA polymerase B gene segment in T cells using cell type-specific gene targeting, *Science,* 265, 103, 1994.

Chapter

The Hippocampal Slice Preparation

Larry G. Stark and Timothy E. Albertson

Contents

0-8493-3362-8/98/$0.00+$.50

I. Introduction

Animal models of epilepsy have been very useful in the preclinical evaluation and development of new antiepileptic drugs and a variety of them exist at the whole animal, organ, tissue, and cellular levels. Fisher[1] has reviewed more than 50 models. The first report of the transverse hippocampal slice preparation maintained *in vitro* appeared in 1971,[2] but the focus of that short report was on the new methodology, rather than as a method applied to epilepsy research. A number of reviews in the past 10 years have focused on the use of the hippocampal slice preparation and adaptations of it for studies of experimental epilepsies, neurotoxicology, and related phenomena in rodents,[3-5] and man.[6]

 Although hippocampal slices have been occasionally used in some long-term tissue culture studies,[7,8] the focus of this review will be on the more common methodologies associated with acute studies of anticonvulsant drugs on hippocampal slices prepared from rats and maintained in a Haas-type superperfusion chamber.

II. Methodology

The advantages offered by an *in vitro* technique such as the hippocampal slice for the study of epileptogenic phenomena are many and include the ability to study a subset of neuronal networks without the influence of the remainder of the brain, the ease with which one can expose brain tissue to drugs and toxicants, the ability to monitor the moment-to-moment changes in neurophysiological properties induced by such exposures, the ability to control local factors such as oxygenation and electrolyte concentrations, and the potential for gaining some insight on possible mechanisms of action since the anatomical and neurophysiological details available about the slice have become more readily available.

A. Preparatory Steps

1.   Rat Selection

Slice preparations have been made from many species and from rats of different ages as well. Surveys of the literature reveal that inbred strains such as Sprague-Dawley and Charles River are most frequently chosen, but Wistar rats have also been used. We have tended to use young rats with weights in the 180 to 220 g weight range.

2.   Methods of Sacrifice

Unlike preparations for many *in vitro* studies, rats used for hippocampal slice preparations are rarely anesthetized prior to decapitation. Anesthetics of course would likely interfere either with normal levels of excitability in the slice or with the actions of drugs to be studied and are to be avoided. Examples of this potential interaction come from Engel et al., who have studied the effect of *in vivo* administration of anesthetics on GABA receptor function.[9]

3.   Methods of Dissection

Following decapitation, the head is moved to a dissection area where the brain is rapidly removed using a combination of scalpel, rongeurs, and a narrow weighing spatula. Some attention to removal of the overlying meninges is useful to make subsequent dissections of the brain easier. Transfer of the whole brain to an ice-cold, well-aerated artificial cerebral spinal fluid (ACSF) solution in a shallow container is the next step. After removal of the cerebellum (Figure 10.1A), the brain is bisected at the midline (Figure 10.1B) and one hemisphere is selected for further dissection. A useful tool for the next stage of preparation is a plastic picnic knife which has been trimmed to a medium point (Figure 10.1C). After slipping the knife just below the hippocampus at the level of the anterior commissure, one carefully slides the blade posteriorly with gentle pressure to separate the whole hippocampus (Figure 10.1E) from the remainder of the brain. Iris scissors facilitate final separation of the hippocampus from the hemisphere. The structure is kept moist with ACSF while it is trimmed of any excess tissues that may be attached. It is then transferred to a moist filter paper on the cutting stage of a slicing device.

4.   Orientation for Cutting

Orientation of the slice relative to the cutting edge of the knife (McIlwain tissue chopper, Brinkman Instruments, Westbury, NY) will depend somewhat on which areas of the hippocampus are of special interest. For studies of the CA1 region, we have found that orienting the slice at about a 30 degree angle with respect to the cutting blade will provide slices with good anatomical definition of the pyramidal cell layers, particularly those chosen from the middle third of the overall slice.

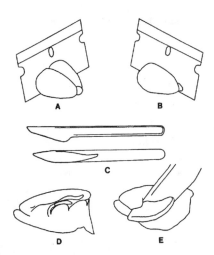

FIGURE 10.1

Drawings represent (A) removal of the cerebellum from the brain with razor blade, (B) separation of the cerebral hemispheres, (C) reshaped plastic picnic knife useful for blunt dissection, (D) a saggital view of the right hemisphere of a rat brain, (E) use of the dissecting knife to separate out the right hippocampus.

Investigators often do not specify the origin of the slices, but may use slices from either end of the hippocampus as well.

5. Slice Thickness

Many studies are done on slices ranging from 200 to 400 μm thick. These slices are translucent when illuminated and make electrode placement somewhat easier. Slices of this thickness also seem to be well nourished by moderate flow rates of bathing solutions and can be maintained for many hours while under study. Thinner slices may tend to disrupt local synaptic networks while thicker ones may not receive adequate nourishment toward their centers.

6. Transfer Techniques

A small-diameter camel hair brush can be used to remove the cut slices from the knife blade of the tissue chopper or to gently tease them from the surface of the hippocampal block as it rests on the chopper platform. Some investigators place the slices in a small amount of chilled buffer (ACSF) while waiting for completion of the slicing process. They then use either a brush, medicine dropper, or pipette to transfer the slices to the recording chamber.[28] We have successfully eliminated this intermediate step by taking the slices directly from the knife to the surface of the recording chamber with a fine bristled brush. We typically place about six slices in the space available within the recording chamber.

Additional slices taken from the remainder of the hippocampus are placed in a well-oxygenated holding chamber (Prechamber, Medical Systems Corp., Greenvale, NY) containing ACSF and are held at room temperature until they are used (Figure 10.2).

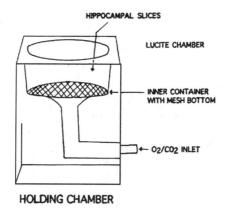

HOLDING CHAMBER

FIGURE 10.2
Depiction of a prechamber for maintaining hippocampal slices in artificial cerebrospinal fluid.

7. Recording Chambers

The recording chamber (Brain Slice Chamber System and Haas Top, Medical Systems Corp., Greenvale, NY) rests on a gimbaled table (High Performance Lab Table, Harvard Apparatus, South Natick, MA) along with the micromanipulators holding both stimulation and recording electrodes. A representative drawing of the recording chamber (Figure 10.3) shows the elements necessary for perfusion of the hippocampal slices. The slices rest on fine nylon mesh in the chamber and are surrounded by a sheet of plastic with oblong openings to create a well for the fluids that are to bathe the slices. A thermistor for temperature monitoring and a reference electrode lie in grooves just below and ahead of the slices so that the temperature of perfusing fluids can be maintained constant. The water-jacketed chamber below the recording chamber contains a heating element controlled by feedback from the thermistor probe and the tubing that supplies the perfusion fluid is wrapped around inside.

8. Composition of Perfusion Fluids

The composition of perfusion fluids used to maintain the slice varies slightly from laboratory to laboratory, but generally represents a balanced, modified Ringer's solution which is often heated prior to exposing the tissues to it. It may be modified further for special studies where ion composition is known to influence the electrophysiological events within the slice (e.g., low or zero magnesium concentrations). A composite formula for ACSF taken from examples in the literature is shown in Table 10.1. Our own formulation for experiments with rats is also shown (Table 10.1). Additional examples for specific species may be found in a report by Alger et al.[10]

9. Temperature

The range of temperatures used to maintain slices for study typically is quite narrow, lows being near room temperature (25°C) and highs closer to the temperature of

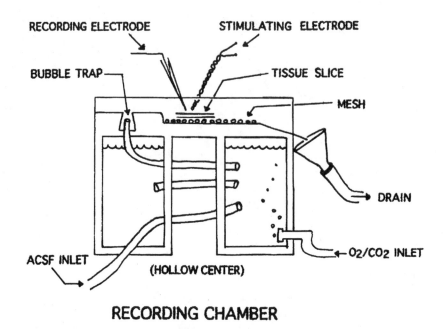

RECORDING CHAMBER

FIGURE 10.3
Cross-sectional view of a Haas-type recording chamber used for brain hippocampal slice studies.

TABLE 10.1
Comparison Between a Composite Formula for Artificial Cerebrospinal Fluid (ACSF) Taken From Examples in the Literature and the One Used by the Authors

	Composition of Artificial Cerebrospinal Fluid (ACSF) in m*M*							
	Na	K	Mg	Ca	Cl	HCO$_3$	Glu	Ref.
Range	143–152	2.5–6.4	1.3–2.4	1.5–2.5	127–136	24–26	4–10	10
Authors	150	3.75	2	2.5	132	26	10	15

small rodents (38 to 39°C). Our laboratory transfers the slices to the chamber at room temperature and permits them to equilibrate with the new environment for 20 min prior to slowly warming them. Many of our studies have been done at 35°C for the duration of the experiment. This is the temperature suggested by Conners and Gutnick[11] to avoid possible metabolic problems in the slice, such as deficits of either oxygen or glucose, essential nutrients for brain tissue. Slice temperature is an important variable and the use of a reliable thermocouple and heater such as the temperature controller available from Medical Systems Corp., Greenvale, NY (Temperature Controller, Model TC-102) has proven reliable and easy to maintain.

10. Humidity Control

Humidity is kept near the saturation point by bubbling 95% O_2/5% CO_2 through the water-jacketed chamber below the slice and providing constant flow into the chamber throughout the experiment. We have found it necessary to limit the area of the working opening of the chamber using thin sheets of plastic to avoid the tendency for the slices to become dry on the surface.

11. Tissue Perfusion

Oxygenated ACSF perfuses the recording chamber from one end and flows over, around, and under the slices, not only providing them with glucose and oxygen, but also a medium through which slices can ultimately be exposed to drugs. Reservoirs (50 cc plastic syringes) containing the ACSF solution are connected to a miniperistaltic pump (Harvard Apparatus, Model 55-4147, South Naticky, MA) which supplies the chamber with ACSF through tubing that is interposed between the outer and inner walls of the water-jacketed base of the recording chamber. Fluids are warmed to a preset temperature as they flow through the tubing to reach the tissue slices.

The actual rate of perfusion required to maintain the slice should be chosen to provide adequate flow around the slices without submersing them or permitting them to float above the nylon mesh on which they are supported. Our laboratory typically uses a flow rate of about 3 ml/min for ACSF perfusions. The perfusion rate is an important variable in experiments of this type, since the rate of drug delivery to the slice will ultimately affect the time to equlibrium between fluid and tissue. Rates that are too high may tend to move the slice upward or away from its location on the mesh. Rates that are too low will cause the slice to dry out and perhaps become compromised from the lack of delivery of nutrients and buffers in the ACSF.

12. Equilibration Times

A period of 60 to 120 min of equilibration with the system without stimulation seems to be required before electrophysiological recordings are begun. We have found empirically that this is so since stable and reproducible control responses are harder to find until 1 to 2 h have elapsed.

13. Types of Electrodes

Although bipolar stimulating electrodes are commercially available (WPI Laboratory Equipment, Sarasota, FL), it is also possible to construct suitable electrodes using small-diameter metal tubing for support and attaching two nichrome Teflon-coated wires which are exposed only at the tip. We have found it useful to bend the end of the support tube slightly so that it is easier to place the electrode tips on the tissue slices from nearly a vertical position while avoiding contact with the chamber and all but surface contact with the slice.

Glass microelectrodes filled with ACSF act as recording electrodes when placed on the slice. These are prepared from glass capillary tubes which have been pulled to a fine point using a commercial electrode puller (Sutter Instrument Co., Model P87, Novato, CA). Resistance of the electrodes we have used measures about 4 mΩ. Some studies have been done using patch clamp techniques in the slice,[12] but those are beyond the scope of this review.

14. Electrode Placement

A depiction of the general placement of the stimulating and recording electrodes on the slices in the recording chamber is shown in Figure 10.4. Stimulating electrodes are typically placed on the Schaeffer collaterals (SC) to the pyramidal cell layer (Figure 10.5) in order to study area CA1 of the slice. The cell layer is quite visible by naked eye or with low-power (dissecting microscope) magnification. In slices taken from the middle third of the hippocampus as described above, the pyramidal cell layer appears as a translucent line a few millimeters below the cortical surface. The tips of the stimulating electrodes are generally placed close to the pyramidal cell layer, usually perpendicular to the orientation of it.

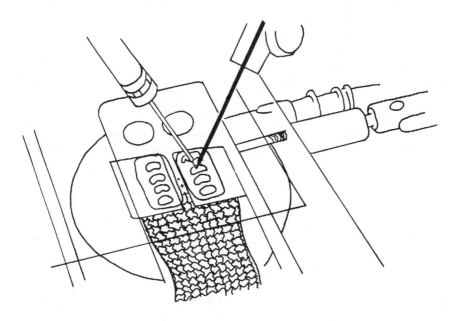

FIGURE 10.4
General placement of electrodes during an experiment on the hippocampal slice.

The recording electrode placement for studying population spikes elicited by SC stimulation also works well if it is placed near the pyramidal cell layer rather than too near the cortical surface (Figure 10.5). Studies of extracellular excitatory postsynaptic potentials (EPSPs)[13] and stimulus input to the slice require placement just below the visible layer. While EPSPs are recorded typically from the stratum radiatum, population spikes (PS)[13] are most easily examined by recording from the stratum pyramidale. Representative potentials recorded from these two locations are

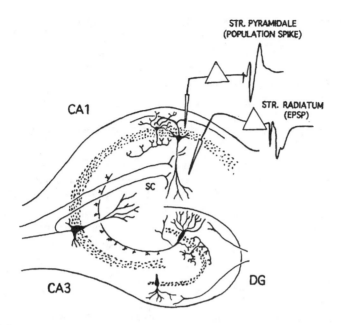

FIGURE 10.5

Top view of a hippocampal slice illustrating the relative locations of cell populations studied in slice experiments. Stimuli delivered to the Schaffer collaterals (SC) evoke excitatory postsynaptic potentials (EPSPs) from the apical dendrites of pyramidal cells in the stratum radiatum while population spikes are evoked in the stratum pyramidale. Typical waveforms are illustrated.

also shown in Figure 10.5. A general reference[13] for explanations of these potentials and the associated terminology may be useful for anyone unfamiliar with the basic neurophysiology.

15. Stimulation Parameters

During the search for electrophysiological responses to stimulation of the slice, stimuli are presented more frequently than during the remainder of the experiment. This enables one to find suitable responses in a shorter time. A 10 s interval between stimuli is useful until a suitable response is found. A 15 s interval between stimuli appears to work well for the remainder of the experiment.

A range of stimulation intensities can be chosen that will evoke variable amplitudes of response from threshold-evoked EPSP levels to a maximum height population spike. Pulse width may vary from 20 to 50 msec. For the study of paired-pulse population spike responses, interpulse intervals for stimulation vary from 15 msec to several hundreds of milliseconds. These pairs are usually delivered at 15 s intervals to avoid interactions between pairs of stimuli.

16. Criteria for Slice Selection/Acceptance

Subjective evaluation of acceptable slices for study usually depend on the surface appearance of the slice, uniform thickness of the slice, and ease of viewing the pyramidal cell layer. With respect to electrophysiological criteria, a stable evoked response, i.e., maintained population spike height and shape for 5 to 10 min under control conditions, is required before initiation of the experiment. If the degree of feedback inhibition is to be studied, the presence or absence and stability of paired-pulse inhibition at 15 msec needs to be quantified.

B. Hippocampal Slice and Epilepsy Research

The neuronal circuitry inherent in the slice makes it suitable for the study of several parameters associated with levels of excitability and inhibition, under both control and hyperexcitable conditions. Changes in evoked EPSP threshold, population spike threshold, waveforms, and the degree of feedback inhibition, as well as paired-pulse facilitation, all represent potential end points for measurement of changes due to experimental manipulation. In addition, changes in spontaneous electrical activity can be measured.

1. Models of Hyperexcitability

Many different methods for inducing hyperexcitability in the slice are available. These offer approaches to the study of epileptogenic phenomena which are devoid of influence from other portions of the brain under these conditions and can provide model systems in which to study the action of both proconvulsant and anticonvulsant chemicals.

a. Electrical Stimulation

The level of stimulus intensity applied to the slice can be varied to provide gradually increasing levels of hyperexcitable responses. Some have used a relatively high combination of current and frequency of stimulation, a "kindling-like" stimulus,[14] which drives the neuronal population to fire at higher frequencies which can be quantified for further study. Clearly these changes can be used as control data against which to compare the effects of anticonvulsant drugs.

b. Chemical Stimulation

Conventional and nonconventional stimulants and convulsants can be perfused over the slices, which ultimately increase the hyperexcitability of the response. Pentyl-enetetrazol, penicillin, bicuculline, and picrotoxin have been used alone or in combination with simultaneous and continuing electrical stimulation, but neurotoxicants such as lindane will also produce increases in hyperexcitability measured by several criteria. Controls and adjustments for changes in electrolyte levels in the perfusion fluid must accompany protocols using these agents. For example, drugs are sometimes dissolved in solvents such as dimethylsulfoxide (DMSO) and the effects of this substance on the slice would have to be evaluated separately from the drug itself.

Convulsants we have used[15] and their concentrations in the perfusion fluid are shown in Table 10.2.

TABLE 10.2
Chemical Convulsants and Concentrations Used in
the Perfusion Fluid

Convulsive agent	Concentration in perfusion fluid
Pentylenetetrazol	4 mM
Picrotoxin	1–2 μM
Lindane	50–75 μM

c.   Variable Magnesium Concentrations

Slices switched to low-magnesium environments via the perfusion fluid become hyperexcitable and begin to fire spontaneously at high rates. These responses can be quantified and used for further study as well.[16] Ashton and Willems have used a low-magnesium model to induce a second population spike to a single stimulus and have tested the ability of an anticonvulsant to abolish it.[17]

d.   Pretreated Animals

Slices taken from animals previously treated either acutely or chronically with other agents or electrical stimulation protocols have also been used to study subsequently the effects on levels of hyperexcitability or diminished inhibition in the slice. These treatments have included exposure to neurotoxicants (lead,[5] alcohol[18,19]) and electrical kindling[20] prior to use of slices in the chamber.

2.   Quantification of Hyperexcitability

a.   Changes in Baseline Measures

Input-output relationships between voltage and EPSP amplitude, population spike amplitude, and ratios among them[21] offer one method for establishing changes in hyperexcitability that have occurred as a function of the experimental treatment. Proconvulsant substances[15] and other treatments typically increase the amplitude of the responses[17] for any given voltage or current applied and some can be shown by paired-pulse analysis of short interval stimulations (15 msec interpulse intervals) to have influenced feedback inhibition of pyramidal cell populations.

b.   Other Measures

If the treatment has increased spontaneous spike discharge rates in the slice, measurement of the new rate can be compared to control values to quantify hyperexcitability. Proconvulsant substances will elicit multiple population spikes visualized in the evoked responses to a single stimulus[15,17] or to the second of two stimuli[15] presented at longer interpulse intervals (e.g., 100 msec). Methods for quantification of these changes in waveform usually involve some analysis of the area under the waveforms.[15,22] Rempke et al.[23] have defined and quantified hyperresponsiveness in neurons within slices after pretreating rats in the self-sustaining limbic status epilepticus model. The criterion for hyperresponsiveness was three or more population spikes in response to a single stimulus adequate to induce a maximum height population spike.[23]

3.   Antiepileptic Drug Evaluation

The general approach to the study of antiepileptic substances in the slice has been one of inducing some quantifiable form of hyperexcitability, which is then followed by exposure of the slice to the drug in the perfusion fluid and additional quantification of the responses. For example, those using a typical slice method in the pharmaceutical industry first measure levels of neuronal discharge following exposure to penicillin and then again following 30 min of exposure to test compounds, quantitating any change in spike discharge rate as a measure of potential antiepileptic activity.[24]

a.   ACSF Controls

Since evaluation of drug exposure in the slice may require prolonged perfusion and time for subsequent data collection periods, it is essential that any changes in function of the slice over time in the absence of any treatment whatsoever be evaluated. Any decrements of function so quantified should be taken into consideration during interpretation of results obtained with other treatments added to the protocol.

b.   Methods of Exposure

Since the experimental apparatus requires a constant flow of ACSF over the slice, it is easy to add convulsant or anticonvulsant chemicals to the perfusion fluid for a predetermined length of time. We attach two 50 cc plastic syringes to a common manifold and turn one supply off (e.g., ACSF + convulsant) while quickly turning on the other (e.g., ACSF + convulsant + anticonvulsant). Figure 10.6 shows this simple arrangement for supplying the perfusion fluids during an experiment. This method of treatment enables one to vary the level of exposure (dose), but requires prolonged perfusion (30 min or more) with most agents to guarantee some level of equilibration with the slices themselves.

Alternatively, known concentrations of agents could be added directly to the slices a drop at a time in the perfused chamber, but this method would have the disadvantage of having variable and rapidly changing drug concentrations with the agent being flushed away fairly quickly at normal perfusion rates. Ionophoretic application of test substances has also been employed.[25]

c. Vehicles

The choice of a solvent for dissolving the chemical under study may be problematic unless the agent dissolves easily in aqueous solvents. Agents such as DMSO (Merck, Rahway, NJ) or surface active agents such as tweens or spans (Aldrich Chemical, Milwaukee, WI) or even alcohol may be required to get the agent into solution, only to have it precipitate back out when it reaches the ACSF perfusing the slice. Appropriate control experiments must be performed examining the effects of all vehicles before interpreting any experimental findings following their use.

MANIFOLD FOR DELIVERY OF PERFUSION FLUIDS

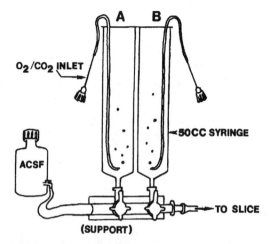

A = ACSF WITH CONVULSANT
B = ACSF, CONVULSANT AND TEST ANTICONVULSANT

FIGURE 10.6
Side view of two syringes (50 cc each) mounted on a common manifold through which perfusing fluids reach slices in the recording chamber. Control experiments are done with artificial cerebrospinal fluid (ACSF) alone. The fluid from chamber A then perfuses the slices for 30 min in order to induce hyperexcitability (epileptiform events) while additional measurements are made. Finally, the solution from chamber B is used to test for anticonvulsant actions produced by the test substance. Note that the convulsant continues to be included in the solution while evaluating the test substance.

d.   End Points for Evaluation

Antiepileptic effects of drug treatments may be evaluated based on any of the chosen end points or measurements discussed. Any distortions or changes in baseline measures induced by convulsants, for example, can be altered back towards normal values by anticonvulsant drug exposure in the slice.[15] Restoration of levels of feedback inhibition after exposure to anticonvulsants may also indicate some potentially useful antiepileptogenic properties of the test agent.[15] Drug treatment may diminish levels of hyperexcitability, i.e., decreasing the number of multiple population spikes following exposure to convulsants.[15]

e.   Reevaluation or Time Course

We have measured the time course of drug effects in the hippocampal slice by reevaluating all measures at 10, 30, and 60 min after continuous drug exposure via the perfusion fluid. Full interpretation of data collected from such a protocol would again require similar studies using only the vehicle for the same treatment period.

III. Interpretation/Evaluation of Results

A. Summary Overview

1. Advantages

Hippocampal anatomy is easy to visualize

The neuronal circuitry has been well characterized

Precise electrode placement is easy

Absence of input from the remainder of the whole brain simplifies variables

Quick alteration of responses and rapid evaluation of drugs is possible

Provides the ability to follow onset and offset of changes in activity in an acute preparation

2. Disadvantages

The absence of input from the remainder of the whole brain may make findings difficult to generalize to the larger system

Difficulty in determining relevance of *in vitro* exposure levels to intact levels of exposure

Variability between slices and responses evoked by treatments makes statistical evaluation difficult

3. Additional Perspectives

The use of any animal "model" of epilepsy inherently raises certain questions about the relevance of findings obtained while using it. In the case of the hippocampal slice preparation, one must also ask questions about the overall involvement of the hippocampus itself in various forms of epilepsy. For example, animals prepared without a hippocampus have been shown to exhibit a fully developed maximal electroshock response.[26] There is also evidence that animals can develop and express a fully kindled response in the amygdala without participation by the hippocampus.[27]

Additional questions arise when evaluating and interpreting data obtained following the use of antiepileptic drugs in the slice preparation. Does any demonstrated anticonvulsant action (by any criterion) accurately predict or correlate with ultimate efficacy in any specific form of human epilepsy? If so, which variants of epilepsy are best modeled by the slice? While the answers to these questions are available for other models, such as maximal electroshock seizures, electrical kindling, and those induced by pentylenetetrazol, the answers are far from clear in the case of the hippocampal slice preparation. It may be some time before the presence or absence of any correlation between anticonvulsant activity in the slice and some variant of human epilepsy is known, simply because there has been no "standard slice" preparation under prolonged study by a large number of neuroscientists.

While the hippocampal slice preparation does offer a convenient *in vitro* method for the study of abnormal and epileptogenic phenomena, it has not yet been critically evaluated with respect to all of these important considerations. Given the need for finding and evaluating new antiepileptic drugs, study of the slice preparation should continue until the overall relevance of its role in preclinical drug testing becomes clear.

References

1. Fisher, R. J., Animal models of epilepsies, *Brain Res. Brain Res. Rev.*, 14, 245, 1989.
2. Skrede, K. K. and Westgaard, R. H., The transverse hippocampal slice: a well defined cortical structure maintained in vitro, *Brain Res.*, 35, 589, 1971.
3. Armstrong, D. L., The hippocampal tissue slice in animal models of CNS disorders, *Neurosci. Neurobehav. Rev.*, 15, 79, 1991.
4. Sarvey, J. M., Burgard, E. C., and Decker, G., Long-term potentiation studies in the hippocampal slice, *J. Neurosci. Methods*, 28, 109, 1989.
5. Wiegand, H. and Altmann, L., Neurophysiological aspects of hippocampal neurotoxicity, *Neurotoxicology*, 15, 451, 1994.
6. Schwartzkroin, P. A., Cellular electrophysiology of human epilepsy, *Epilepsy Res.*, 17, 185, 1994.
7. Bahr, B. A., Long-term hippocampal slices: a model system for investigating synaptic mechanisms and pathologic processes, *J. Neurosci. Res.*, 42, 294, 1995.
8. Heimrich, B. and Frotscher, M., Slice cultures as a model to study entorhinal-hippocampal interaction, *Hippocampus*, Spec. No. 3, 11, 1993.
9. Engel, S. R., Gaudet, E. A., Jackson, K. A., and Allan, A. M., Effect of in vivo administration of anesthetics on GABA-A receptor function, *Lab. Anim. Sci.*, 46, 425, 1996.
10. Alger, B. E., Dhanjal, S. S., Dingledine, R., Garthwaite, J., Henderson, G., King, G. L., Lipton, P., North, A., Schwartzkroin, P. A., Sears, T. A., Segal, M., Whittingham, T. S., and Williams, J., Brain slice methods, in *Brain Slices*, Dingledine, R., Ed., Plenum Press, New York, 1984, 381.
11. Connors, B. W. and Gutnick, M. J., Neocortex: cellular properties and intrinsic circuitry, in *Brain Slices*, Dingledine, R., Ed., Plenum Press, New York, 1984, 313.
12. Isokawa, M., Decrement of GABA-A receptor-mediated inhibitory postsynaptic currents in dentate granule cells in epileptic hippocampus, *J. Neurophysiol.*, 75, 1901, 1996.
13. Shepherd, G. M., *Neurobiology*, Oxford University Press, New York, 1988, 122.
14. Stasheff, S. F., Hines, M., and Wilson, W. A., Axon terminal hyperexcitability associated with epileptogenesis in vitro, *J. Neurophysiol.*, 70, 961, 1993.
15. Stark, L. G., Joy, R. M., Walby, W. F., and Albertson, T. E., Interactions between convulsants at low-dose and phenobarbital in the hippocampal slice preparation, in *Kindling 5*, Corcoran, M. and Moshe, S., Eds., Plenum Press, 1997, in press.
16. Anderson, W. W., Lewis, D. V., Swartzwelder, H. S., and Wilson, W. A., Magnesiuim-free medium activates seizure-like events in the rat hippocampal slice, *Brain Res.*, 398, 215, 1986.
17. Ashton, D. and Willems, R., In vitro studies on the broad spectrum anticonvulsant loreclezole in the hippocampus, *Epilepsy Res.*, 11, 75, 1992.

18. Chepkova, A. N., Doreulee, N. V., Trofimov, S. S., Gudasheva, T. A., Orstrovskaya, R. U., and Skrebitsky, V. G., Nootropic compound L-pyroglutamyl-D-alanine-amide restores hippocampal long-term potentiation impaired by exposure to alcohol in rats, *Neurosci. Lett.*, 188, 163, 1995.

19. Diao, L. and Dunwiddie, T. V., Interactions between ethanol, endogenous adenosine and adenosine uptake in hippocampal brain slices, *J. Pharmacol. Exp. Ther.*, 278, 542, 1996.

20. Behr, J., Gloveli, T., Gutierrez, R., and Heinemann, U., Spread of low Mg^{2+} induced epileptiform activity from the rat entorhinal cortex to the hippocampus after kindling studied in vitro, *Neurosci. Lett.*, 216, 41, 1996.

21. Joy, R. M., Walby, W. F., Stark, L. G., and Albertson, T. E., Lindane blocks GABA-A-mediated inhibition and modulates pyramidal cell excitability in the rat hippocampal slice, *Neurotoxicology*, 16, 217, 1995.

22. Balestrino, M., Aitken, P. G., and Somjen, G. G., The effects of moderate changes in extracellular K^+ and Ca^{++} on synaptic and neural function in the CA 1 region of the hippocampal slice, *Brain Res.*, 377, 229, 1986.

23. Rempke, D. A., Mangan, P. S., and Lothman, E. W., Regional heterogeneity of pathophysiological alterations in CA1 and dentate gyrus in a chronic model of temporal lobe epilepsy, *J. Neurophysiol.*, 74, 816, 1995.

24. Garske, G. E., Palmer, G. C., Napier, J. J., Griffith, R. C., Freedman, L. R., Harris, E. W., Ray, R., McCreedy, S. A., Blosser, J. C., Woodhead, J. H., White, H. S., and Swinyard, E. A., Preclinical profile of the anticonvulsant remacemide and its enantiomers in the rat, *Epilepsy Res.*, 9, 161, 1991.

25. Collins, D. R. and Davies, S. N., Potentiation of synaptic transmission in the rat hippocampal slice by exogenous L-glutamate and selective L-glutamate receptor subtype agonists, *Neuropharmacology*, 33, 1055, 1994.

26. Browning, R. A. and Nelson, D. K., Modification of electroshock and pentylenetetrazol seizure patterns in rats after precollicular transections, *Exp. Neurol.*, 93, 546, 1986.

27. Racine, R. J., Paxinos, G., Mosher, J. M., and Kairiss, E. W., The effects of various lesions and knife-cuts on septal and amygdala kindling in the rat, *Brain Res.*, 454, 64, 1988.

28. Rondouin, G., personal communication.

Chapter

Microdialysis Techniques for Epilepsy Research

John W. Dailey and Pravin K. Mishra

Contents

I. Introduction

Dialysis is a process of passive diffusion of solute across a semipermeable membrane. *In vivo* microdialysis is essentially a dialysis procedure, but miniaturized to

0-8493-3362-8/98/$0.00+$.50

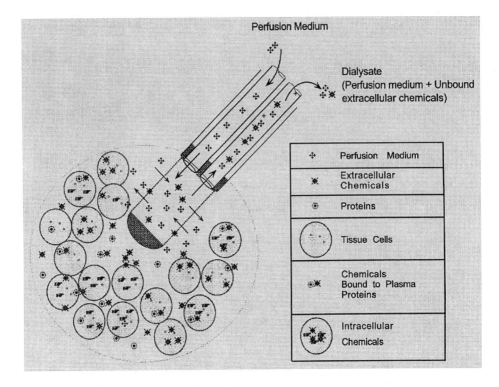

FIGURE 11.1

Principle of microdialysis as depicted by a schematic microdialysis probe placed in a tissue. An artificial extracellular fluid perfusion medium such as artificial cerebrospinal fluid is delivered through the microdialysis probe inlet tube. As the perfusion medium passes the active dialysis area, solute moves in both directions across the dialysis membrane. The perfusion medium collects extracellular solute via the process of dialysis and the dialysate containing the solute of interest is delivered from the outlet tube.

allow continuous sampling of molecules from extracellular fluid of animals and man. The principle of microdialysis is illustrated in Figure 11.1. A microdialysis probe is composed of a tube-like dialysis membrane or fiber attached to two rigid or semirigid transport tubes. Both the inlet and outlet transport tubes, as well as the dialysis fiber, are hollow. Although dialysis probes are quite small (200 to 600 μm in diameter), they can accommodate passage of several microliters of liquids through their lumen each minute. When used in experiments, the membrane portion of the probe is placed in the extracellular space of a tissue or tissue matrix and a liquid perfusion medium is pumped continuously through the dialysis fiber lumen. This perfusion medium usually has an ionic and osmotic composition similar to the extracellular fluid, but lacks the chemicals that are intended to be sampled. This way, molecules that can travel across the membrane diffuse into the medium from the extracellular space when the perfusion medium passes through the fiber. The perfusion medium at this point is termed dialysate as it contains the molecules from the extracellular space by virtue of dialysis. In some situations, the perfusion medium may contain an excess of certain chemicals such as a drug which can diffuse out into the extracellular space.

There are several types of dialysis membranes appropriate for use in microdialysis. All membranes are semipermeable and allow passive entry of molecules that can readily diffuse through its complex porous maze. The solutes with molecular weight lower than the permeability allowed by the membrane are the only substances that can diffuse through the membrane. While microdialysis fibers have been used to sample the chemistry of the extracellular fluid of many organs and tissues, most applications of microdialysis techniques to studies of epilepsy have involved placement of the microdialysis fiber into specific brain regions or specific nuclei.

When used to dialyze brain extracellular fluid, the perfusion medium usually has an ionic and osmotic composition that is similar to cerebrospinal fluid. Most brain dialysis experiments involve sampling the extracellular fluid for neurotransmitter concentrations or for concentrations of a drug. For these applications, the contents of the artificial cerebrospinal fluid (ACSF) dialysate usually are analyzed by high-performance liquid chromatography (HPLC).

An advantage of microdialysis for studies of brain chemistry is that the technique allows continuous sampling of low-molecular-weight substances in the extracellular fluid. Neurons store relatively large concentrations of neurotransmitters intracellularly. This stored neurotransmitter does not have functional significance until it is released as part of the process of chemical neurotransmission. Neurons release neurotransmitter into the extracellular or synaptic space when they are stimulated. This extraneuronal neurotransmitter interacts with receptors on other cells to produce a response. Since microdialysis allows sampling of the extracellular neurotransmitter, it provides a direct estimate of the neurotransmitter available to interact with cellular receptors. Earlier techniques for estimating extracellular or functional neurotransmitter (e.g., turnover rate or isotopic dilution measurements) required large numbers of animals and allowed only a single estimate of functional neurotransmitter release per experimental animal. In some ways these earlier techniques provided data analogous to a photograph of an event. Microdialysis allows continuous sampling of the extracellular fluid in an individual animal so that it is more analogous to a videotape of an event as it takes place.

In studies of epilepsy, microdialysis can be used to sample extracellular neurotransmitters, cellular substrates, or metabolites such as glucose or lactate and ions such as calcium or potassium. This sampling can take place in the preictal phase, during the seizure, and postictally. Thus, microdialysis allows an evaluation of the effects of a seizure on the solutes of interest. Microdialysis also can be used for studies of the distribution and metabolism of drugs. Several groups have carried out detailed pharmacokinetic studies of antiepileptic drugs in brain extracellular fluid while, in some cases, simultaneously measuring drug concentration in peripheral compartments such as blood or subcutaneous fluids. This technique allows a complete pharmacokinetic profile to be generated in a single subject.

Many laboratories around the world have used microdialysis for studies in both animals and humans. As might be anticipated the bulk of these investigations have taken place in animals, but an increasing number of microdialysis studies are being carried out in humans.[1-14] Microdialysis technique as it is used today was first described in 1966[15] and in 1969,[16] but it became a routine research tool only in the last decade and credit for its popularity is due to Ungerstedt and Hallstrom.[17] Since then, over 4000 reports have been published using a variety of microdialysis applications in humans, animals, and many kinds of plant. The increasing popularity and utilization of the technique is largely due to many improvements in commercially available materials, the availability of more sensitive analytical techniques, and constant refinements in the technique itself.[18] While these improvements definitely have made microdialysis easier to use, it remains a technically demanding procedure. Individuals who wish to learn microdialysis to apply it to studies in their laboratories would be well advised to enroll in one of the microdialysis courses offered by equipment manufacturers or scientific societies. Alternatively, a visit to a laboratory with an ongoing microdialysis setup is worthwhile.

II. Methodology

Microdialysis sampling can be accomplished in many ways. Just like any other research technique, the standard procedures may have to be optimized and/or modified in order to incorporate the requirements of a given experiment. Several fine textbooks and review articles are available on microdialysis.[19-22] This section summarizes the common practices and discusses the pros and cons of common variations. It details some aspects of methodological refinements used in our laboratory. Lastly, some discussion on how this technique can be implemented in epilepsy research is provided.

One could opt for a commercially available system, as it can save substantial time required to configure a new system. Although fabricating the probes is not complicated, making them work with all the connecting tubes, characterizing all the parameters of the system, and validating the results obtained from a newly configured system before using it can take several years. In commercially available systems, most of these parameters are already established.

A. Probe Design

More than half of the published studies use "homemade probes" or probes that were made by investigators. This was common because initially the commercially available probes were prohibitively expensive and not as robust as they are now. Also, many probes are wasted during the learning period. However, once an investigator has enough experience, probes can be utilized in a very efficient manner, sometimes even the same probe over many experiments. Thus, the probes and supplies are a very small fraction of the total cost of microdialysis setup.

In the early years of microdialysis use, some investigators[23-27] used "transcranial probes" or probes that passed straight through the brain (Figure 11.2B). However,

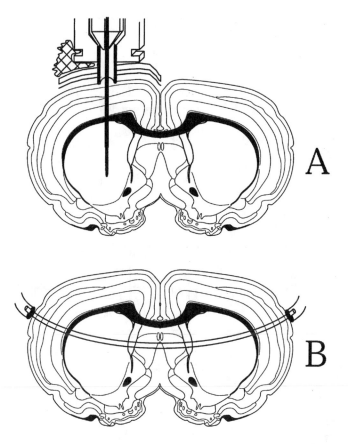

FIGURE 11.2
Panels A and B depict coronal brain sections showing the placement of a radial-type microdialysis probe (panel A) or a transcranial-type microdialysis probe (panel B).

the popularity of this approach has diminished over time. First, transcranial probes are difficult to put in place and could not be used easily in freely moving animals. Second, recent refinements in analytical techniques allow adequate analysis samples from relatively smaller probes that can be precisely inserted through the dorsal surface of the skull. These probes are most popular now and are called "radial probes" (Figure 11.2A). Both "homemade" and commercially available radial probes employ one of three configurations: "loop-type," "concentric," or "side-by-side" (Figure 11.3). The loop-type probes have the fiber bent in a "V" or a "U" shape and have a wire inside the loop to prevent the fiber from kinking (Figure 11.3). Both concentric and side-by-side probes have a fiber with one end permanently sealed with water-insoluble glue and are quite similar in appearance and functionality. In the concentric probe, one transport tube is arranged inside the lumen of the other, whereas side-by-side probes have the inlet and outlet tubes arranged next to each other (Figure 11.3).

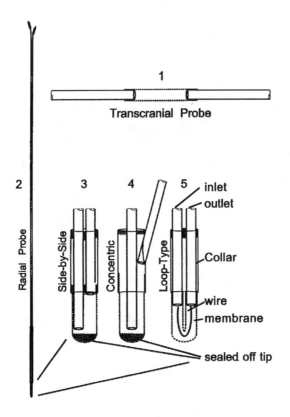

FIGURE 11.3

This figure depicts a schematic drawing of a transcranial probe (1) and a radial probe (2). Radial probe dialysis areas and inlet and outlet tubes can be side-by-side (3), concentric (4), or loop-type (5) designs. Dashed lines represent the active dialysis area while the support structures are represented by solid lines.

B. Materials

Many fiber-shaped dialysis materials have been successfully used in fabrication of microdialysis probes. Popular ones are regenerated cellulose, cellulose acetate, cellulose ester, polysulfone, etc.[28] These membranes vary in their permeability limits, diameter, wall thickness, etc. When choosing a dialysis membrane, the primary consideration should be the permeability limit and its performance *in vivo*. Certain membrane materials (such as polysulfone) offer excellent *in vitro* recovery numbers for a given surface area, but may perform poorly *in vivo*, and others are vice versa. The regenerated cellulose membranes are usually rugged enough to sustain the handling required under common laboratory conditions. If the membrane must be wet in order to stay patent, threading it inside the plumbing tubes and gluing it can be tricky. Flaccidity and handling of the membrane in dry and wet conditions should be considered. The membrane should be firm enough that it can be conveniently placed inside the target tissue. Also important factors are availability of diameters that are suitable for the target tissue, and their ability to withhold internal fluid pressures. Such information is usually available from the membrane manufacturer.

The material and dimensions for the inlet and outlet connecting tubing can vary depending upon the material and dimensions of the membrane. The choice of material and diameter of connecting tubing also depends upon the flow rate to which the finished probe is going to be subjected. The fluid pressure inside the dialysis area depends upon the flow rate, diameter, and length of the tubes connected to it, as well as viscosity of the fluid. The relationship between the fluid pressure P inside extremely narrow bore cylindrical tubing is given by the following equation:

$$P = 8VLQ/\pi R^4$$

where $8/\pi$ is a constant of proportionality, V is the viscosity of the fluid, L is the length of the tubing, Q is the flow rate, and R is the radius of the tubing. Physiological fluids, such as ACSF and Ringer's solution, have similar viscosity. Therefore, the length and diameter are the only factors one can manipulate. Ideally, one should have access to the dialysate as soon as it leaves the fiber but this is practically impossible. Thus, one should choose tubes with the shortest length and smallest diameter. According to the above equation, the pressure inside the fiber can be estimated and some theoretical numbers are calculated in Table 11.1.

TABLE 11.1
Pressure (mmHg) Inside the Fiber Due to the Outlet Tube

Diameter (μm)	Length (cm)	Flow Rate (μl/min)					
		0.05	1	2	3	4	5
25	100	6,532	13,064	26,129	39,193	52,258	65,322
50	100	408	817	1,633	2,450	3,266	4,083
75	100	81	161	323	484	645	806
100	100	26	51	102	153	204	255
200	100	2	3	6	10	13	16

While membranes can be engineered to withstand very high internal fluid pressure, there are other reasons why it is important to use combinations of outlet tubing material to keep the internal fluid pressure in the optimum range. Because the membrane is porous, high internal fluid pressure can push fluid outside the fiber, thereby causing a sweating effect. Moreover, at high pressures there is an increase in the osmotic pressure inside the membrane, which can reduce recovery of extracellular chemicals into the dialysate. Good results can be obtained if pressure inside the fiber does not exceed 500 mmHg. Also, for the tubing inert material such as Teflon, PEEK, fused silica, and polyimide are ideal (see Table 11.2 for sources of materials). Other plastic materials can offer similar results but some (such as polyvinyl chloride or PVC) can leak plasticizer compounds into the dialysate, which can seriously affect the analysis. Metal or other reactive material (such as ordinary plastic or glass) should be avoided in the plumbing, as well as in the collection tube, because

TABLE 11.2
Microdialysis Supply Vendors

ESA, Inc.	Bedford, MA
Bioanalytic Systems, Inc.	West Lafayette, IN
Harvard Apparatus	Holliston, MA
CMA/Microdialysis	Solna, Sweden

of reactivity with the perfusion medium or dialysate and the potential for trapping the sampled substances.

The injury to the tissue at the site where the probe is inserted for prolonged periods depends upon the material used in constructing of the probe and the duration for which the probe invades the tissue. The brain reacts very little to the actual fiber.[29,30] The stiffness of the probe, particularly probes that are constructed with stainless steel, can increase trauma to the brain. It is notable that all earlier studies that reported excessive gliosis around the inserted portion of the probe after prolonged placement[31,32] were conducted when the microdialysis technique was in its infancy and probes were made out of stainless steel material. In recent reports, probes made out of thin fused silica tubes without a rigid support structure have been reported to induce far less damage to the brain, as seen from postmortem histology.[29,33] Furthermore, recent observations indicate that there is no significant difference in the basal levels of dopamine if the probe is reinserted in the same area after 1 week.[25]

Figures 11.4 and 11.5 depict some of the types of microdialysis probes, guide cannulae, and stylets that are currently in use for sampling from animals. Besides the variations in the tip configuration of radial probes (loop-type, concentric, and side-by-side types discussed above), there are numerous ways in which the fiber tip is joined to the connecting tubes and subsequently plumbed to the rest of the system. In some probes the inlet and outlet tubes are ported out via a rigid plastic mount, which also serves as a probe holder. In contrast, other probes (homemade, as well as some commercial ones) consist only of a slender tube but as an option can be secured to a probe holder by the user (Figure 11.4A, B, C). The probe holder in either configuration supports the delicate body of the probe and is used to secure it to the guide cannula (Figure 11.4E). Moreover, if the probe is used with anesthetized animals, the probe holder secures the probe to the arm of the stereotaxic equipment. The fixed length probe holder offers an advantage in that it does not require any configuration prior to use. However, this fixed probe length design suffers from a lack of flexibility, as it does not allow the investigator to adjust the length of the probe projecting from it. Commonly available probes are designed to fit into a guide with a 1 cm guide tube (Figure 11.4D, E and Figure 11.5, left), which is usually sufficient for most targets in a rat brain. Such fixed-length matching guides and probes work very well for targets located far ventral to the dorsal surface of the skull. But for those targets that are located near the dorsal surface of the skull, such a guide cannula may be too long. On the other hand, there are probes that can be adjusted in the probe holder to allow their use with short guide cannulae that do not penetrate the dura (Figure 11.5, right). Although the nonpenetrating guide cannulae

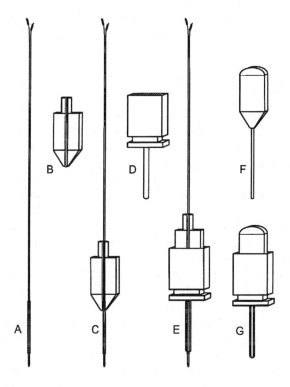

FIGURE 11.4
Various designs of microdialysis probes, probe holders, and guides. (A) Radial dialysis probe; (B) probe holder; (C) radial dialysis probe affixed to a probe holder; (D) probe guide; (E) radial dialysis probe affixed to the probe holder, with the probe and holder placed in the probe guide; (F) stylet; (G) stylet in probe guide.

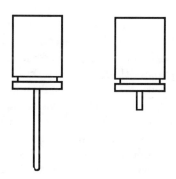

FIGURE 11.5
Commonly used microdialysis guide cannulae. The cannula on the left has the long, penetrating guide tube projecting from the body. The guide is designed to penetrate the dura with the end of the guide tube being placed immediately above the area to be dialyzed. The cannula on the right has the short, nonpenetrating guide tube. The short guide tube is positioned above the dura such that the dura must be punctured before the dialysis probe is inserted.

offer the flexibility of adjusting the ventral position, they require an additional step of precise affixation to the probe holder.

C. Perfusion Media and Procedure

The ideal perfusion medium is a solution that is identical to the chemical and physiological characteristics of the tissue where the dialysis fiber is placed. If the fluid perfused through the probe is identical to the fluid outside the fiber, there will be no net exchange of chemicals. Therefore, a good perfusion solution is one that is physiologically identical to the extracellular fluid but it either is devoid of the substance that is intended to be sampled or has excess of the substance intended to be delivered. Normally, ACSF is used for most sampling applications in the central nervous system. The ACSF is a preferred perfusion medium when microdialysis sampling is done from the brain, as it approximates ionic concentration of the extracellular fluid and contains isomolar calcium. It has been shown that the neurotransmitter release is dependent on calcium and if the perfusion medium is devoid of Ca^{2+} ions, the local extracellular calcium can be depleted and in turn the neurotransmitter release is adversely affected.[34] As a matter of fact, a calcium-free medium is applied in order to evaluate if the release of a neurotransmitter is neuronal. An excess of K^+ in the perfusion medium causes membrane depolarization, which results in neurotransmitter release.[35] In such formulations, Na^+ is substituted with varying concentrations of K^+ so that the medium is isotonic. The substitution of the ACSF to a higher K^+ (50 to -150 mM) solution after a baseline of neurotransmitter release is achieved is a common practice for testing the neuronal release capacity of a brain area. The method for preparing the ACSF and its variations appears in Table 11.3. The purity of chemicals and water is important in formulation of ACSF, as it may interfere with analysis. Other media used for perfusion are Ringer's solution and deionized water. Since only small-molecular-weight substances can pass through the fiber, and the perfusion media are generally aqueous, only water-soluble drugs within the permeability limits of the fiber can be administered or sampled.

The most common way of accomplishing a continuous perfusion of the medium for microdialysis is using perfusion pumps. Most commercially available microprocessor-controlled syringe pumps deliver the fluid precisely at desired flow rates. The important parameters to consider are the ability of the pumps to consistently perform over the entire duration of the experiment and whether their motors can sustain the load that is required to pump in narrow bore tubes. Among the syringes that are used in the perfusion pumps, those that have inert material pistons and glass bodies are usually better than the plastic disposable ones. The plasticizers used in plastic disposable syringes with rubber pistons may contaminate the perfusion medium. Many investigators have successfully utilized other types of pumps, such as peristaltic and osmotic pumps.

TABLE 11. 3
A Solution that Closely Matches the Electrolytic Concentrations of the Cerebrospinal Fluid (CSF) can be Prepared by Combining Equal Volumes of Solutions A and B

A: Make a 500 ml solution using sterile deionized water and ultra:		B: Make a 500 ml solution using sterile deionized water:	
Chemical	Amount (g)	Chemical	Amount (g)
NaCl	8.66	$Na_2HPO_4 \cdot H_2O$	0.214
KCl	0.224	$NaH_2PO_4 \cdot H_2O$	0.027
$CaCl_2 \cdot 2H_2O$	0.206		
$MgCl_2 \cdot 6H_2O$	0.163		

Note: Final ionic concentrations (mM): Na, 150; K, 3.0; Ca, 1.4; Mg, 0.8; P, 1.0; Cl, 155.

Variations: Keeping the overall molarity the same, the ionic concentration of K+ can be raised by substituting with Na⁺. The variations of solution A are indicated below. Solution B remains the same. The four commonly used solutions are

	NaCl	KCl	$CaCl_2 \cdot 2H_2O$	$MgCl_2 \cdot 6H_2O$
ACSF	8.66	0.224	0.206	0.163
40 mM K⁺	6.52	2.99	0.206	0.163
100 mM K⁺	3.07	7.47	0.206	0.163
150 mM K⁺	0.175	11.2	0.206	0.163

D. Implanting the Probes

Microdialysis probes are placed in the tissue of choice using various techniques. Their placement in the brain is quite similar to that of electrodes or injection cannulae. If the sampling is intended to be from anesthetized animals, a bare probe mounted on the arm of a stereotaxic instrument can be adequate. However, if the sampling is intended to be from an awake and behaving animal, it is required that either a guide cannula or the probe itself is surgically implanted beforehand. See Chapter 3 for a detailed description of intracranial implantation surgery.

Surgical implantation of microdialysis probes is a debatable approach. This was the only approach in early microdialysis experiments. Both transcranial and radial probes have been used after direct implantation in awake animals. But a chronic placement of the probes for more than 3 to 4 d reduces physiological response at the probe site due to gliosis.[36,37] The early probes used were made of stainless steel tubes which did not move in relation to the brain, which may have been the cause for the excessive gliosis. Newer, more flexible probe designs allow movement of the probes in relation to the brain and may cause relatively less gliosis.

Most experiments are conducted by inserting probes through preimplanted guide cannulae. Commonly used guide cannulae are depicted in Figure 11.5. While designs can vary, guide cannulae have two main parts: the body and the guide tube. A guide tube is the distal slender portion (Figure 11.5), which is inserted through a drilled hole in the skull. The proximal portion or the body accommodates the probe holder (Figure 11.5), or the stylet (to keep the guide tube from clogging after surgery until the experiment, Figure 11.4F, G). While the body of the guide cannula varies in design to accommodate specific types of probes, the guide tube is either short (nonpenetrating type, Figure 11.5, right) or long (deep penetrating type, Figure 11.5, left). When inserted into the guide cannula, the dialysis membrane protrudes beyond the end of the guide cannulae into previously undisturbed tissue (Figure 11.4E). Many prefer that the guide tube, when inserted through the skull, should terminate at the dorsal edge of the target area for probe placement as it allows more precise placement of the probe. Such users prefer the long or deep penetrating type of guide tube. While it may be appealing to some investigators by giving a perception of more precise placement, the deep penetrating-type guides results in a greater insult to the brain for a more prolonged period. Others opt for a nonpenetrating cannula, which is implanted above the dura mater and therefore does not cause any insult to the brain. The extent of insult to the brain tissue is limited to the dimensions of the probe and duration of actual probe placement. However, when using the nonpenetrating-type guides, an additional step of puncturing the dura mater prior to the probe placement is required because the probe tips are not hard enough to pierce the dura.

E.   Sample Collection and Analysis

The concentration of analytes collected in the dialysate usually is 1% to 30% of those of extracellular fluid. While increasing surface area of the dialysis fiber or reduction in flow rate can increase the relative recovery of the probes this also results in an increase in probe size and fluid handling problems, respectively. The flow rate and the size of fiber have to be optimized in a manner that the available analytical techniques can reliably detect the substance of interest in the dialysate fluid. In other words, the flow rate, probe dimensions, and collection intervals are largely dictated by the requirements of the analytical technique.

F.   Fluid Handling

At low concentrations, many substances are subject to a rapid degradation or binding to the surface of the material they come in contact with. At the flow rates of 0.5 to 5.0 µl/min, all fluid handling tubes and collection gear has to be adequate to reliably handle microliter quantities of the dialysate. The smallest possible diameter inlet tube, outlet tube, and connectors with low dead volume swivels allow access to the fluid in the shortest possible time after it has passed through the fiber. In a typical setup, using a 75 μm inner diameter fused silica tube which has 0.038 μl/cm dead volume allows access to the dialysate in <2 min. By using shorter collection tubes in anesthetized animals it is possible to collect the fluid right after it leaves the outlet of the probe. It is ideal to collect the fluid samples directly into the sampling valve of the analytical system because it eliminates the need for handling it. The material of each component in the fluid pathway should be carefully chosen so that the analytes are not quenched or bound, and do not react with it. Generally, Teflon, PEEK, polyimide, and fused silica are sufficiently inert to provide adequate preservation of the analytes and are available in a variety of inner and outer diameter choices so that they can be customized for use in most situations.

G. Sample Preservation

Enzymes that metabolize extracellular chemicals do not permeate the dialysis fiber, because of their high molecular weight. This offers an advantage, as the postdialysis metabolism of the analytes is largely eliminated. Thus, the microdialysis dialysate is a chemically fixed sample from a range of sample collection times which is something a tissue biopsy or a blood/plasma sample cannot provide. Despite the lack of enzymes in the dialysate some analytes are susceptible to degradation due to oxidation and binding to the surface of the collection vial. The monoamines can degrade rapidly at natural pH but if they are preserved in an acidified environment, they maintain their concentration in the dialysate long enough to be analyzed. Figure 11.6 shows monoamine stability over a 4-h period. The biogenic amino acids, on the other hand, are quite stable in aqueous solution and therefore require no special preservation. Other drugs or chemicals, depending upon their stability and reactivity, may require a fixative/stabilizing agent to be added immediately upon collection. A variety of methods are employed for sample preservation, which include removal of oxygen by bubbling inert nitrogen in the perfusion medium prior to the experiment or adding antioxidants such as ascorbic acid. Some investigators measure the degradation and recovery by adding to the perfusion medium or to dialysate an internal standard that has similar physical properties to the analyte.[38] Adding fixative or an internal standard can be achieved either by collecting in vials preloaded with the fixative or by connecting the outflow tube of a second syringe on the same pump to the microdialysis outlet using a Y connector.

H. Analysis of Dialysate

The biggest challenge in microdialysis is to quantify the analytes that are sampled. The concentration of these analytes typically is in femtomolar or at best picomolar range and the quantity of available volume is in microliters. In fact, the microdialysis technique established its utility more than two decades after its conception because the analytical techniques became available only during the last decade. Dopamine

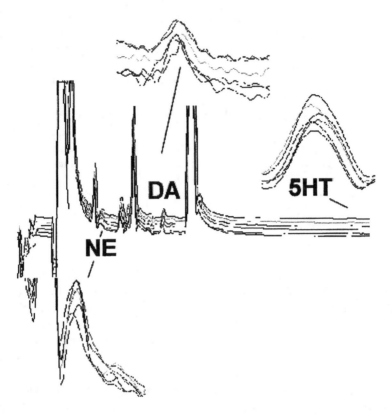

FIGURE 11.6

Stability of monoamine neurotransmitters in dialysate. The HPLC chromatograms depict six repeated analyses of the same dialysate sample over a 4-h period and show that there is no significant degradation of the samples.

is probably the most sampled analyte in dialysate because its abundance in the striatum made it very convenient to sample and analyze. But in the last 5 years, the HPLC systems have evolved considerably and now allow a simultaneous detection of norepinephrine, dopamine, and serotonin in <1 min samples from rat brain (Figure 11.7). Commercially available microbore systems allow routine analysis of 1 to 2 fM samples. Amino acid analysis systems are not as advanced, but it is possible to accurately quantify 5 to 10 min samples of neurotransmitter amino acids. Chemicals of the cholinergic system (acetylcholine and choline) as well as drugs, electrolytes, and arachidonic acid metabolites have been successfully analyzed in the dialysates.

I.   Data Analysis

Microdialysis offers many advantages over the techniques it replaces. For example, in pharmacokinetic studies, precise timed samples can be obtained simultaneously from various sites of the same animal. In pharmacodynamic and physiologic studies

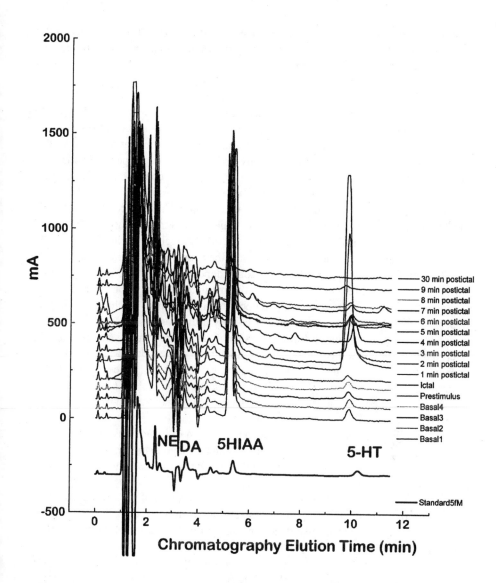

FIGURE 11.7

Analysis of dialysates collected over a 1-min period in the preictal, ictal, and postictal phases of a class 9 seizure in a genetically epilepsy-prone rat. NE (norepinephrine), 5HIAA (5-hydroxyindoleacetic acid), and 5-HT (5-hydroxytryptamine, serotonin) increased in the immediate postictal phase and returned to near normal at 30 min postictal, while DA (dopamine) did not change during or after the seizure.

in which an endogenous chemical such as a neurotransmitter is measured, the same animal and same implantation site data prior to drug or physiological challenge can serve as its own control.

In the current state of technology, it is relatively easy and reliable to compare the concentration from pretreatment (or baseline) levels and post-treatment levels and interpret them as a ratio of basal levels. However, care should be taken in interpreting the numbers of relative change. While the estimation of extracellular concentration of an endogenous chemical requires independent experiments and thus cannot be combined with other experiments, their importance in interpreting data is well established. There are several methods in practice for estimation of true *in vivo* recovery (an extracellular concentration). One such method is the "no net flux method" which assumes that, if the concentration inside and outside the fiber is the same, there will be no net exchange of a given chemical. In this method, the substance of interest is added to the perfusion medium in varying concentrations. The values at which there is no difference between analyzed concentrations in perfusion medium and dialysate is considered to be the true extracellular concentration.[39] Another method is the "zero flow extrapolation method" which assumes that if the flow rate was nil, the dialysate will equilibrate to the extracellular fluid concentration. In this method the flow rate is reduced sequentially which allows increasing periods of time for the fluid inside the fiber to equilibrate with the extracellular fluid. The flow rate vs. concentration data are plotted to extrapolate the concentration for zero flow rate.[40]

The relative recovery (or efflux) of probes depends upon many factors.[41] While surface area of the fiber and flow rate of the medium can be precisely controlled, the recovery of the dialysate can depend upon the tissue matrix in which the dialysis fiber is placed.[42] *In vivo* recovery of probes may also change in relation to change in concentration, for example following cocaine and amphetamine administration, microdialysis reports a greater increase in dopamine release than was observed by conventional methods, which needs to be carefully interpreted.[43] These factors make accurate generalization difficult. In uniform tissue matrices such as blood and brain, it may be possible to use a predetermined factor of the ratio of *in vivo* and *in vitro* recovery.[44]

When it is desirable to measure the concentration of exogenously administered substances such as drugs, an internal standard can be added to the perfusion medium. Since chemical efflux from and influx into the perfusion medium are equivalent passive processes, loss of the internal standard from the perfusion medium and influx of the drug of interest into the dialysate are equal. Thus, loss of the internal standard predicts recovery of the drug of interest. An elegant example of the use of an internal standard for pharmacokinetic studies is provided by the studies of carbamazepine pharmacokinetics carried out by Van Belle et al.[45,46]

When establishing a new application of microdialysis or setting up a new laboratory, repeating published work and performing a number of physiological and pharmacological procedures can validate the system.[34,35] If the microdialysis application is in the brain, a potassium challenge (replacing the infusion medium with 100 to 50 mM K$^+$ CSF, according to Table 11.3, after baseline has been established) would reveal a rapid neuronal release of all neurotransmitters.[47] Using calcium-free medium should reveal a reduction in levels of neurotransmitters that have a calcium-dependent release.[35] Pharmacological validation of the identity of an analyte can be done with reuptake blockers in the perfusion medium, which should increase the analyte in the dialysate. For example, addition of fluoxetine to the perfusion medium causes blockade of serotonin reuptake and an increase in the extracellular serotonin.[48] Addition of tetrodotoxin in the perfusion medium should successfully reduce the extracellular neurotransmitter levels.[34,35]

J.   Microdialysis Resources

Microdialysis courses are offered by all leading vendors of microdialysis equipment on a regular basis. A list of vendors appears in Table 11.2. These courses offer something for all levels of skill. While they are useful for beginners, investigators who have started some microdialysis typically learn the most from them. A hands-on laboratory is included in these 2- to 4-d courses which are adequate for starters. Another alternative for learning microdialysis is visiting a microdialysis laboratory. Even though the literature and manuals give a lot of information on different aspects of microdialysis, actual visualization of the technique simplifies the fine aspects. Besides publications, a vast amount of resources is available on the internet and the world wide web.

III.   Interpretation

A.   Brain Chemistry and Seizures

There have been many studies of changes in brain chemistry associated with seizures. Generally seizures are very brief and sometimes are accompanied by violent convulsive movements. Because of the brief and violent nature of many seizures, it often is difficult to obtain an analytical sample large enough to allow accurate assessment of neurotransmitter changes that occur during the ictal phase. Although there are a few published reports of neurotransmitter changes during the ictal event,[49-51] most of these studies have involved evaluation of neurotransmitters in the preictal or postictal phase.[48,52-76] Neurotransmitters that have been evaluated by microdialysis include the monoamines (norepinephrine, dopamine, histamine, and serotonin), acetylcholine, and amino acids (glutamate, aspartate, GABA, and glycine). Other epilepsy-related studies include analysis of free radicals such as nitric oxide[71,77,78] and reactive oxygen species.[79]

B.   Pharmacokinetic Studies

Another type of study increasingly of interest to epileptologists is pharmacokinetic studies in which tissue distribution of antiepileptic drugs is evaluated with the aid of microdialysis.[5,6,8,13,45,48,80-86] This procedure allows continuous sampling of the body compartments to which drugs are distributed so that a complete pharmacokinetic profile can be determined in one experimental subject. Many drugs are extensively bound to plasma proteins or are sequestered in lipids. It is axiomatic in pharmacology that only the unbound or free drug is available for interaction with the tissue receptors that produce the pharmacologic response for nearly all drugs. Because microdialysis depends on passage of solutes across a semipermeable dialysis membrane, only the free drug is collected for analysis. Thus, some scientists view microdialysis as a means of improving the relevance of pharmacokinetic data to pharmacologic responses.

C. Pharmacodynamic Studies

Many studies of neurotransmitter changes in seizure models have involved evaluation of the response of a neurotransmitter to administration of an anticonvulsant or convulsant drug. These studies allow understanding of neurochemical correlates of therapeutic as well as toxic effects of the anticonvulsant drug. Microdialysis studies with carbamazepine,[62,87] valproate,[88,89] zonisamide,[90,91] loreclezole, and antiepilepsirine[92] have revealed a better understanding of their possible therapeutic mechanisms. In such studies, drugs can be administered either systemically or directly through the microdialysis probe.[48] Systemic administration of a drug allows evaluation of the integrated effects of that drug on neurotransmitters in the specific brain region being sampled by the microdialysis probe. Administration of a drug through the microdialysis probe produces pharmacologically relevant concentrations of the drug in a limited area of tissue surrounding the probe. Probes deliver solute to and recover solute from the same sized area. Thus, if the drug delivered through the probe produces an effect on neurotransmitters, the neurotransmitters are sampled from the same region to which effective concentrations of the drug had been delivered. In many ways, studies in which drugs are delivered focally to a brain region are analogous to studies of neurotransmitter function in a brain slice. In a slice, drugs can be superfused over the neurons in the slice, while in microdialysis focally delivered drugs are delivered to a cylinder of tissue surrounding the probe's active dialysis area.

D. Conclusion

Microdialysis is a very valuable and increasingly utilized technique for studying epilepsy. It allows continuous sampling of extracellular fluid for solutes, including neurotransmitters, cellular metabolites, ions, and drugs. Much of what has been learned about the effects of seizures and drugs on brain chemistry has been learned through the use of microdialysis. Although the technique is technically demanding, most individuals skilled in careful laboratory techniques can master microdialysis and use it effectively in studies of epilepsy.

References

1. Ronneengstrom, E., Hillered, L., Flink, R., Spannare, B., Ungerstedt, U., and Carlson, H., Intracerebral microdialysis of extracellular amino acids in the human epileptic focus, *J. Cereb. Blood Flow Metab.*, 12, 873, 1992.

2. Hillered, L., Persson, L., Ponten, U., and Ungerstedt U., Neurometabolic monitoring of the ischaemic human brain using microdialysis, *Acta. Neurochir. Wien.*, 102, 91, 1990.

3. Meyerson, B. A., Linderoth, B., Karlsson, H., and Ungerstedt, U., Microdialysis in the human brain: extracellular measurements in the thalamus of parkinsonian patients, *Life. Sci.*, 46, 301, 1990.

4. During, M. J., Fried, I., Leone, P., Katz, A., and Spencer, D. D., Direct measurement of extracellular lactate in the human hippocampus during spontaneous seizures, *J. Neurochem.*, 62, 2356, 1994.

5. Scheyer, R. D., During, M. J., Cramer, J. A., Toftness, B. R., Hochholzer, J. M., and Mattson, R. H., Simultaneous HPLC analysis of carbamazepine and carbamazepine epoxide in human brain microdialysate, *J. Liquid Chromatogr.*, 17, 1567, 1994.

6. Scheyer, R. D., During, M. J., Spencer, D. D., Cramer, J. A., and Mattson, R. H., Measurement of carbamazepine and carbamazepine epoxide in the human brain using *in vivo* microdialysis, *Neurology*, 44, 1469, 1994.

7. During, M. J. and Spencer, D. D., Extracellular hippocampal glutamate and spontaneous seizure in the conscious human brain, *Lancet*, 341, 1607, 1993.

8. Scheyer, R. D., During, M. J., Hochholzer, J. M., Spencer, D. D., Cramer, J. A., and Mattson, R. H., Phenytoin concentrations in the human brain — an *in vivo* microdialysis study, *Epilepsy Res.*, 18, 227, 1994.

9. Kanthan, R., Shuaib, A., Griebel, R., and Miyashita, H., Intracerebral human microdialysis — *in vivo* study of an acute focal ischemic model of the human brain, *Stroke*, 26, 870, 1995.

10. During, M. J., Ryder, K. M., and Spencer, D. D., Hippocampal GABA transporter function in temporal-lobe epilepsy, *Nature*, 376, 174, 1995, [see comments].

11. Mizuno, T. and Kimura, F., Medial septal injection of naloxone elevates acetylcholine release in the hippocampus and induces behavioral seizures in rats, *Brain Res.*, 713, 1, 1996.

12. Kanthan, R., Shuaib, A., Griebel, R., el-Alazounni, H., Miyashita, H., and Kalra, J., Evaluation of monoaminergic neurotransmitters in the acute focal ischemic human brain model by intracerebral *in vivo* microdialysis, *Neurochem. Res.*, 21, 563, 1996.

13. Stahle, L., Alm, C., Ekquist, B., Lundquist, B., and Tomson, T., Monitoring free extracellular valproic acid by microdialysis in epileptic patients, *Ther. Drug Monit.*, 18, 14, 1996.

14. Kanthan, R., Shuaib, A., Goplen, G., and Miyashita, H., A new method of *in vivo* microdialysis of the human brain, *J. Neurosci. Methods*, 60, 151, 1995.

15. Bito, L., Davson, H., Levin, E., Murray, M., and Snider, N., The concentrations of free amino acids and other electrolytes in cerebrospinal fluid, *in vivo* dialysate of brain, and blood plasma of the dog, *J. Neurochem.*, 13, 1057, 1966.

16. Delgado, J. M. R., United States Patent Appl. 869170, 1969, Fluid-conducting instrument insertable in living organism, United States Patent #3,640,269, 1972.

17. Ungerstedt, U. and Hallstrom, A., *In vivo* microdialysis — a new approach to the analysis of neurotransmitters in the brain, *Life. Sci.*, 41, 861, 1987.

18. Benveniste, H. and Huttemeier, P. C., Microdialysis — theory and application, *Prog. Neurobiol.*, 35, 195, 1990.

19. Robinson, T. E. and Justice, J. B., *Microdialysis in the Neurosciences*, Elsevier, Amsterdam, 1991.

20. Lindefors, N., Amberg, G., and Ungerstedt, U., Intracerebral microdialysis. I. Experimental studies of diffusion kinetics, *J. Pharmacol. Methods,* 22, 141, 1989.

21. Amberg, G. and Lindefors, N., Intracerebral microdialysis: II. Mathematical studies of diffusion kinetics, *J. Pharmacol. Methods,* 22, 157, 1989.

22. Ungerstedt, U., Microdialysis — principles and applications for studies in animals and man, *J. Intern. Med.,* 230, 365, 1991.

23. Day, J. and Fibiger, H. C., Dopaminergic regulation of cortical acetylcholine release, *Synapse,* 12, 281, 1992.

24. Miu, P., Karoum, F., Toffano, G., and Commissiong, J. W., Regulatory aspects of nigrostriatal dopaminergic neurons, *Exp. Brain Res.,* 91, 489, 1992.

25. Camp, D. M. and Robinson, T. E., On the use of multiple probe insertions at the same site for repeated intracerebral microdialysis experiments in the nigrostriatal dopamine system of rats, *J. Neurochem.,* 58, 1706, 1992.

26. Martel, P. and Fantino, M., Mesolimbic dopaminergic system activity as a function of food reward — a microdialysis study, *Pharmacol. Biochem. Behav.,* 53, 221, 1996.

27. DiChiara, G., Carboni, E., Morelli, M., Cozzolino, A., Tanda, G. L., Pinna, A., Russi, G., and Consolo, S., Stimulation of dopamine transmission in the dorsal caudate nucleus by pargyline as demonstrated by dopamine and acetylcholine microdialysis and for immunohistochemistry, *Neuroscience,* 55, 451, 1993.

28. Buttler, T., Nilsson, C., Gorton, L., Markovarga, G., and Laurell, T., Membrane characterisation and performance of microdialysis probes intended for use as bioprocess sampling units, *J. Chromatogr. A,* 725, 41, 1996.

29. Gerin, C. and Privat, A., Evaluation of the function of microdialysis probes permanently implanted into the rat cns and coupled to an on-line hplc system of analysis, *J. Neurosci. Methods,* 66, 81, 1996.

30. Gale, K., Olson, D., Mihali, M., Keough, L., Gunderson, V., and Dubach, M., *Soc. Neurosci. Abstr.,* 21, 204, 1995

31. Robinson, T. E. and Camp, D. M., The effects of 4 days of continuous striatal microdialysis on indices of dopamine and serotonin neurotransmission in rats, *J. Neurosci. Methods,* 40, 211, 1991.

32. Georgieva, J., Luthman, J., Mohringe, B., and Magnusson, O., Tissue and microdialysate changes after repeated and permanent robe implantation in the striatum of freely moving rats, *Brain Res. Bull.,* 31, 463, 1993.

33. Bossi, S. R., Yu, J., Acworth, I. N., and Maher, T. J., An histological approach to identify viability of microdialysis probe usage in the striatum of freely moving rats, *Soc. Neurosci. Abstr.,* 20, 704, 1994.

34. Sharp, T., Bramwell, S. R., Clark, D., and Grahame Smith, D. G., *In vivo* measurement of extracellular 5-hydroxytryptamine in hippocampus of the anaesthetized rat using microdialysis: changes in relation to 5-hydroxytryptaminergic neuronal activity, *J. Neurochem.,* 53, 234, 1989.

35. Westerink, B. H. and De Vries, J. B., Characterization of *in vivo* dopamine release as determined by brain microdialysis after acute and subchronic implantations: methodological aspects, *J. Neurochem.,* 51, 683, 1988.

36. Santiago, M. and Westerink, B. H., Characterization of the *in vivo* release of dopamine as recorded by different types of intracerebral microdialysis probes, *Naunyn Schmiedebergs Arch. Pharmacol.,* 342, 407, 1990.

37. Wang, J., Lieberman, D., Tabubo, H., Finberg, J. P. M., Oldfield, E. H., and Bankiewicz, K. S., Effects of gliosis on dopamine metabolism in rat striatum, *Brain Res.,* 663, 199, 1994.

38. Scheller, D. and Kolb, J., The internal reference technique in microdialysis — a practical approach to monitoring dialysis efficiency and to calculating tissue concentration from dialysate samples, *J. Neurosci. Methods,* 40, 31, 1991.

39. Crippens, D., Camp, D. M., and Robinson, T. E., Basal extracellular dopamine in the nucleus accumbens during amphetamine withdrawal — a no net flux microdialysis study, *Neurosci. Lett.,* 164, 145, 1993.

40. Menacherry, S., Hubert, W., and Justice, J. B., *In vivo* calibration of microdialysis probes for exogenous compounds, *Anal. Chem.,* 64, 577, 1992.

41. Linhares, M. C. and Kissinger, P. T., *In vivo* sampling using loop microdialysis probes coupled to a liquid chromatograph, *J. Chromatogr. B Biomed. Appl.,* 578, 157, 1992.

42. Le Quellec, A., Dupin, S., Genissel, P., Saivin, S., Marchand, B., and Houin, G., Microdialysis probes calibration — gradient and tissue dependent changes in no net flux and reverse dialysis methods, *J. Pharmacol. Toxicol. Methods,* 33, 11, 1995.

43. Olson, R. J. and Justice, J. B., Quantitative microdialysis under transient conditions, *Anal. Chem.,* 65, 1017, 1993.

44. Sarre, S., Van Belle, K., Smolders, I., Krieken, G., and Michotte, Y., The use of microdialysis for the determination of plasma protein binding of drugs, *J. Pharm. Biomed. Anal.,* 10, 735, 1992.

45. Van Belle, K., Dzeka, T., Sarre, S., Ebinger, G., and Michotte, Y., *In vitro* and *in vivo* microdialysis calibration for the measurement of carbamazepine and its metabolites in rat brain tissue using the internal reference technique, *J. Neurosci. Methods,* 49, 167, 1993.

46. Van Belle, K., Sarre, S., Ebinger, G., and Michotte, Y., Quantitative microdialysis for the *in vivo* measurement of carbamazepine, oxcarbazepine and their major metabolites in rat brain and liver tissue and in blood using the internal standard technique, *Eur. J. Pharm. Sci.,* 3, 273, 1995.

47. Kalen, P., Kokaia, M., Lindvall, O., and Bjorklund, A., Basic characteristics of noradrenaline release in the hippocampus of intact and 6-hydroxydopamine-lesioned rats as studied by *in vivo* microdialysis, *Brain Res.,* 474, 374, 1988.

48. Dailey, J. W., Yan, Q. S., Mishra, P. K., Burger, R. L., and Jobe, P. C., Effects of fluoxetine on convulsions and on brain serotonin as detected by microdialysis in genetically epilepsy-prone rats, *J. Pharmacol. Exp. Ther.,* 260, 533, 1992.

49. Mishra, P. K., Burger, R. L., Cullison, J. K., and Jobe, P. C., Simultaneous measurement of subfemtomole amounts of norepinephrine, dopamine and serotonin in microdialysis samples using HPLC-EC, *Soc. Neurosci. Abstr.,* 21, 1137, 1995.

50. Jobe, P. C., Ko, K. H., and Mishra, P. K., Norepinephrine release during seizures in GEPR and nonepileptic rats: analysis of one minute microdialysis samples, *Epilepsia,* 36, 18, 1995.

51. Jobe, P. C., Deoskar, V. U., Burger, R. L., Dailey, J. W., Ko, K. H., and Mishra, P. K., Norepinephrine and serotonin release during seizures in GEPRs and nonepileptic rats: analysis of one minute microdialysis samples, *Soc. Neurosci. Abstr.,* 21, 1966, 1995.

52. Ludvig, N., Mishra, P. K., Yan, Q. S., Lasley, S. M., Burger, R. L., and Jobe, P. C., The combined EEG-intracerebral microdialysis technique: a new tool for neuropharmacological studies on freely behaving animals, *J. Neurosci. Methods,* 43, 129, 1992.

53. Yan, Q. S., Jobe, P. C., and Dailey, J. W., Thalamic deficiency in norepinephrine release detected via intracerebral microdialysis: a synaptic determinant of seizure predisposition in the genetically epilepsy-prone rat, *Epilepsy Res.,* 14, 229, 1993.

54. Cavalheiro, E. A., Fernandes, M. J., Turski, L., Mazzacoratti, M. G. N., Neurochemical changes in the hippocampus of rats with spontaneous recurrent seizures, *Epilepsy Res. Suppl.,* 9, 239, 1992.

55. During, M. J., Craig, J. S., Hernandez, T. D., Anderson, G. M., and Gallager, D. W., Effect of amygdala kindling on the *in vivo* release of GABA and 5-HT in the dorsal raphe nucleus in freely moving rats, *Brain Res.,* 584, 36, 1992.

56. Minamoto, Y., Itano, T., Tokuda, M., Matsui, H., Janjua, N. A., Hosokawa, K., Okada, Y., Murakami, T. H., Negi, T., and Hatase, O., *In vivo* microdialysis of amino acid neurotransmitters in the hippocampus in amygdaloid kindled rat, *Brain Res.,* 573, 345, 1992.

57. Nishikawa, Y., Takahashi, T., and Ogawa, K., Redistribution of glutamate and GABA in the cerebral neocortex and hippocampus of the mongolian gerbil after transient ischemia — an immunocytochemical study, *Mol. Chem. Neuropathol.,* 22, 25, 1994.

58. Doretto, M. C., Burger, R. L., Mishra, P. K., Garciacairasco, N., Dailey, J. W., and Jobe, P. C., A microdialysis study of amino acid concentrations in the extracellular fluid of the substantia nigra of freely behaving GEPR-9s — relationship to seizure predisposition, *Epilepsy Res.,* 17, 157, 1994.

59. Yan, Q. S., Jobe, P. C., Cheong, J. H., Ko, K. H., and Dailey, J. W., Role of serotonin in the anticonvulsant effect of fluoxetine in genetically epilepsy-prone rats, *Naunyn Schmiedebergs Arch. Pharmacol.,* 350, 149, 1994.

60. Yan, Q. S., Jobe, P. C., and Dailey, J. W., Further evidence of anticonvulsant role for 5-hydroxytryptamine in genetically epilepsy-prone rats, *Br. J. Pharmacol.,* 115, 1314, 1995.

61. Yan, Q. S., Jobe, P. C., and Dailey, J. W., Noradrenergic mechanisms for the anticonvulsant effects of desipramine and yohimbine in genetically epilepsy-prone rats — studies with microdialysis, *Brain Res.,* 610, 24, 1993.

62. Dailey, J. W., Reith, M. E. A., Yan, Q. S., Li, M. Y., and Jobe, P. C., Anticonvulsant doses of carbamazepine increase hippocampal extracellular serotonin in genetically epilepsy-prone rats: dose response relationships, *Neurosci. Lett.,* 227, 13, 1997.

63. Lasley, S. M. and Yan, Q. S., Diminished potassium-stimulated GABA release *in vivo* in genetically epilepsy-prone rats, *Neurosci. Lett.,* 175, 145, 1994.

64. Yan, Q. S., Jobe, P. C., and Dailey, J. W., Evidence that a serotonergic mechanism is involved in the anticonvulsant effect of fluoxetine in genetically epilepsy-prone rats, *Eur. J. Pharmacol.,* 252, 105, 1994.

65. Dailey, J. W., Seo, D., Yan, Q. S., Ko, K., Jo, M., and Jobe, P. C., The anticonvulsant effect of the broad spectrum anticonvulsant loreclezole may be mediated in part by serotonin in rats: a microdialysis study, *Neurosci. Lett.,* 178, 179, 1994.

66. Yan, Q. S., Mishra, P. K., Burger, R. L., Bettendorf, A. F., Jobe, P. C., and Dailey, J. W., Evidence that carbamazepine and antiepilepsirine may produce a component of their anticonvulsant effects by activating serotonergic neurons in genetically epilepsy-prone rats, *J. Pharmacol. Exp. Ther.*, 261, 652, 1992.

67. Waldmeier, P. C., Martin, P., Stocklin, K., Portet, C., and Schmutz, M., Effect of carbamazepine, oxcarbazepine and lamotrigine on the increase in extracellular glutamate elicited by veratridine in rat cortex and striatum, *Naunyn Schmiedebergs Arch. Pharmacol.*, 354, 164, 1996.

68. Kanda, T., Kurokawa, M., Tamura, S., Nakamura, J., Ishii, A., Kuwana, Y., Serikawa, T., Yamada, J., Ishihara, K., and Sasa, M., Topiramate reduces abnormally high extracellular levels of glutamate and aspartate in the hippocampus of spontaneously epileptic rats (SER), *Life Sci.*, 59, 1607, 1996.

69. Wu, H. Q. and Schwarcz, R., Seizure activity causes elevation of endogenous extracellular kynurenic acid in the rat brain, *Brain Res. Bull.*, 39, 155, 1996.

70. Richards, D. A., Lemos, T., Whitton, P. S., and Bowery, N. G., Extracellular GABA in the ventrolateral thalamus of rats exhibiting spontaneous absence epilepsy: a microdialysis study, *J. Neurochem.*, 65, 1674, 1995.

71. Rigaud-Monnet, A. S., Heron, A., Seylaz, J., and Pinard, E., Effect of inhibiting NO synthesis on hippocampal extracellular glutamate concentration in seizures induced by kainic acid, *Brain Res.*, 673, 297, 1995.

72. Rowley, H. L., Marsden, C. A., and Martin, K. F., Differential effects of phenytoin and sodium valproate on seizure-induced changes in gamma-aminobutyric acid and glutamate release *in vivo, Eur. J. Pharmacol.*, 294, 541, 1995.

73. Obrenovitch, T. P., Urenjak, J., and Zilkha, E., Evidence disputing the link between seizure activity and high extracellular glutamate, *J. Neurochem.*, 66, 2446, 1996.

74. Takazawa, A., Murashima, Y. L., Minatogawa, Y., Kojima, T., Tanaka, K., and Yamauchi, T., *In vivo* microdialysis monitoring for extracellular glutamate and GABA in the ventral hippocampus of the awake rat during kainate-induced seizures, *Psychiatry Clin. Neurosci.*, 49, S275, 1995.

75. Rowley, H. L., Martin, K. F., and Marsden, C. A., Decreased GABA release following tonic-clonic seizures is associated with an increase in extracellular glutamate in rat hippocampus *in vivo, Neuroscience*, 68, 415, 1995.

76. Sayin, U., Timmerman, W., and Westerink, B. H. C., The significance of extracellular gaba in the substantia nigra of the rat during seizures and anticonvulsant treatments, *Brain Res.*, 669, 67, 1995.

77. Rauca, C. and Ruthrich, H. L., Moderate hypoxia reduces pentylenetetrazol-induced seizures, *Naunyn Schmiedebergs Arch. Pharmacol.*, 351, 261, 1995.

78. Rigaudmonnet, A. S., Heron, A., Seylaz, J., and Pinard, E., Effect of inhibiting no synthesis on hippocampal extracellular glutamate concentration in seizures induced by kainic acid, *Brain Res.*, 673, 297, 1995.

79. Layton, M. E., Pazdernik, T. L., Nelson, S. R., and Samson, F. E., Reactive oxygen species detected by microdialysis in brain extracellular fluid increase during seizures, *Soc. Neurosci. Abstr.*, 17, 1464, 1991.

80. Muller, M., Schmid, R., Georgopoulos, A., Buxbaum, A., Wasicek, C., and Eichler, H. G., Application of microdialysis to clinical pharmacokinetics in humans, *Clin. Pharmacol. Ther.*, 57, 371, 1995.

81. Sato, Y., Shibanoki, S., Sugahara, M., and Ishikawa, K., Measurement and pharmaco-kinetic analysis of imipramine and its metabolite by brain microdialysis, *Br. J. Pharmacol.*, 112, 625, 1994.

82. Delange, E. C. M., Danhof, M., Deboer, A. G., and Breimer, D. D., Critical factors of intracerebral microdialysis as a technique to determined the pharmacokinetics of drugs in rat brain, *Brain Res.*, 666, 1, 1994.

83. Walker, M. C., Alavijeh, M. S., Shorvon, S. D., and Patsalos, P. N., Microdialysis study of the neuropharmacokinetics of phenytoin in rat hippocampus and frontal cortex, *Epilepsia*, 37, 421, 1996.

84. Yan, Q. S., Dailey, J. W., Steenbergen, J. L., and Jobe, P. C., Anticonvulsant effect of enhancement of noradrenergic transmission in the superior colliculus in genetically epilepsy-prone rats (GEPRs): a microinjection study, *Brain Res.*, 765, 149, 1997.

85. Takahashi, R., Hagiwara, M., Watabe, M., Kan, R., and Takahashi, Y., Carbamazepine and carbamazepine-10, 11-epoxide concentrations in rat brain and blood evaluated by *in vivo* microdialysis, *Jpn. J. Psychiatry Neurol.*, 47, 293, 1993.

86. Van Belle, K., Sarre, S., Ebinger, G., and Michotte, Y., Brain, liver and blood distribution kinetics of carbamazepine and its metabolic interaction with cloripramine in rats — a quantitative microdialysis study, *J. Pharmacol. Exp. Ther.*, 272, 1217, 1995.

87. Dailey, J. W., Li, M. Y., Yan, Q. S., Reith, M. E. A., and Jobe, P. C., Carbamazepine anticonvulsant effect: systemic or focal administration releases serotonin in the hippocampus, *Epilepsia*, 38, 45, 1997.

88. Löscher, W., Effects of the antiepileptic drug valproate on metabolism and function of inhibitory and excitatory amino acids in the brain, *Neurochem. Res.*, 18, 485, 1993.

89. Whitton, P. S. and Fowler, L. J., The effect of valproic acid on 5-hydroxytryptamine and 5-hydroxyindoleacetic acid concentration in hippocampal dialysates *in vivo*, *Eur. J. Pharmacol.*, 200, 167, 1991.

90. Okada, M., Kaneko, S., Hirano, T., Ishida, M., Kondo, T., Otani, K., and Fukushima, Y., Effects of zonisamide on extracellular levels of monoamine and its metabolite, and on Ca^{2+} dependent dopamine release, *Epilepsy Res.*, 13, 113, 1992.

91. Okada, M., Kaneko, S., Hirano, T., Mizuno, K., Kondo, T., Otani, K., and Fukushima, Y., Effects of zonisamide on dopaminergic system, *Epilepsy Res.*, 22, 193, 1995.

92. Dailey, J. W., Yan, Q. S., Adams-Curtis, L. E., Ryu, J. R., Ko, K. H., Mishra, P. K., and Jobe, P. C., Neurochemical correlates of antiepileptic drugs in the genetically epilepsy-prone rat (GEPR), *Life Sci.*, 58, 259, 1996.

Chapter **12**

Methodologies for Determining Rhythmic Expression of Seizures

Thomas H. Champney

Contents

I.   Introduction

Virtually all physiological processes have a rhythmic component. When examining pathological changes in a physiological process, investigators often fail to realize that a disruption in the rhythmic nature of the process could be involved in the pathology. Epilepsy, defined as an unexpected increase in neuronal firing, is a specific pathological disorder that has well-characterized rhythmic components. Two approaches can be used when examining the nature and control of rhythmic seizures produced by epilepsy. Investigators can attempt to eliminate the rhythmic nature of the seizures as a variable or they can explore the rhythmic nature of the seizures as a means of understanding the fundamental aspects of epilepsy.

A.   Definitions of Rhythms

A rhythm is characterized as a repeatable change in a measurement that occurs over time. Rhythms are classified based on the frequency of the rhythm.[1-3] Those rhythms that occur less than once per day are ultradian rhythms and include the electroencephalographic (EEG) rhythm, heart rate, and hormonal pulses (Figure 12.1). Those rhythms that occur on a daily basis and are endogenously generated are circadian rhythms. Conversely, those rhythms that occur on a daily basis but are maintained by other factors (such as the light/dark cycle) are diurnal rhythms. Examples of circadian or diurnal rhythms include the body temperature rhythm and the sleep/wake cycle. Rhythms that occur approximately once a month are circalunar rhythms and those that occur on a yearly basis are circannual rhythms.

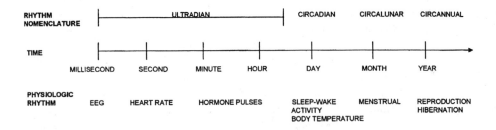

FIGURE 12.1
A time scale depiction of the nomenclature used when describing rhythmic processes. All rhythms shorter than 24 h are classified as ultradian rhythms and include heart rate, hormone pulses, and electroencephalographic (EEG) recordings. Rhythms with a period of approximately 24 h can be diurnal or circadian. Diurnal rhythms are controlled by another factor (such as the light:dark cycle) and will not be observed if the controlling factor is removed. Circadian rhythms are endogenously generated and will continue even in constant conditions with external sources merely acting as entrainment cues. Examples of circadian rhythms include the body temperature rhythm and general activity. Rhythms that occur on a monthly basis are termed circalunar, while those that occur on a yearly basis are circannual.

B. Structures/Compounds that Control Rhythms

Rhythms can be generated by exogenous sources or can be produced endogenously by a physiological "clock".[2-6] The daily light/dark cycle is the best-known exogenous source that can set and regulate daily rhythms. This cycle is primarily conveyed to the nervous system by the retina. The site of the biological "clock" is the suprachiasmatic nucleus of the hypothalamus[6] and this nucleus receives input from the retina to synchronize the endogenous rhythms with the exogenous light/dark cycle (Figure 12.2). Another site that contributes to timing mechanisms is the pineal gland. In mammals, this gland receives rhythmic neural information from the suprachiasmatic nucleus and produces a hormone, melatonin, in response. Melatonin (5-methoxy-*N*-acetyltryptamine) can modulate the biological "clock" as well as provide a hormonal signal to other cells to produce coordinated rhythmic responses.[7] Melatonin is currently used to reset circadian rhythms as a jet lag treatment and is used to induce sleep. It is also being used experimentally as an anticonvulsant to control or modify the rhythmic nature of seizures.[8,9]

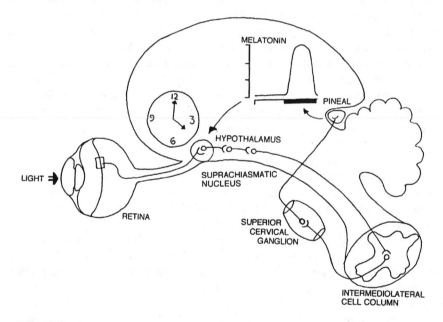

FIGURE 12.2

Artistic depiction of the neuroanatomic pathway that conducts photic information from the retina to the suprachiasmatic nucleus (the biologic clock) and the multisynaptic pathway that conveys this information to the pineal gland. The pineal gland releases the hormone, melatonin, on a rhythmic basis, with the peak occurring during the dark period. Melatonin is able to act on the suprachiasmatic nucleus and regulate "clock" function.

C. Generalized Studies of Rhythms

When investigating physiological processes that have a rhythmic component, it is important to consider the time of day the experiment takes place, the current and previous lighting schedule, and other factors that could impact on the expression of rhythms. Specific factors to control when studying rhythmic processes include (1) performing the experiments at the same time of day and (2) interspersing the control and experimental groups throughout the sampling period, so that any rhythmic differences will be dispersed throughout each group. The previous and current photoperiod exposure can also have a major impact on rhythmic expression.[10] It is recommended that experimental animals be exposed to a consistent photoperiod for 2 weeks prior to experimentation[10] and that human subjects maintain a consistent daily schedule.[11] Concerning animal studies, the use of 12L:12D (12 hours of light and 12 hours of darkness) photoperiods (a very common photoperiod) is discouraged, since many species have difficulty interpreting this photoperiod and greater variability of results can occur.[12] The use of a standard long photoperiod, such as 14L:10D (14 hours of light and 10 hours of darkness), allows the animals to synchronize their rhythms, resulting in less variability within groups. The choice of photoperiod exposure should be considered carefully. Most rodent species are long day breeders, so long photoperiods (14 hours of light per day) will maintain reproductive competence and provide a less variable animal model. However, some large animals (e.g., sheep) are short day breeders and, if these animals are kept in long photoperiods, they will have a reduction in reproductive competence that could unfavorably impact on the measures under study. Therefore, it is important to maintain research animals in controlled environments that will not confound the research protocol.

As with all experiments, consistency in treatment of the subjects is vital to gathering reliable data. This consistency needs to be extended to variables such as duration and intensity of photoperiod exposure, handling and maintenance of the animals, and other environmental cues the animals could use to synchronize their rhythms (such as ambient temperature and noise).[3,10]

When examining rhythmic processes, many researchers attempt to determine if the rhythm is endogenous or exogenous, and whether the rhythm is circadian or diurnal. These results are important in determining the nature of the rhythm, but they may not provide insight into the process being investigated. Likewise, researchers interested in rhythms may choose a rhythmic physiologic process as an output measure, but these researchers may not have an interest in the physiologic process, only in the rhythm inherent in the process. Therefore, when examining studies that report rhythmic processes, such as epilepsy, it should be determined whether the researchers are investigating the rhythm or the process. The main emphasis of this chapter is to provide investigators interested in the process (epilepsy), the means to either control or analyze the rhythmic nature of this process.

D. Previous Research on Rhythms in Epilepsy

Previous research on rhythms in epilepsy has occurred on two major research fronts, the rhythmic expression of seizures and the chronopharmacokinetics of anticonvulsants.[13] The rhythmic expression of seizures has been investigated, with daily, monthly, and seasonal rhythms observed.[14-18] A daily difference in seizure susceptibility in animals has been found in virtually all convulsant methodologies tested.[19-23] For example, there appears to be a daily rhythm in kindling-induced seizures,[20] but no seasonal rhythm in kindling.[17] However, a seasonal rhythm in pentylenetetrazol-induced seizures has been observed,[18] suggesting that convulsant methodologies may have seasonal as well as daily rhythms. Daily rhythms in electroconvulsive shock-induced seizures,[23] in generalized spike wave activity,[22] and in absence epilepsy have also been observed.[16] The observance of daily and seasonal rhythms in models of epilepsy raises important issues on control and interpretation of experiments, such as the value of performing experiments at the same time of day and having controls that are sampled at the same time as the experimental groups.

These rhythms in the epileptic models may be endogenously generated and maintained, but the majority of these rhythms appear to be linked to other physiological processes, such as the EEG, sleep/wake cycle, or the menstrual rhythm.[19-23] Numerous books and reviews have examined this topic, although much of the research in this area is still descriptive and has not determined the mechanism(s) involved.[16-18,24-29]

The other prominent area of investigation is the chronopharmacokinetics of anticonvulsants.[13,30-36] By providing anticonvulsants at specific times throughout the day, control of seizures can be increased while side effects are reduced. For example, the use of carbamazepine as an anticonvulsant is much more effective when administered in a controlled-release preparation that reduces fluctuations in plasma levels.[30-32] This is a clinically relevant area of investigation that has yielded beneficial results for patients on anticonvulsant therapy. This area of investigation is also descriptive and the rhythmic mechanism(s) of action of many of the anticonvulsants is still unknown.

II. Methodology

The methodologies available for the study of rhythms can be subdivided into those methods available for animal use (primarily rodents) and those methods available for use with human subjects. In general, the same methods used to study other rhythmic processes can be used to study the rhythmic nature of seizures. Conversely, those mechanisms used to investigate seizures can be examined for a rhythmic component. For example, the seizure inductive methodologies described elsewhere in this book could be examined for rhythmic components by producing the seizures at different times of the day and then examining for rhythmicity in seizure severity.

To further characterize the rhythm, an investigator could place the subjects in constant conditions and determine if the rhythm still exists and whether it continues with a periodicity of its own (Figure 12.3), i.e., whether the rhythm is circadian or diurnal.

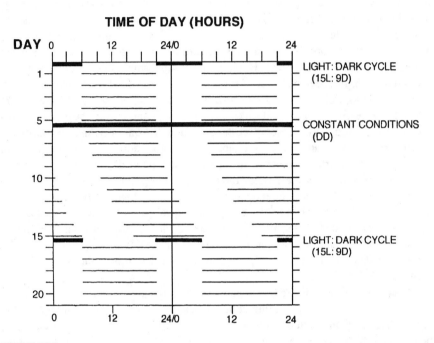

FIGURE 12.3
Artistic depiction of a standard "double plotted" circadian rhythm with the peak of the rhythm occurring in the dark and the nadir occurring in the light. This type of figure is generated by plotting the 24-h rhythm of interest on a linear scale and then duplicating the result to the right of the first plot. The next day's measurement of the rhythm is plotted beneath the first day with the remaining days of the experiment following below. This allows the figure to be "read" like a book, from left to right and top to bottom. Experimental treatments are usually placed to the right of the double plotted figure at the time they are begun. Under an external light:dark cycle (days 1 to 5), the rhythm is fixed to the onset and offset of darkness. When the subject is placed in constant conditions (days 6 to 15), the rhythm begins to "free run" following its own endogenous period (tau) which in this case is slightly longer than 24 h. When the subject is placed back into a light:dark cycle (days 16 to 20), the rhythm is entrained to the new photoperiod.

In general, if an investigator wants to determine if a specific process is rhythmic, at least six time points throughout the cycle of interest are necessary. This number of time points allows for the statistical interpretation of the rhythm and the generation of a curve depicting the rhythm. Specific mathematical treatments available for characterizing and interpreting rhythms include cosinor analysis, power spectrum analysis (fast Fourier transformations), and periodograms.[37] These mathematical formulations are available in specific software programs developed by companies specializing in rhythms research (Table 12.1). Each of these methods has advantages

TABLE 12.1
Companies Providing Hardware or Software that Collect or Analyze Rhythmic Processes

Company name and address	Products available
Ambulatory Monitoring, Inc. 731 Saw Mill River Road P.O. Box 609 Ardsley, NY 10502-0609	Actillume®, Mini-Motionlogger Actigraph®, and other monitoring devices for measurement of rhythms in human subjects
Circadian Technologies, Inc. 125 Cambridge Park Drive Cambridge, MA 02140	Techniques available for treatment of shift work, jet lag and other circadian disorders in human subjects
Data Sciences International 4211 Lexington Avenue North St. Paul, MN 55126-6164	Temperature and activity monitoring: implantable transmitters for animal research
Harvard Apparatus 22 Pleasant Street South Natick, MA 01760	Feeding/drinking monitors; activity monitoring in animals
Mini-Mitter Company P.O. Box 3386 Sunriver, OR 97707	Implantable transmitters; temperature and activity monitoring in animals; Tau® software for analyzing rhythms
Nalgene Company P.O. Box 20365 Rochester, NY 14602	Wheel running apparatus for rodents
Stanford Software Systems 574 Sims Road Santa Cruz, CA 95060	Collect® software for analyzing rhythms
Stoelting Company 620 Wheat Lane Wood Dale, IL 60191	Activity monitoring apparatus for animals

Note: The majority of these companies have Internet sites that can be accessed using available search engines.

and disadvantages. Cosinor analysis seeks to impose a cosine rhythm on the data set. This assumes that the actual data set is rhythmic and that it is biologically expressed as a cosine waveform. Power spectrum analysis also imposes rhythms on the data set, but uses more sophisticated approaches to rhythm interpretation (both sine and cosine values). An advantage of the previous two methods is that the "power" or strength of the rhythm can be determined (i.e., how well the data set fits the hypothetical curve generated). Periodograms, on the other hand, do not assume any inherent rhythmic nature to the data set and, therefore, may provide a more realistic description of the rhythm. However, the power of the rhythm generated by a periodogram cannot be determined because periodograms do not have an inherent rhythm to be compared against. For a more complete description of these methods, Enright provides a detailed discussion.[37]

A. Animal Experiments

1. Rhythms that Can be Measured

The rhythmic nature of seizures can be directly determined by analyzing seizure severity throughout a cycle of interest (day, month, year) either under normal living conditions or under controlled living conditions. In this manner, ultradian, daily,[15,25,26] and seasonal[17,18] rhythms of seizure severity can be examined. This information can be important to investigators, so that they produce consistent levels of seizures throughout their experimental protocols. Both acute and chronic models of epilepsy can be examined for rhythmic expression of seizures. Acute models require multiple sampling points and multiple groups of animals to measure a rhythm of seizure severity. Chronic models, on the other hand, can be continuously sampled over the time period of interest, providing rhythmic information from individual subjects.

Besides analyzing the rhythmic nature of seizures, investigators can also examine other well-characterized rhythms and compare these rhythms to the seizure rhythm. Examples of other rhythms studied include locomotor activity (wheel running), body temperature, heart rate, feeding, and drinking. In the past, the easiest rhythms to measure were wheel running activity or feeding and drinking rhythms. However, with the advent of implantable transmitters (Table 12.1), generalized activity, body temperature, and heart rate can be measured in the same animal with relative ease. More difficult rhythms to determine include plasma levels of cortisol or melatonin, although some investigators have routinely measured these hormones with the use of indwelling venous catheters.

To measure wheel running activity, a running wheel is placed in the rodent's cage with the wheel having a magnet attached to the spoke. At the same distance from the axle, a small magnetometer is attached to the support structure of the wheel that can measure each time the magnet passes (one wheel revolution). This information is transmitted to a dedicated computer that accumulates the number of revolutions in a quantity of time (a bin, usually 10 to 15 min). The number of revolutions per bin is plotted on the ordinate of a graph with the time of day plotted on the abscissa over a range of 24 h. This 24-h plot is duplicated and placed beside the original to create a 48-h plot. Each subsequent 48-h plot is placed below the previous 48-h plot, so that a continuous reading of an individual animal's rhythm can be observed from left to right and top to bottom much like reading a book (Figure 12.3). This is the conventional manner in which rhythms are displayed, so that alterations in rhythms can be observed clearly.[2,3]

Alterations in rhythms include changes in free running rhythm (tau) and phase shifts.[1,3] The free running rhythm is that period of time (usually close to 24 h in length) that the rhythm of interest displays when the subject is placed in constant conditions (usually constant darkness). Different species display different free running rhythms, with mice having rhythms that are usually shorter than 24 h, while humans have rhythms that are usually longer than 24 h.[3] This period can be determined by software that analyzes the rhythms or by a simple calculation of the time of activity onset from one day to the next. For example, if an individual has a period that is longer than 24 h and it is plotted on a 24-h basis, the onset of activity will occur later each day and a line that connects each time of activity onset will slope to the right (Figure 12.3). After specific treatments, the period can be determined to assess whether the treatment has altered clock function by altering period length. For a rhythm in seizure severity, it would be important to determine if an anticonvulsant was as effective throughout the rhythm and if the period of sensitivity was altered by the treatment.

Phase shifts are determined by placing a subject in constant conditions (constant darkness, variable feed, water and cage changes; no synchronizing cues) and then exposing the individual to a small duration of an entraining stimulus (usually a short pulse [15 min] of light) at different times throughout the circadian cycle. The individual will either advance, delay, or not shift the rhythm of interest, depending on the time that the stimulus is presented. A curve of responsiveness (phase response curve) can be constructed with the period of the cycle on the abscissa and phase advances plotted in the positive direction of the ordinate and phase delays plotted in the negative direction of the ordinate (Figure 12.4).[2,3] This curve indicates when the stimulus is most and least effective at shifting the rhythm of interest and provides information about the function of the circadian clock. Researchers interested in rhythms use phase response curves to predict when and how specific stimuli will alter circadian clock function.

Similarly, a response curve can be determined for the effectiveness of an anticonvulsant. In this case, an anticonvulsant could be given once at various times throughout the day/night cycle and the ability of the dose of anticonvulsant to prevent seizures could be plotted. This response curve would provide the optimal time for the administration of the anticonvulsant, allowing the lowest effective dose to be used. In addition, the severity of the anticonvulsant side effects can be plotted and compared with the plot of effectiveness. Utilizing both response curves, a time for administration of the anticonvulsant can be chosen when seizure prevention is high and side effects are low.

Many of the measurements that circadian biologists gather during a course of study can be important for determining how the circadian clock functions, but may not be applicable for control of rhythmic seizures. However, without some knowledge of the rhythmic aspects of the seizures, it is difficult to determine the cause of the disturbance and to provide proper and timely treatment.

2. Equipment and Controls Required

The equipment that is necessary for these type of studies can be obtained from numerous companies that specialize in rhythms research (Table 12.1). Also, information on the current state of affairs in rhythms research can be found by searching the internet (Table 12.2). When determining the type of equipment that will be used in an experiment, two major considerations are involved, the actual rhythms to be measured and the mathematical treatment of the data. The rhythms to be measured are at the discretion of the investigator, although choosing a rhythm that could have impact on seizure severity would have advantages (e.g., the sleep/wake rhythm could provide more correlative information on seizure rhythmicity than the heart rate rhythm). The type of equipment used to gather the rhythmic data can be important. Some measures of rhythms are gathered continuously and then captured or reported

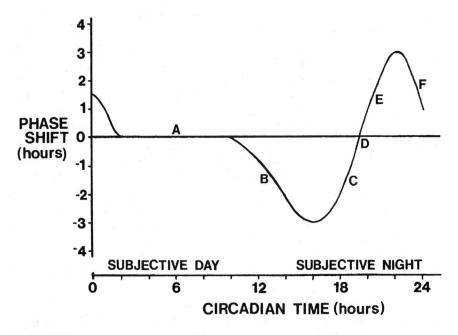

FIGURE 12.4
Artistic depiction of a typical phase response curve (PRC). A PRC is produced by placing subjects in
constant conditions (no external cues available for synchronizing) and exposing the subjects to a stimulus
(usually light) for a short period of time (usually 15 min). The subjects are monitored for the next few
days for changes in phase produced by the stimulus and this phase change is plotted on the ordinate of
a graph with the abscissa as the circadian time of stimulus treatment. Phase advances are plotted on the
positive scale of the ordinate while phase delays are plotted on the negative scale of the ordinate. The
subjective day and subjective night are listed on the abscissa in relation to the subject's previous
photoperiod. Most PRCs have a "dead zone" (A) where stimuli do not produce any changes in phase
and this usually is found during the subjective day. At the onset of the subjective night, a phase delay is
typically produced by stimulus (B). Throughout the subjective night, stimuli will produce a maximum
phase delay, a reduction in phase delay response (C), pass through a point where neither phase delays
or advances occur (D, "singularity") and then enter a region of phase advances (E, F). At the transition
of subjective night to subjective day, a phase advance is typically produced by the stimulus (F).

as number of events per unit time (wheel running) while other rhythms may be
marked by an onset or offset with little other information gathered (sleep/wake).

The mathematical treatment of the data is also an important consideration. Some
of the mathematical treatments available begin with the assumption that a rhythm
exists and "fit" the data to a specific rhythmic expression (i.e., cosinor analysis).
This may not be an appropriate treatment for the data. Other means of examining
the data do not begin with these assumptions and may provide a more realistic
portrayal of the rhythmic process (i.e., periodograms).

3.   General Methodologies Utilized
For animal experiments, the first choice an investigator must make is the experimental animal model. The majority of epilepsy studies are conducted in rodents,

TABLE 12.2
Selected Organizations and University Research Centers Devoted to Rhythms Research

Name and address	Brief description
Society for Research into Biological Rhythms University of Virginia Department of Biology Gilmer Hall Charlottesville, VA 22903	Major society organized to investigate research on rhythms from unicellular organisms to humans; *Journal of Biological Rhythms* is a bimonthly publication of the society
Society for Light Treatment and Biological Rhythms 10200 W. 44th Avenue, Suite 304 Wheat Ridge, CO 80033-2840	Society primarily devoted to understanding rhythms research in humans
Biological Rhythm Research Department of Physiology University of Leiden P.O. Box 9604 2300 RC Leiden The Netherlands	*Journal for European Society for Chronobiology*
San Diego Sleep and Rhythms Society Research Service 151 Veterans Affairs Medical Center 3350 La Jolla Village Drive San Diego, CA 92161	Local society to coordinate rhythms research in the San Diego area
Chronotherapeutics Searle Corporation 5200 Old Orchard Road Skokie, IL 60077	Information on circadian effectiveness of pharmaceuticals
NSF Center for Biological Timing Department of Biology University of Virginia Charlottesville, VA 22903	A regional center established by the National Science Foundation for the study of biological rhythms
Center for Circadian Biology & Medicine Northwestern University Evanston, IL 60208	Midwestern center for the study of circadian biology
Medical Chronobiology Lab Stratton Veteran's Medical Center 113 Holland Avenue Albany, NY 12208	Veteran's center chronobiology group
Chronobiology Research Department of Biology Texas A&M University College Station, TX 77840	Chronobiology research at Texas A&M University is highlighted

Note: The majority of these groups have internet sites that can be accessed using available search engines.

with rats and gerbils predominating.[18,38,39] Other species, including cats and primates, are also used in epilepsy studies.[40] When the specific animal model is chosen, the animals are acquired and allowed to acclimate to the housing conditions. As mentioned earlier, the housing conditions must remain constant throughout any experiment that attempts to investigate rhythmic processes. If a rhythm in seizure severity is under investigation, the researchers would place all of the animals in the same conditions and produce the seizures at specific times throughout the light:dark cycle, usually at intervals of 4 h or less.[38] The time points that occur in the dark portion of the light:dark cycle are conducted in the dark with the aid of a dim red light (Kodak, Safelight, Rochester, NY) or the use of infrared night vision goggles (IR Goggles, Electrophysics, Nutley, NJ). The investigators must determine whether the same animals can receive multiple seizures throughout a 24-h period (a within-animal design) or whether different animals should be used for each seizure induction (a between-animal design). It is recommended that a between-animal design be used, due to the long-term postictal refractory period that may affect subsequent seizure responses. The results from studies of this type would determine if a diurnal rhythm in seizure severity exists, but would not indicate whether this rhythm is endogenous to the animal or is controlled by exogenous factors, such as the light:dark cycle.

A second experiment is necessary to determine if the rhythm is endogenous and circadian. The same species of animals would be placed in constant darkness and another rhythm, such as activity, would be measured to determine that the subjects were not entrained to any outside influences, but were free running. After a free running rhythm is established, the animals would receive the seizure stimulus at 4-h intervals throughout a 24- to 48-h period. Again, a 24-h curve to seizure induction can be generated. The rhythm in seizure severity found in both experiments can be compared and correlated to determine if the seizure rhythm is endogenous and circadian.

Utilizing the rhythm of seizure severity, the time of peak seizure production can be determined. This time point could be of interest to researchers exploring cellular mechanisms of seizure production and propagation by providing the proper time to examine these mechanisms (providing the largest window of response to experimental treatments). Conversely, the time of least seizure production could also be used to investigate the endogenous processes that limit seizure severity.

This same methodology can be used to investigate diurnal/circadian differences in anticonvulsant effectiveness. The results from studies of this type allow the administration of anticonvulsant compounds at the proper time for optimal control of seizures. It should be stressed that any treatments that increase or decrease seizure expression may have rhythmic components and that conclusions reached on the efficacy of these treatments may only be true at specific times of the day.

B. Human Experiments

The use of human subjects in epilepsy research is well established, especially in the search for effective anticonvulsants. Although many humans have epileptic seizures at specific times of day, the study of rhythmic seizures in humans is limited.[25,28,41,42] Also, specific anticonvulsants are known to have markedly different levels of effectiveness, dependent on the time of day they are administered.[30-33,43] Therefore, both of these areas of research have rhythmic processes that can be explored.

1.   Rhythms that Can be Measured

Virtually all of the rhythms that can be determined in animals can also be determined in humans. However, the methods for collecting these rhythms are quite different. General activity can be determined with the use of a "wrist watch" sized monitor (Ambulatory Monitoring, Inc., Ardsley, NY) (Table 12.1). Body temperature and heart rate can be assessed by placement of specific transceivers that relay the information to dedicated computers. Subjects can also have sequential blood samples collected for determination of hormonal rhythms, such as cortisol or melatonin.[11]

2.   Equipment and Controls Required

Most of the equipment used to study rhythmic seizures in humans is the same as would be used to study any seizure response. The only difference is that the measurements would be taken over time in each individual subject. Using these methods, investigators can determine if a rhythm of seizure severity is present in specific individuals by utilizing rhythm analysis software (Table 12.1). Also, the effectiveness of anticonvulsants can be measured over a 24-h period to determine the optimal time for administration.

It is much more difficult to place human subjects into constant conditions for monitoring circadian rhythms, although experiments of this type have been conducted in the past.[44] Therefore, most rhythm studies in humans occur with the subject exposed to normal photoperiods and other environmental cues, but some studies use more controlled situations, such as constant routine conditions.[11] In constant routine conditions, subjects are placed in a semirecumbent position with few options for activity. The subjects are kept awake for the entire protocol and are provided small meals hourly to prevent any feeding cues. These conditions can be maintained for up to 40 h, allowing daily rhythms to be clearly observed.

3.   General Methodologies Utilized

In general, patients with epilepsy can be monitored for aspects of seizures throughout the 24-h cycle. Usually the patients have continuous EEG recordings and video monitoring that provide an indication of the onset of seizure activity. These studies will frequently take place in a controlled hospital environment that limits the patient's exposure to other variables.[11] Constant routine conditions could be considered to examine the endogenous nature of the seizures, but may not be appropriate for determination of optimal anticonvulsant timing since the non-natural aspect of the conditions may change the anticonvulsant response curve.

III. Interpretation

A. Awareness of Rhythmic Nature of Seizures

Determining the rhythmic components of seizure activity has many advantages. If anticonvulsants are provided at (or just prior to) the maximal seizure response, then the compounds can have their greatest effectiveness. In addition, anecdotal evidence suggests that the severity of seizures may be reduced if consistent daily rhythms are maintained that provide strong cues for synchronization of the circadian system. By making patients aware of the rhythmic nature of seizures, they may be able to gain more control of their seizures and not feel as dominated by the seizure events.

For researchers investigating epilepsy, it is important to maintain consistent conditions for their human or animal subjects, so that rhythmic variables can be reduced or eliminated. If the rhythmic nature of the seizures cannot be eliminated by consistent treatments, then the experiments must be designed so that sampling from all of the groups occurs throughout the rhythmic period. This prevents one group from being sampled at the nadir of the rhythm while another group is sampled at the peak of the rhythm, leading to an incorrect claim that the treatment was effective.

B.   Chronopharmacokinetics of Anticonvulsants

Finally, the administration of anticonvulsants at specific times of day can reduce the side effects of the anticonvulsants while maintaining their effectiveness in preventing seizures. This could be especially useful in patients who have rhythmic seizures, by linking the rhythmic application of anticonvulsants to the rhythmic appearance of seizures. Although there may be well-known rhythmic differences in effectiveness of specific anticonvulsants, it is important to determine the rhythmic nature in each individual patient.

Epilepsy is a major neurological deficit that has tremendous impact on patient's lives. By determining the rhythmic nature of the seizures associated with epilepsy and the optimal time of day for administration of anticonvulsants, better and more consistent control of seizures can be accomplished.

References

1. Moore-Ede, M. C., Sulzman, F. M., and Fuller, C. A., *The Clocks That Time Us,* Harvard University Press, Cambridge, 1982.
2. Binkley, S. A., *The Clockwork Sparrow,* Prentice-Hall, Englewood Cliffs, NJ, 1990.
3. Aschoff, J., *Handbook of Behavioral Neurobiology, Vol. 4, Biological Rhythms,* Plenum Press, New York, 1981.
4. Murphy, P. J. and Campbell, S. S., Physiology of the circadian system in animals and humans, *J. Clin. Neurophysiol.,* 13, 2, 1996.
5. Ralph, M. R., Circadian rhythms — mammalian aspects, *Semin. Cell Dev. Biol.,* 7, 821, 1996.
6. Reuss, S., Components and connections of the circadian timing system in mammals, *Cell Tissue Res.,* 285, 353, 1996.
7. Yu, H. S. and Reiter, R. J., Eds., *Melatonin. Biosynthesis, Physiological Effects and Clinical Applications,* CRC Press, Boca Raton, FL, 1993.

8. Champney, T. H., Hanneman, W. H., Legare, M. E., and Appel, K., Acute and chronic effects of melatonin as an anticonvulsant in male gerbils, *J. Pineal Res.*, 20, 79, 1996.

9. Fauteck, J. D., Bockmann, J., Bockers, T. M., Wittkowski, W., Kohling, R., Lucke, A., Straub, H., Speckmann, E. J., Tuxhorn, I., Wolf, P., Pannek, H., and Oppel, F., Melatonin reduces low Mg^{2+} epileptiform activity in human temporal slices, *Exp. Brain Res.*, 107, 321, 1995.

10. Hoffman, R. A. and Corth, R., Photic environments: physical and biological considerations, *Pineal Res. Rev.*, 6, 95, 1988.

11. el-Hajj Fuleihan, G., Klerman, E. B., Brown, E. N., Choe, Y., Brown, E. M., and Czeisler, C. A., The parathyroid hormone circadian rhythm is truly endogenous — a general clinical research center study, *J. Clin. Endocine Metab.*, 82, 281, 1997.

12. Alleva, J. J., The biological clock and pineal gland: how they control seasonal fertility in the golden hamster, *Pineal Res. Rev.*, 5, 95, 1987.

13. Battino, D., Estienne, M., and Avanzini, G., Clinical pharmacokinetics of antiepileptic drugs in paediatric patients. I. Phenobarbital, primidone, valproic acid, ethosuximide and mesuximide, *Clin. Pharmacol.*, 29, 257, 1995.

14. Bulau, P. and Clarenbach, P., Interaction of epileptic seizures and biological rhythms, *Wien. Med. Wochenschr.*, 145, 448, 1995.

15. Shouse, M. N., Dasilva, A. M., and Sammaritano, M., Circadian rhythm, sleep, and epilepsy, *J. Clin. Neurophysiol.*, 13, 32, 1996.

16. Champney, T. H. and Peterson, S. L., Circadian, seasonal, pineal, and melatonin influences on epilepsy, in *Melatonin: Biosynthesis, Physiological Effects and Clinical Applications*, Yu, H. S. and Reiter, R. J., Eds., CRC Press, Boca Raton, FL, 1993, 477.

17. Wláz, P. and Löscher, W., The role of technical, biological, and pharmacological factors in the laboratory evaluation of anticonvulsant drugs. V. Lack of seasonal influences on amygdala kindling in rats, *Epilepsy Res.*, 16, 131, 1993.

18. Löscher, W. and Fiedler, M., The role of technical, biological and pharmacological factors in the laboratory evaluation of anticonvulsant drugs. VI. Seasonal influences on maximal electroshock and pentylenetetrazol seizure thresholds, *Epilepsy Res.*, 25, 3, 1996.

19. Rajna, P. and Veres, J., Correlations between night sleep duration and seizure frequency in temporal lobe epilepsy, *Epilepsia*, 34, 574, 1993.

20. Weiss, G., Lucero, K., Fernandez, M., Karnaze, D., and Castillo, N., The effect of adrenalectomy on the circadian variation in the rate of kindled seizure development, *Brain Res.*, 612, 354, 1993.

21. Balish, M., Albert, P. S., and Theodore, W. H., Seizure frequency in intractable partial epilepsy: a statistical analysis, *Epilepsia*, 32, 642, 1991.

22. Burr, W., Korner, E., and Stefan, H., Circadian distribution of generalized spike-wave activity in relation to sleep, *Epilepsy Res. Suppl.*, 2, 121, 1991.

23. Oliverio, A., Castellano, C., Puglisi-Allegra, S., and Renzi, P., Diurnal variations in electroconvulsive shock-induced seizures: involvement of endogenous opioids, *Neurosci. Lett.*, 57, 237, 1985.

24. Yehuda, S. and Mostofsky, D. I., Circadian effects of beta-endorphin, melatonin, DSIP, and amphetamine on pentylenetetrazol-induced seizures, *Peptides*, 14, 203, 1993.

25. Ellis, C. R., Chronobiological aspects of epileptic phenomena: a literature review, implications for nursing and suggestions for research, *J. Neurosci. Nurs.*, 24, 335, 1992.

26. Poirel, C. and Ennaji, M., Circadian aspects of epileptic behavior in comparative psychophysiology, *Psychol. Rep.,* 68, 783, 1991.
27. Poirel, C., Circadian chronobiology of epilepsy: murine models of seizure susceptibility and theoretical perspectives for neurology, *Chronobiologia,* 18, 49, 1991.
28. Dreifuss, F. E., Meinardi, H., and Stefan, H., *Chronopharmacology in Therapy of the Epilepsies,* Raven Press, New York, 1990.
29. Schachter, S. C., Neuroendocrine aspects of epilepsy, in *Epilepsy and Behavior,* Devinsky, O. and Theodore, W. H., Eds., Wiley-Liss, New York, 1991, 303.
30. Bareggi, S. R., Tata, M. R., Guizzaro, A., Pirola, R., Parisi, A., and Monza, C. G., Daily fluctuation of plasma levels with conventional and controlled-release carbamazepine: correlation with adverse effects, *Int. Clin. Psychopharmacol.,* 9, 9, 1994.
31. Haefeli, W. E., Meyer, P. G., and Luscher, T. F., Circadian carbamazepine toxicity, *Epilepsia,* 35, 400, 1994.
32. Bonneton, J., Iliadis, A., Genton, P., Dravet, C., Viallat, D., and Mesdjian, E., Steady state pharmacokinetics of conventional versus controlled-release carbamazepine in patients with epilepsy, *Epilepsy Res.,* 14, 257, 1993.
33. Petker, M. A. and Morton, D. J., Comparison of the effectiveness of two oral phenytoin products and chronopharmacokinetics of phenytoin, *J. Clin. Pharmacol. Ther.,* 18, 213, 1993.
34. Cloyd, J., Pharmacokinetic pitfalls of present antiepileptic medications, *Epilepsia,* 32 (Suppl. 5), S53, 1991.
35. Hartley, R., Forsythe, W. I., McLain, B., Ng, P. C., and Lucock, M. D., Daily variations in steady-state plasma concentrations of carbamazepine and its metabolites in epileptic children, *Clin. Pharmacol.,* 20, 237, 1991.
36. Rogawski, M. A. and Porter, R. J., Antiepileptic drugs: pharmacologic mechanisms and clinical efficacy with consideration of promising developmental stage compounds, *Pharmacol. Rev.,* 42, 223, 1990.
37. Enright, J. T., Data analysis, in *Handbook of Behavioral Neurobiology, Vol. 4, Biological Rhythms,* Aschoff, J., Ed., Plenum Press, New York, 1981, 21.
38. Masukawa, T. and Nakanishi, K., Circadian variation in enoxacin-induced convulsions in mice coadministered with fenbufen, *Jpn. J. Pharmacol.,* 73, 175, 1997.
39. Löscher, W. and Schmidt, D., Which animal models should be used in the search for new antiepileptic drugs? A proposal based on experimental and clinical observations, *Epilepsy Res.,* 2, 145, 1988.
40. Fisher, R. S., Animal models of the epilepsies, *Brain Res. Rev.,* 14, 245, 1989.
41. Gallerani, M., Manfredini, R., and Fersini, C., Chronoepidemiology in human diseases, *Ann. Istit. Super. Sanita,* 29, 569, 1993.
42. Autret, A., Lucas, B., Laffont, F., Bertrand, P., Degiovanni, E., and De Toffol, B., Two distinct classifications of adult epilepsies: by time of seizures and by sensitivity of the interictal paroxysmal activities to sleep and waking, *Electroencephalogr. Clin. Neurophysiol.,* 66, 211, 1987.

43. Brouwer, O. F., Pieters, M. S., Edelbroek, P. M., Bakker, A. M., van Geel, A. A., Stijnen, T., Jennekens-Schinkel, A., Lanser, J. B., and Peters, A. C., Conventional and controlled release valproate in children with epilepsy: a cross-over study comparing plasma levels and cognitive performances, *Epilepsy Res.*, 13, 245, 1992.
44. Aschoff, J. and Wever, R., The circadian system of man, in *Handbook of Behavioral Neurobiology, Vol. 4, Biological Rhythms,* Aschoff, J., Ed., Plenum Press, New York, 1981, 311.

Index